T0189672

International Explorations in Outdoor and Environmental Education

Volume 7

This series focuses on contemporary trends and issues in outdoor and environmental education, two key fields that are strongly associated with education for sustainability and its associated environmental, social and economic dimensions. It also has an international focus to encourage dialogue across cultures and perspectives. The scope of the series includes formal, nonformal and informal education and the need for different approaches to educational policy and action in the twenty first century. Research is a particular focus of the volumes, reflecting a diversity of approaches to outdoor and environmental education research and their underlying epistemological and ontological positions through leading edge scholarship. The scope is also be both global and local, with various volumes exploring the issues arising in different cultural, geographical and political contexts. As such, the series aims to counter the predominantly "white" Western character of current research in both fields and enable cross-cultural and transnational comparisons of educational policy, practice, project development and research. The purpose of the series is to give voice to leading researchers (and emerging leaders) in these fields from different cultural contexts to stimulate discussion and further research and scholarship to advance the fields through influencing policy and practices in educational settings. The volumes in the series are directed at active and potential researchers and policy makers in the fields. Book proposals for this series may be submitted to the Publishing Editor: Claudia Acuna E-mail: Claudia.Acuna@springer.com

More information about this series at http://www.springer.com/series/11799

Karen Haydock • Abhijit Sambhaji Bansode
Gurinder Singh • Kalpana Sangale

Learning and Sustaining Agricultural Practices

The Dialectics of Cultivating Cultivation in Rural India

 Springer

Karen Haydock
Homi Bhabha Centre for Science Education
Tata Institute of Fundamental Research
Mumbai, Maharashtra, India

Gurinder Singh
Homi Bhabha Centre for Science Education
Tata Institute of Fundamental Research
Mumbai, Maharashtra, India

Abhijit Sambhaji Bansode
Tata Institute of Social Sciences
Mumbai, Maharashtra, India

Kalpana Sangale
Homi Bhabha Centre for Science Education
Tata Institute of Fundamental Research
Mumbai, Maharashtra, India

ISSN 2214-4218 ISSN 2214-4226 (electronic)
International Explorations in Outdoor and Environmental Education
ISBN 978-3-030-64067-5 ISBN 978-3-030-64065-1 (eBook)
https://doi.org/10.1007/978-3-030-64065-1

This Springer imprint is published by the registered company Springer Nature Switzerland AG
The registered company address is: Gewerbestrasse 11, 6330 Cham, Switzerland

We dedicate this book to the farming family:
Namdev, Prabhavati, Bhimraj, Smita, Pratik,
and Pranay.

Series Editors' Foreword

Passage O soul to India!...
Not you alone, proud truths of the world!
Nor you alone, ye facts of modern science!
But myths and fables of eld—Asia's, Africa's fables
The far-darting beams of the spirit, the unloos'd dreams,
The deep diving bibles and legends,
The daring plots of the poets, the elder religions;

(Walt Whitman, 1870)

This book invites readers to undertake a metaphorical 'Passage to India' – not so much the 'passage' of E.M. Forster's (1924) celebrated novel, but rather of the Walt Whitman poem that inspired it. In 1869, Whitman was himself inspired by two history-making events, namely the completion of the American transcontinental railroad, connecting the USA from East to West, and the opening of the Suez Canal joining the Mediterranean Sea to the Red Sea, which allowed transportation and trade between Europe and Asia in record time, without navigating around Africa. The canal promised to change the face of world commerce, but it also extended the possibilities of cultural exchange between nations (see Harsharan S. Ahluwalia, 1983).

In 'Passage to India', Whitman celebrates the canal's construction as both a feat of engineering and a triumph of the human imagination. The 'facts of modern science' alone are not enough to explain the project's completion. By directly addressing the 'proud truths' and 'fables' in parallel with the 'far-darting beams of the spirit', 'deep diving bibles and legends' and 'the daring plots of the poets', Whitman brings 'modern science' into perspective with the 'elder religions' and expresses his admiration for both.

Learning and Sustaining Agricultural Practices: The dialectics of cultivating cultivation in rural India relates a very different 'Passage to India'. This volume discusses the authors' understanding of historical dialectical materialist science and explains how and why their observations, learning, and work on the farm led them to see science as a process of doing and working, even as many educationists see science as a 'body of knowledge'. But it is much more than this. It is very interdisciplinary. It is an example of a participatory case study. And, as the authors note, 'different parts of the book will be useful to people who have more specialised interests, which may range from rural school education, or science education, to skill training, agricultural development and sustainability, indigenous knowledge and multicultural education, philosophies of science, and political economy'.

We also see this book as epitomising the changes in orientations from environmental education to education for sustainable development. While this volume is about education and relationships with the environment, it is much more.

Early conceptions of environmental education were focused on the quality of the environment and quality of life for humans. For example, the Belgrade Charter stated that:

> ...the foundations must be laid for a world-wide environmental education programme that will make it possible to develop new knowledge and skills, values and attitudes, in a drive towards a better quality of environment and, indeed, towards a higher quality of life for present and future generations living within that environment. (UNESCO 1975, p. 2)

In 1987 the report of the World Commission on Environment and Development, also known as *Our Common Future* or the Brundtland Report, included what is now frequently quoted as the standard definition of sustainable development: 'Sustainable development is development that meets the needs of the present without compromising the ability of future generations to meet their own needs.' (World Commission on Environment and Development 1987, p. 43). This report marked a turning point for the environment movement and environmental education in that subsequent United Nations' meetings moved to reorient environmental education towards education for sustainable development. Fast forward to 2015 and the United Nations' adoption of the 2030 agenda for sustainable development that included 17 Sustainable Development Goals. The goals address issues confronting the majority world such as poverty, health, income, agricultural sustainability, food security, educational opportunity and achievement – all of which are addressed (and problematised) in this volume.

We are especially pleased to introduce *Learning and Sustaining Agricultural Practices* as the first volume in our series that wholly arises from a 'majority world' context. As David Cheruiyot & Raul Ferrer-Conill (2020, p. 9) write:

> 'Minority World Countries' (North America, Europe and Australasia) and 'Majority World Countries' (Africa, Asia, Latin America and the Middle East) are terms increasingly used in place of the misleading 'Global North/South' or reductionist 'West/Rest'.

We also prefer 'majority world' to the largely inaccurate, outdated and/or non-descriptive terms 'developing' nations and the offensive 'third world'. Since the early 1990s, the communications cooperative New Internationalist (www.newint. org) has used 'majority world' to describe this global community by reference to what it *is*, rather than what it lacks, and also to draw attention to the disproportionate impact that the largest economies in the world (variously known as the Group of Seven/Eight countries), which represent a relatively small fraction of humankind, have on the majority of the world's peoples.

'Majority world' perspectives have been included in previous volumes in this series. In *Green Schools Globally*, accounts of green school movements were provided from China (Yu and Lee 2002), Hong Kong (Tsang et al. 2020), India (Sharma & Kanaujia 2020), Israel (Tal 2020), Kenya (Otieno et al. 2020), Mexico (González-Gaudiano et al. 2020), South Africa (Rosenberg 2020), Taiwan (Wang et al. 2020), Turkey (Taşar 2020) and the Western Indian Ocean (Copsey 2020). In *Education and Climate Change: The Role of Universities*, David Rhodes and Margaret Wang (2021) discuss how to develop leadership capacities that help students in Israel and Palestine address climate change, Lina Lopez Lalinde and Carrie Maierhofer (2021) look at 'Creating a Culture of Shared Responsibility for Climate Action in Guatemala through Education', Ashley Bazin and Christelle Saintis (2021) discuss building climate change resilience in Haiti through educational radio programming, and Natasha Japanwal (2021) discusses a climate change curriculum for out-of-school children in Pakistan.

We hope this volume will encourage others from majority world perspectives to contribute to this series. We also hope that this volume will reach those for whom the research is important – those in rural school education, or science education, skill training, agricultural development and sustainability, indigenous knowledge and multicultural education, philosophies of science, political economy, and those engaging in ethnographic and case study–based research. We can all learn from this work as we struggle with understanding sustainable development in the majority world.

RMIT University   Annette Gough
Melbourne, VIC, Australia

La Trobe University   Noel Gough
Melbourne, VIC, Australia

References

Ahluwalia, Harsharan S. 1983. A reading of Whitman's 'Passage to India'. *Walt Whitman Quarterly Review 1*(1): 9–17. https://doi.org/10.13008/2153-3695.1002.

Bazin, Ashley, and Christelle Saintis. 2021. Rezistans Kimatik. Building climate change resilience in Haiti through educational radio programming. In *Education and Climate Change: The Role of Universities,* ed. Fernando Reimers, 113–136. Cham: Springer.

Cheruiyot, David, and Raul Ferrer-Conill. 2020. Pathway outta pigeonhole? De-contextualizing majority world countries. *Media, Culture & Society* 0163443720960907. https://doi.org/10.1177/0163443720960907.

Copsey, Olivia. 2020. A regional approach to eco-schools in the Western Indian Ocean. In *Green Schools Globally: Stories of Impact on Education for Sustainable Development*, eds. Annette Gough, John Chi-Kin Lee, & Eric Po Keung Tsang, 403–418. Cham: Springer. https://doi.org/10.1007/978-3-030-46820-0_22.

Forster, E.M. 1924. *A Passage to India*. London: Edward Arnold.

González-Gaudiano, Edgar, Pablo Á. Meira-Cartea, and José M. Gutiérrez-Bastida. 2020. Green schools in Mexico and Spain: Trends and critical perspectives. In *Green Schools Globally: Stories of Impact on Education for Sustainable Development*, eds. Annette

Gough, John Chi-Kin Lee, & Eric Po Keung Tsang, 269–287. Cham: Springer. https://doi.org/10.1007/978-3-030-46820-0_15.

Japanwal, Natasha. 2021. Adaptation, migration, advocacy. A climate change curriculum for out-of-school children in Badin, Sindh. In *Education and Climate Change: The Role of Universities*, ed. Fernando Reimers, 137–152. Cham: Springer.

Lopez Lalinde, Lina, and Carrie Maierhofer. 2021. Creating a culture of shared responsibility for climate action in Guatemala through education. In *Education and Climate Change: The Role of Universities*, ed. Fernando Reimers, 85–112. Cham: Springer.

Otieno, Dorcas, David Wandabi, and Lorraine Dixon. 2020. Eco-schools Kenya: Practising education for green economy and sustainability. In *Green Schools Globally: Stories of Impact on Education for Sustainable Development*, eds. Annette Gough, John Chi-Kin Lee, & Eric Po Keung Tsang, 245–267. Cham: Springer. https://doi.org/10.1007/978-3-030-46820-0_14

Rosenberg, Eureta. 2020. Eco-schools as education for sustainable development in rural South Africa. In *Green Schools Globally: Stories of Impact on Education for Sustainable Development*, eds. Annette Gough, John Chi-Kin Lee, & Eric Po Keung Tsang, 289–307. Cham: Springer. https://doi.org/10.1007/978-3-030-46820-0_16.

Sharma, Pramod Kumar, and Preeti Rawat Kanaujia. 2020. Journey of green schools in India. In *Green Schools Globally: Stories of Impact on Education for Sustainable Development*, eds. Annette Gough, John Chi-Kin Lee, & Eric Po Keung Tsang, 203–226. Cham: Springer. https://doi.org/10.1007/978-3-030-46820-0_12.

Tal, Tali. 2020. Green schools in Israel: Multiple rationales and multiple action plans. In *Green Schools Globally: Stories of Impact on Education for Sustainable Development*, eds. Annette Gough, John Chi-Kin Lee, & Eric Po Keung Tsang, 227–244. Cham: Springer. https://doi.org/10.1007/978-3-030-46820-0_13.

Taşar, Mehmet Fatih. 2020. Impact of the eco-schools program on "Education for sustainable development" in Turkey. In *Green Schools Globally: Stories of Impact on Education for Sustainable Development*, eds. Annette Gough, John Chi-Kin Lee, & Eric Po Keung Tsang, 345–363. Cham: Springer. https://doi.org/10.1007/978-3-030-46820-0_19.

Tsang, Eric Po Keung, John Chi-Kin Lee, Sai Kit Eddie Yip, and Annette Gough. 2020. The green school award in Hong Kong: Development and impact in the school sector. In *Green Schools Globally: Stories of Impact on Education for Sustainable Development*, eds. Annette Gough, John Chi-Kin Lee, & Eric Po Keung Tsang, 189–201. Cham: Springer. https://doi.org/10.1007/978-3-030-46820-0_11.

UNESCO. 1975. *The Belgrade Charter: A Global Framework for Environmental Education.* http://unesdoc.unesco.org/images/0001/000177/017772eb.pdf.

United Nations. 2015. *Transforming Our World: The 2030 Agenda for Sustainable Development.* Resolution adopted by the General Assembly on 25 September 2015. Retrieved from www.un.org/ga/search/view_doc.asp?symbol=A/RES/70/1&Lang=E.

Rhodes, David, and Margaret Wangs. 2021. Learn to lead: Developing curricula that foster climate change leaders. In *Education and Climate Change: The Role of Universities*, ed. Fernando Reimers, 45–83. Cham: Springer.

Wang, Shun-Mei, John Chi-Kin Lee, and Sin-Jao Ho. 2020. The development of greenschools in Taiwan: Current situation, obstacles and prospects. In *Green Schools Globally: Stories of Impact on Education for Sustainable Development*, eds. Annette Gough, John Chi-Kin Lee, & Eric Po Keung Tsang, 333–344. Cham: Springer. https://doi.org/10.1007/978-3-030-46820-0_18.

Whitman, Walt. 1870. Passage to India. In *Leaves of Grass* (Comprehensive Reader's ed., 411–421), eds. Harold W. Blodgett, & Sculley Bradley (1965). New York: New York University Press.

World Commission on Environment and Development. (WCED). 1987. *Our Common Future.* Oxford: Oxford University Press.

Yu, Huang, and John Chi-Kin Lee. 2020. The past, present and future of mainland China's green schools. In *Green Schools Globally: Stories of Impact on Education for Sustainable Development*, eds. Annette Gough, John Chi-Kin Lee, & Eric Po Keung Tsang, 125–146. Cham: Springer. https://doi.org/10.1007/978-3-030-46820-0_8.

Preface

The title of this book refers to two different meanings of the word 'cultivate'. The word comes from the Latin word *cultura*, which originally meant inhabiting, caring for, cultivating, or doing agriculture. It also came to mean cultivating the mind: a kind of intellectual learning. But in English, the word 'culture' came to be more of a noun – an orderly state of being – rather than a dynamic verb – a continuing process of doing. This is indicative of the dialectical nature of cultivating cultivation. And it points to the main questions we ask in this book: Do the paddy cultivators learn and develop a dynamic, ever-changing process of cultivation, or does each generation somehow learn a rather static method of cultivation from the previous generations? Is their cultivation a 'body of knowledge'? Or is cultivating a creative process of doing and changing? Is it manual work, distinct from, and subservient to, mental work? How and why is their process of doing cultivation similar to or different from the process of doing science?

Our aim is not to generalise, but to understand how one family in one village cultivates – passes on, learns, and/or develops – methods of paddy cultivation. The family lives in the village of Rudravali, in Raigad District, in the state of Maharashtra, in India.

Ironically, although it is almost the smallest unit of social organisation, a village is commonly thought to represent the nation – the 'real' India – and even provide a generalisation of the social processes in South Asia. 'Traditional' villages are counterpoised to 'modern, westernised' towns and cities. These are oversimplifications, but they persist, despite the huge diversity of cultures, languages, and complex, highly stratified communities throughout the region. As with other forms of identity, the identity of the 'traditional village' is imposed from outside, not by the villagers deciding to define themselves. In this book, we hope that the voices and actions of one village family will be heard.

Mahatma Gandhi is remembered as having advocated the revival of the spirit of traditional village life. He wanted India to 'recover its lost self and attain true freedom' by constructing 'a harmonious and self-contained village, uncorrupted by the modern life of the city and western technology' (Jodhka 2002). He recognised the high levels of alienation that characterise modern development and are much reduced in village life. However, claims that Gandhi was very 'anti–western science' are exaggerated – his dreams of reconstructing village life did not exclude all modern technology. Gandhi had also written about how the British colonial rule impaired the creativity of the villager who had previously used the mind and body to produce all daily needs. This is in contrast to the views of the other two most important leaders of the Nationalist movement, BR Ambedkar and Jawaharlal Nehru. Nehru viewed a village as a community characterised by class division, ignorance, and backwardness. He thought villagers needed to be motivated to learn modern methods and adopt the new technologies of modern agriculture. Ambedkar, who represents the downtrodden, and is now more influential than either Gandhi or Nehru, saw and experienced traditional village living as an oppressive, undemocratic structure dominated by caste-based inequality, which is inherent to Hinduism.

As we shall see, these conflicts between the so-called 'traditional' and 'modern' continue to define and characterise the process of cultivating cultivation. These conflicts give rise to a number of questions that will keep appearing throughout the book: Are the family's methods of paddy cultivation traditional? Are they harmonious and unchanging? Is the family able to sustain itself by producing its own food and other needs? Does the family use modern 'western' technology? If so, how do they learn to use it? If not, do they want to? Does the family develop their own methods of cultivation? Are the family members ignorant and backward? How do oppressive social structures affect the family's learning, teaching, and developing of paddy cultivation? And for all of these questions, we will ask additional 'Why?' questions.

Readers who are not very familiar with villages of India, where most people in the country still live, may have some difficulty understanding the context of our study. We have found that it is not uncommon for English-speaking intellectuals outside India, and even some inside India, to have certain misconceptions about the basic situation and the vast diversities in India. Many of these misconceptions are generalisations that have been spread deliberately, so it is easy to be misinformed. For example, at the time of this study (2015–2018) the following generalisations about India are all false: (1) Due to neoliberalism and India's very rapid rate of development, the lives of its common people have improved tremendously in recent times. (2) These days people are no longer dying due to not having enough food. (3) Everyone in India has a phone, and/or most people have smartphones with internet. (4) Most people know English. (5) All children in India go to school. (6) Caste, untouchability, and Brahmanism are no longer problems. (7) India is a Hindu country. (8) Inequality is decreasing as poverty is decreasing. (9) Unemployment is decreasing and is not a huge problem. (10) Farming is mechanised, and fields are ploughed by tractors rather than oxen throughout the country.

We will not provide evidence or try to explain any of these misconceptions – that is beyond the scope of this book. However, clearing up such basic misconceptions will help readers to understand the context of our study. Otherwise, some of what you read here may sound too unbelievable. Readers may refer to the following sources: *India after Naxalbari*, by Bernard D'Mello (2018); *Resisting Dispossession*, by Ranjana Padhi and Nigamananda Sadangi (2020); *Women and Work in Rural*

India, edited by Madhura Swaminathan, Shruti Nagbhushan, and V.K. Ramachandran (2020); *Dispossession, Deprivation and Development*, edited by Arindam Banerjee and C.P. Chandrasekhar (2018); and *Is This Azaadi?*, by Anand Chakravarti (2018).

We begin, in Chap. 2, with a discussion of the superficial appearances of what might be called 'traditional' paddy cultivation. When we take a closer look at the process of cultivation, in Chap. 3, we find a number of dialectical conflicts, and it is no longer clear that the traditional is really traditional. In Chap. 4, we discuss dialectical conflicts that we find in the process of learning cultivation in the field, as well as in schools.

We then explore the historical development of education related to cultivation, in Chap. 5, and the historical development of paddy cultivation, in Chap. 6. By considering general historical, social, and economic commonalities, we are able to make more sense of the particularities that we observed on the farm. In Chap. 7, we discuss how this application of historical development leads us to understand the process of cultivating (learning and passing on) paddy cultivation by the family. We come to a better understanding of what is 'traditional', and we find that the family is, in some ways, actually learning by doing science.

Taking into account the historical development of education related to cultivation allows us to analyse the roles of formal education and informal learning and their relation to paddy cultivation by the family, in Chap. 8.

Finally, in Chap. 9, we summarise our findings and discuss implications for the possible future of education related to cultivation.

The nature of this book is that it is very interdisciplinary. As some of you read the book, this interdisciplinarity and the interdependencies between learning, teaching, cultivating, and doing science will be your chief interest. We have drawn upon all these areas in crafting an investigation in which we keep trying to go beyond initial appearances, interrelating different observations and understandings. We keep finding contradictions in our observations, and new questions keep arising. We keep revising our analysis each time we reconsider and expand our observations. In other words, as we proceed, we keep asking questions and questioning our answers. Thus, we find that the traditional is not traditional, school science is not science, education is not learning, and capitalist cultivation prevents cultivators from cultivating cultivation. As you read the book, we hope you find the same contradictions we found, as well as some that we did not find. We hope the book will lead you to come up with your own questions.

Different parts of the book will be useful to people who have more specialised interests, which may range from rural school education, or science education, to skill training, agricultural development and sustainability, indigenous knowledge and multicultural education, philosophies of science, and political economy.

If you are an ethnographic researcher, you may find the book especially useful as an example of a participatory case study, which has the advantage of going into more depth and exploring a broader range of issues than larger studies that rely on surveys and interviews. We made audio and video recordings of the family and ourselves as the family did their work in the paddy field, and we tried to learn by working alongside them. Our methods can be understood by reading our accounts and analysis, particularly in Chap. 3. As researchers, we tried to analyse ourselves as well as the family members. The family members also discussed and analysed along with us.

For those of you who are interested in understanding education, Chaps. 4, 5, and 8 will be most relevant. In Chap. 4, we focus on understanding the learning that we saw and experienced in the field, as compared to the classroom. In Chap. 5, we give an overview of the historical development of education related to agriculture in India. We discuss how education in India developed to cater to a colonial and capitalist structure in which work with the mind was separated from work with the hands, which was devalued (Sect. 5.1). This leads to a discussion of the historical rationale for including skill training as part of formal school education (Sect. 5.3). It is only after we also consider the history of agricultural development and the political economy in India (in Chaps. 6 and 7) that we investigate, in Chap. 8, what formal education actually means to the family we are studying. As you will see, the family led us to become disillusioned with the current schemes to increase vocational education and skill certification related to agriculture. Thus, if you are looking for a specific curriculum or program for reformed education regarding cultivation, you may be disappointed. Our effort has been aimed more at understanding, comparing, and analysing how people learn in the field and in formal education, rather than at designing new curricula, methods, or materials. As our study progressed, we became disillusioned from our initial aim of making school education more relevant to children from cultivating families by including learning by doing cultivation in the curricula. The book documents how and why our aims changed. Nevertheless, based on our findings, we do give some suggestions for new directions in formal school-level education in Chap. 9.

Chapter 6 will be particularly relevant to those of you who are interested in the history of rice and the history of paddy cultivation in India. Rice is the main food of the majority of people in South Asia, and the importance of understanding how its cultivation has been developed is obvious. A historical perspective is needed in order to understand the meaning of 'traditional' paddy cultivation and to understand how and why the family cultivates as they do. We try to explain the social, political, and economic significance of the current agricultural crisis and why it is occurring. This requires an overview of the history of agricultural development in India, and the failure to maintain or devise a system of sustainable agriculture, which is becoming all the more alarming as the environmental crisis develops. We relate these problems to the appalling increase in hunger and malnutrition, and try to understand how and why people who cultivate food for others do not have enough for themselves.

If you are a student of political economy and social sciences, you may be particularly interested in our historical dialectical materialist framework and methodology. We have summarised our framework, and how it developed from our observations and analysis, in Chap. 7. We have used this framework to provide concrete evidence for theories of economic agricultural development.

Based on our observations of the family doing cultivation, we have developed a definition of science as a process of using both the hands and mind to do historical dialectical materialist science (Sect. 7.2.7). If you are particularly interested in the philosophy of science, you may be curious about this and other discussions of the nature of science throughout the book. In Chap. 3, we have discussed how the family engages in several aspects of a dialectical process of doing science as they work in the field. These aspects are summarised in Chap. 7 (Sect. 7.2.1). This leads us to a definition of science in Sect. 7.2.1.9. We compare the sorts of science that we observed them doing with how different members of the family defined science (Sect. 7.2.2). We discuss questions of truth and objectivity in Sect. 7.2.3. Based on what the family did and said regarding religion and ritual (Sect. 3.3), we discuss our conclusions regarding the relation between practising rituals and doing science in the process of paddy cultivation (Sect. 7.2.1.2). We discuss how the science the family was doing compares to capitalist science (Sect. 7.2.6). Finally, this leads us to define historical dialectical materialist science (Sect. 7.2.7). We explain how and why our observations, learning, and work on the farm led us to see science as a process of doing and working, even as many educationists see science as a 'body of knowledge' (Sect. 7.3). This is one of the main differences we have with researchers who study 'traditional knowledge' and advocate for multicultural science education.

Mumbai, Maharashtra, India   Karen Haydock

Mumbai, Maharashtra, India   Abhijit Sambhaji Bansode

Mumbai, Maharashtra, India   Gurinder Singh

Mumbai, Maharashtra, India   Kalpana Sangale

References

Banerjee, Arindam, and C.P. Chandrasekhar, eds. 2018. *Dispossession Deprivation and Development: Essays for Utsa Patnaik.* First edition. New Delhi: Tulika Books.

Chakravarti, Anand. 2018. *Is This Azaadi? Everyday Lives of Dalit Agricultural Labourers in a Bihar Village.* First edition. New Delhi: Tulika Books.

D'Mello, Bernard. 2018. *India After Naxalbari.* New Delhi: Aakar Books.

Jodhka, Surinder S. 2002. Nation and village: Images of rural India in Gandhi, Nehru and Ambedkar. *Economic and Political Weekly* 37(32): 3343–3352.

Padhi, Ranjana, and Nigamananda Sadangi. 2020. *Resisting Dispossession: The Odisha Story.* New Delhi: Aakar Books.

Swaminathan, Madhura, Shruti Nagbhushan, and V.K. Ramachandran, eds. 2020. *Women and Work in Rural India.* Chennai: Tulika Book.

Acknowledgements

We would like to thank everyone who was involved in this research project, as well as in other projects from which it sprang. We are especially indebted to the work initiated by Rosemary Varkey on farmers' understandings of their own knowledge in her village in Arpookara in central Kerala. She investigated farmers' understandings of science and the differences and similarities between modern science and their own traditional knowledge and processes of learning about agriculture. Rosemary conducted semi-structured interviews and completed case studies of five farmers, including her own father. Through interviews, she explored whether the farmers said they did science, what their conceptions of science were, how they obtained traditional knowledge, and how they evaluated the worthiness of scientific and traditional knowledge. Her study is what got us interested in working with a few farmers in Madhya Pradesh and Maharashtra, where the initial work was done by Subeer Kangsabanik, Kranti Patil, and Nikhil Narkar.

We had many useful discussions with Rossi D'Souza, Jeenath Rahaman, Jaikishan Advani, and MC Arunan. Rafikh Shaikh and Shweta Naik helped in translations and analysis.

Note that all photographs and drawings are by the authors, unless otherwise noted.

Parts of Chap. 7 are adapted from our previously published work (Singh, G., Shaikh, R., Haydock, K. 2019). The discussion on the 'Balance of Nature' in Chap. 6 is adapted from Haydock and Srivastava (2019).

This project would not have been possible without the support of many other members, staff, and contract workers of the Homi Bhabha Centre for Science Education, especially the Director, Jayashree Ramadas. We also acknowledge the support provided by Nandini Manjrekar and Disha Nawani at the School of Education, Tata Institute of Social Sciences, where Karen was provided support through a Senior Fellowship from the Indian Council of Social Science Research for 2017–2019.

The main people who were also involved in this project, in addition to the family in Rudravali and the authors, were Ankita Sawant and Swapnaja Patil. Our work would not have been possible without the difficult labour by the people who drove us to and from the village, especially Dhananjay. We also thank the family of

Kalpana, who took care of us in Rudravali and fed us very special 'traditional' food. Karen acknowledges the important support from Shana, who asks, and forces her to ask, some of the most uncomfortable questions.

References

Singh, Gurinder, Rafikh Shaikh, and Karen Haydock. 2019. Understanding student questioning. *Cultural Studies of Science Education* 14(3): 643–79.
Haydock, Karen, and Himanshu Srivastava. 2019. Environmental philosophies underlying the teaching of environmental education: A case study in India. *Environmental Education Research* 25(7): 1038–65.

About the Book

This book describes a participatory case study of a small family farm in Maharashtra, India. It is a dialectical study of cultivating cultivation: how paddy cultivation is learnt and taught, and why it is the way it is. The paddy cultivation that the family is doing at first appears to be 'traditional'. But by observing and working along with the family, the authors have found that they are engaging in a dynamic process in which they are confronted by conflicts, which lead them to engage in a process of questioning, investigating, and learning by doing the work. The authors compare this to the process of doing science, and to the sort of learning that occurs in formal education. Thus, the superficial appearances of 'traditional' cultivation and formal education mask dialectical contradictions which are revealed upon closer observation. In order to analyse these contradictions, we relate them to the historical development of paddy cultivation and agricultural education in India. The book presents evidence that paddy cultivation has always been varying and evolving through chance and necessity, experimentation, and economic contingencies. Through the example of one farm, the book provides a critique of current attempts to sustain agriculture and an understanding of the ongoing agricultural crisis.

Contents

About the Authors

Karen Haydock (PhD, Biophysics) recently retired from the faculty of the Homi Bhabha Centre for Science Education (Tata Institute of Fundamental Research), Mumbai, where the research described in this book began. She then continued the research as a Senior Fellow of the Indian Council of Social Science Research at the Tata Institute of Social Sciences, Mumbai. Her research focuses on understanding school-level learning/teaching of natural and social sciences and art (particularly evolution, agriculture, and environmental science) in a few places in urban and rural India. She has also been developing teaching materials and methods and teaching both children and teachers. She is also an artist and has illustrated and written numerous books for children.

Abhijit Sambhaji Bansode is a PhD student in Development Studies at the Tata Institute of Social Sciences, Mumbai. His PhD research is on the Nutritional Status of Tribal Children in the Palghar District of Maharashtra. He completed his MPhil in Development Studies and Masters in Social Work in Dalit and Tribal Studies and Action at the same institute. He obtained his BA in Political Science (First Class) from the University of Mumbai.

Gurinder Singh recently completed his PhD in Science Education at the Homi Bhabha Centre for Science Education (TIFR), Mumbai. His thesis is on understanding the process and dynamics of student questioning and its role in doing science. He is interested in studying student talk/discussion/argument and investigations initiated by students' authentic questions (questions for which they do not know the answers). He also explores how such an understanding can help in constructing meaningful contexts for classrooms in which students' questions acquire a central role. Before joining the PhD program, he taught Physics for about 8 years at secondary and senior secondary schools in Ludhiana, Punjab (which is his hometown), and Mathematics briefly in the UK. He has a Bachelor of Education and Master of Science in Physics. He has recently been working with Navnirmiti, a Mumbai-based NGO, for projects on teaching of science and mathematics to middle-school students of government schools. In these projects, he is mainly involved in teacher professional development and the development of science modules.

Kalpana Sangale was a Project Scientific Officer at the Homi Bhabha Centre for Science Education (TIFR), Mumbai, when the work described in this book was initiated. She has a BSc in Agriculture and an MSc in Horticulture from Dr. Balasaheb Sawant Konkan Krishi Vidyapeeth, Dapoli, Ratnagiri. She grew up in Rudravali, the same village where the farming family that is the focus of this book resides. She now works in a bank in Mumbai. Her family still resides in Rudravali.

List of Figures

List of Tables

Chapter 1
Introduction: Who Are We, and What Are We Doing?

How can it be that people who work every day growing food do not have enough food to eat?
Where does wealth come from, and why are people who do work in agriculture poor?
How much can a few rich people eat?

These are some simple-minded questions that bother even small children. But some say that these are not the sorts of questions that can be easily answered and they are not to be asked in schools. And they usually are not asked in schools, even if the students reach 16 or 18 years old. In fact, there is surprisingly little study about agriculture anywhere in the Indian school curriculum except for the memorisation of some word definitions and a few 'facts' about where various foods and crops come from. Some of the recent textbooks for very young children do have chapters on farming, but most of them do little more than tell a very simplified happy story of someone producing a nice crop. Hunger is not mentioned. Whether the farmers got the true value of what they produced is not mentioned. Who really got rich from the crop is not mentioned. If students continue in school after Class VIII they may study more about agriculture, but it is usually separated into biology, chemistry, politics, geography, history, and economics, making it difficult to see the big picture or ask the sorts of basic questions we began with. Apparently, these questions are both too basic and too complex.

Thinking that children need to ask these sorts of questions is what first motivated us to do some work related to learning and teaching about agriculture. We are teachers and researchers in India who study, design, write, and illustrate methods, textbooks, and materials for teaching children. We thought that schools should be more concerned with agriculture. This is especially because most people in the country are still involved at least part of the time in doing work in agriculture, and because there is a huge agricultural crisis, combined with the crises in environment and healthcare. When we look around, we see that following along the present path of development under our present political and economic system is not alleviating these crises – they are actually becoming more serious. Are we facing a future in

© The Author(s), under exclusive license to Springer Nature
Switzerland AG 2021
K. Haydock et al., *Learning and Sustaining Agricultural Practices*,
International Explorations in Outdoor and Environmental Education 7,
https://doi.org/10.1007/978-3-030-64065-1_1

which these crises will lead to even less food, water, and basic necessities for the majority of people?[1]

Some people say that the widespread hunger and poverty in India is due to the centuries of stagnation and backwardness that characterise 'traditional' agriculture in the country. They say that the solution lies in the application of modern science, formal education, and in the industrialisation of agriculture, as was begun by the 'Green Revolution' in the 1960s.

Others say that the present agricultural and environmental problems are due to modern science and technology: the growth of agribusiness and development itself. They point to a past way of human life that was in 'balance with nature'. The implication is that we need to return to this balance, and to past, pre-industrial agricultural methods. Here also, formal education could play a role.

We are questioning both of these points of view. We see a need to investigate what 'traditional' cultivation really was and is, and whether or how it actually was or is traditional. In this connection, we need to investigate what is the nature of science, what kind of science (if any) existed or exists in 'traditional' cultivation, and how it compares to modern capitalist science. This will help us figure out a role for science and education in solving the agricultural crisis.

Our stand is to focus "in particular on the needs of the poorest and most vulnerable", which is mentioned in the UN 2030 Agenda for Sustainable Development (United Nations 2015). The 1st goal of this Agenda is to "End poverty in all forms everywhere", and the 2nd goal is to "End hunger, achieve food security and improved nutrition and promote sustainable agriculture." The Agenda does not explain exactly how all this is to be accomplished, but it does mention a commitment to "making fundamental changes in the way that our societies produce and consume goods and services." The 4th goal is to ensure "inclusive and equitable quality education and promote lifelong learning opportunities for all". In this book, we will investigate questions related to all these goals.

What can we do about the agricultural crisis? We believe it is necessary to follow BR Ambedkar's advice; educate, agitate, and organise. But what kind of education do we need – and for whom? Ambedkar was speaking for those who are most oppressed. We agree that, as Richard Shaull wrote in the forward to The Pedagogy of the Oppressed, by Paulo Freire (1970, p. 15), meaningful education must become the practice of freedom – "the means by which people deal critically and creatively with reality and discover how to purposively participate in the transformation of their world". Through education, children need to ask basic questions in these areas and figure out how to work together to find answers, demand change, and implement solutions.

---

[1] After we had finished this book, the corona pandemic broke out and this is exactly what happened. Corona has highlighted the crises that already existed, and made it even more obvious that we cannot continue living in a world like this.

1.1 Our Aims in Investigating the Cultivation of Cultivation

We wanted to figure out how we can address the topic of agriculture in school education, mainly at the level of primary and middle school (for children up to about 12 or 14 years old), because most people in India do not have access to formal education beyond this. But from the beginning, we were also interested in working in informal education related to agriculture for people of all ages.

This study evolved from our initial desire to improve formal school science education by making it more relevant to rural children growing up on small family farms (given that the majority of children in India are growing up on small family farms). We thought that these children needed to be taught more about how to do agriculture. Hearing about the problems of modern agriculture, we wanted to develop activity-based teaching methods so that school children could be taught how to do sustainable 'organic farming' that would be 'greener' and less destructive to the environment. We also thought school should teach them to develop their 'scientific temper', i.e. to use a scientific method to ask questions and search for answers and solve problems throughout their everyday lives. However, our findings caused this entire foundation for our work to be transformed during the process of our study. As we shall see, our suggestions for the objectives of formal education related to agriculture are now very different.

However, we realised that in order to do any work related to learning and teaching agriculture, we ourselves needed to learn more about agriculture. We had resources and privileges that might enable us to write research papers, scholarly books, and reports, and help develop textbooks and teaching methods. But, except for Kalpana, we did not know much about paddy cultivation. The rest of us doubted whether we would be able to understand much about how to cultivate paddy just by discussing, reading about it, and talking to experts. We decided to go to people who actually do the work in the field in order to find out how they do it. By observing them at work, we thought we could learn how paddy is cultivated, and we could also learn how they learned and taught cultivation to each other. We also thought that if they need help in learning and teaching, we need to find out from them what help they need, and what their problems are. Therefore, we decided to ask them to teach us how to cultivate paddy, and to work along with them in the field, in order to find out how they teach us and how we learn. Thus, we would be studying ourselves as well as the cultivators.

Rather than trying to conduct a survey or trying to understand cultivation on a larger scale, we decided to work with a few farms in order to investigate these questions in depth and in detail. We began by contacting a few people growing various food crops in villages near Mumbai. Eventually, we focussed on paddy cultivation in just one small family farm. We decided to focus on paddy because it is the major crop in our area, and rice is the main food. Agriculture in Maharashtra is very diverse, and much has been said about how different farms have been undergoing various kinds of transformation over the past several decades. The present study is focussed on one family and one set of researchers, and it is not suggested to be typical or representative of paddy cultivation in general. Rather, we would like it to be an encouragement to other farmers to communicate their own stories, problems, and needs, and to analyse and compare their experiences.

As soon as we ask how paddy cultivation is learned and taught, we are also confronted by questions regarding why it is learned and taught the way it is. Due to our background as science educationists, we also wonder how the learning/teaching of paddy cultivation is similar and/or different from the sorts of learning and teaching of science that we have previously encountered. We have learned, taught, and done research in basic sciences, and we wondered if people learn to do cultivation in similar ways, or if they use methods that are at all similar to the methods we use to do science research.

The broader questions underlying our research are how learning and teaching paddy cultivation is related to change in paddy cultivation, and how paddy cultivation has been changing. If it has not been changing in particular ways, then why not? And if it has been changing, then how and why?

These questions are related to the claims of some researchers that 'traditional' cultivation is different from modern, 'scientific' cultivation – not only with regard to the technology that is used, but also with regard to the sort of 'knowledge' cultivators have and how this 'knowledge' is created and passed along from generation to generation. Here we are critically examining such claims, and questioning whether or not the family we are working with does some sort of science. This requires us to examine not just the 'body of knowledge' that is commonly considered to be science, but rather the process of doing science. Note that throughout this book we are using the word 'process' not to mean a formal procedure or series of steps, but in the marxist sense, to mean acting, doing, and changing, whether purposively or not. We are also questioning and critically examining the meaning of 'traditional' in the context of the work on the farm. We are investigating how the family knows what to do, and how they teach, learn, solve problems, and make decisions in the process of one season of paddy cultivation. And we are also investigating ourselves: how we (the researchers) learn and understand paddy cultivation – and science.

For all of these questions, historical, social, political, and economic relations are important. By exploring in detail how paddy cultivation is learned and taught – through our own observations and direct experience as well as the historical research of others – we hope to get some insight and a better understanding, and we also hope to find more questions.

In sum, our aim is to investigate the process of developing paddy cultivation – the process of cultivating cultivation – how it is learnt and taught, and why paddy cultivation is the way it is, in its dynamic interconnection to historical and social relations.

1.2 Who We Are

The researchers contacted and began work in April 2015 with a family in Rudravali Village, Maangao Taluka, Raigad District, Maharashtra, because one of the researchers, Kalpana, is from the same village. It is about 130 km from Mumbai, where the researchers were living.

The main family members (those who live in the house on the farm) are: Prabhavati and her husband Namdev. their son Bhimraj and his wife Smita, and their two sons, Pratik (9), and Pranay (6). The researchers are: Kalpana, Ankita, Swapnaja, and Karen (all female), and Abhijit and Gurinder (both male). Except for Karen, who is about the same age as Prabhavati and Namdev, all the other researchers are in their 20s and 30s. The family members communicate in Marathi, and also in Hindi to some extent. All the researchers communicate in Hindi and English. Kalpana, Ankita, Swapnaja, and Abhijit also communicate in Marathi, and Gurinder in Punjabi.

The family has been living in Rudravali for several generations. Their ancestral home is in the main part of the village, which had an official population of 825 in 2011. For the last 40–50 years the family has been living in a house on the other side of the highway that runs by one side of the village.

All the researchers except Abhijit were members of Homi Bhabha Centre for Science Education in Mumbai when the project began. Kalpana, Ankita, and Swapnaja were Project Scientific Officers. Abhijit and Gurinder were PhD Research Scholars. The project was completed when Karen was a Senior Fellow of The Indian Council of Social Science Research at Tata Institute of Social Sciences, Mumbai, where Abhijit was a student. All the researchers accompanied the family in their work in the fields, learning some aspects of paddy cultivation, conducting interviews, videotaping and photographing. All the researchers participated in the analysis, transcription, and translation of the recordings. The family members also participated in the translation and analysis, and gave feedback when we returned to the farm for follow-up visits in which we did additional filming, photographing, and audiotaping, viewed videotapes, and read or discussed parts of the text of this book.

Kalpana has a BSc in Agriculture and an MSc in Horticulture from Krishi Vidypeth, Dapoli, Ratnagari. She grew up in Rudravali, and her parents and other family members still live there. However, she did not know the family very well, having left the village for her education, employment, and marriage. After one year of working on this project, she took up a government job in a bank.

Ankita has an MSc in Biochemistry, and is from Mumbai. After her year of working on this project, she went abroad to pursue an MSc in Neurosciences.

Swapnaja has an MSc in Medical Technology, and is from Sangli, a city about 280 km south of Rudravali. After leaving Homi Bhabha Centre for Science Education she has been working in a few non-governmental organisations (NGOs) in the area.

During this project, Gurinder was a research scholar pursuing his Ph.D. in Science Education on understanding the process and dynamics of student questioning and investigating. He has a Bachelor of Education and an MSc in Physics, and about 8 years teaching experience at the secondary school level. He has limited experience of village life, although his father taught in a Government school in a village near their home in Ludhiana, Punjab.

Abhijit's PhD research is on child nutrition. He was brought up in Mumbai, where his father has been holding a Government job. He has some past experience of visiting villages in Maharashtra in order to visit relatives as well as to do research. He joined this research project after the initial data had already been collected, and he visited the farm to conduct additional interviews and get feedback on the analysis.

Karen has been living in India since 1985, but was born to artist parents in New York City and grew up in Santa Fe, a small town in the southwestern part of the USA. Both she and her two brothers became scientists. Other than some experience of growing vegetable plots at home, she had very limited knowledge or experience of cultivation when this project began. She has a PhD in Biophysics from USA, and used to do research in computational structural biology. In India she has been working as a school teacher, and as a teacher and researcher in science education. She also has been doing artwork and illustrating, designing, and writing children's books.

Thus, the researchers are coming from very diverse backgrounds. The researchers and the family all tried to work together with each other in a cooperative, democratic manner, in which power relations are minimised. However, there are some inescapable hierarchies between them, based on class, caste, creed, gender, ability, age, race, education, language, and many other differences. Our work together was often troubled by these power imbalances, and we often discussed the problems and tried to minimise them. At different times, we were all getting very different amounts of money for the work we did for this study. Was it just a coincidence that the researchers of lowest caste got less (or no) payment? In hindsight we see that we should have tried harder. Karen, was the leader, and that could hardly be denied. The researchers were afraid that they might be interfering too much and making too many demands on the family. In writing this book, we have tried not to deny the conflicts that arose due to these power differences. But the imbalance is particularly evident in the book because Karen is the one who has done most of the writing. She has been particularly troubled by the possibility that she has put words into someone else's mouth. That is why she tried to include as many direct quotations as possible. At least this way, readers can find evidence to disagree with the written interpretations. The work in the field as well as the interpretation and analysis was shared more evenly, although Karen had only minor roles in the transcription and translation. The analysis involved continual referral back to the original tapes as new questions and understandings arose that required more observation and analysis of the recordings. Analysis was done by the family as well as the researchers, with much reconsideration, revision of understandings, changing of opinions, learning, and requestioning of ourselves and each other. Many questions were never resolved. Differences of opinion remained, and will be mentioned in the text.

All the researchers made conscious efforts to avoid treating the farming family in a patronising or demeaning way, which occasionally happened despite our efforts.

However, the family members were not only very open and friendly, but were also very assertive, dignified, and critical, not allowing others to mistreat them. All of us expressed both agreements and disagreements with each other, in friendly discussions, as is shown in the text.

Except for Kalpana, the rest of the researchers were not knowledgeable or experienced in paddy cultivation at the beginning of the project. Since we wanted to find out how people learn and teach cultivation in the field, we avoided engaging in library research on the topic before or during our initial visits to the farm. The researchers wanted to be taught by the family, so we could find out something about how they teach us as well as how they teach each other.

1.3   The Dialectical Organisation of the Book

This book is organised in a dialectical format that to some extent follows the chronology of our investigation.

We begin, in Chap. 2, with the surface appearances of physical reality: what we immediately observe on the farm.

Upon further investigation, we see dialectical conflicts in what we observe as well as in how we interpret what we observe, as discussed in Chap. 3. In Chap. 4, we discuss dialectical conflicts in school education related to agriculture.

In order to further understand learning, teaching and doing paddy cultivation in this small family, we need to relate it to the historical development of education (Chap. 5) and to the historical development of paddy cultivation (Chap. 6). This allows us to get a better understanding of why education is the way it is, why paddy is the way it is, and what is 'traditional' and what is 'science'.

Further analysis and relation to historical development brings us to a deeper understanding of what is happening on the farm, in Chap. 7, and in school, in Chap. 8.

We conclude, in Chap. 9, with a summary and discussion of suggestions for what is to be done.

1.4   The Transcription Notation

Transcription and translation of the dialogues is not straight-forward, partly because the dialogues were colloquial and very informal. We have tried to present the dialogues in a way that makes them easy to read, and gives some hint as to the style of speaking rather than using any more technical notation system.

The time of each utterance is included in order to indicate the pace of the dialogue and overlapping utterances by more than one person. A hyphen (-) at the end of a word or sentence indicates that a speaker seemed to cut off what they were saying (sometimes by another person interrupting). This happened frequently. A hyphen at the beginning of an utterance indicates that they are continuing a previous sentence.

We use commas (,) to indicate short pauses, full stops (.) to indicate longer pauses, question marks (?) for explicit questions (often indicated by intonation), and exclamation marks (!) for emphasis. Inaudible parts are indicated by three dots in square brackets, and deletions by three dots without brackets.

We give all utterances in the language in which they were originally spoken, followed by translations into English in parentheses (). Translations are fairly literal, with explanations added if necessary. In square brackets [] we describe actions, expressions, and gestures.

Reference

Freire, Paulo. 1970. *Pedagogy of the Oppressed.* New York: Seabury Press.

Chapter 2
The Superficial Appearances

2.1 Traditional Cultivation in the Past

If by 'traditional', we simply mean a practice that has occurred relatively unchanged over a long period of time - perhaps hundreds of years, then we can find evidence for whether or not paddy cultivation has remained unchanged or traditional by examining artworks, artefacts, and texts from previous centuries. If we see similar practices over many years, we can consider the practices to be traditional, according to this definition.

2.1.1 Artworks Depicting Cultivation

It is difficult to find artworks that depict paddy cultivation. It is easier to find paintings of rulers, armies, tigers, and elephants than paintings of farmers and oxen pulling ploughs. One reason for this is that paintings are often made to record History, which has usually been defined by the rich and powerful to be about the rich and powerful - and paddy cultivators are not rich or powerful. Artwork often supports and glorifies those in power. Another reason is that paintings were often made for upper castes and upper classes who had no interest in showcasing or acknowledging the physical labour done by ordinary people. Also, artwork is often made for religious purposes, which may not be directly related to everyday labour.

Art is not just done for the sake of art. It is not a luxury - it is a necessity, and it has existed in all human societies (Ernst Fischer 1971). Art may be aimed at imagining and creating a better life, as well as escaping, avoiding, or counteracting reality - in order to provide some solace and hope. It also functions as a memorial. Therefore, everyday life - especially tedious everyday labour - may not be the intended subject of art (or of history), even though it forms its basis. But whatever its original

K. Haydock et al., *Learning and Sustaining Agricultural Practices*, International Explorations in Outdoor and Environmental Education 7, https://doi.org/10.1007/978-3-030-64065-1_2

purpose, and whatever the intentions of the artists, we can use art to learn about how cultivation was done in the past.

Some of the earliest surviving art in India consists of paintings on rocks and cave walls, some dating from as early as the 2nd millennium BCE. We do not know why it was painted. We do not find many scenes from farming depicted in rock art - they more likely show scenes of people encountering or killing large animals, fighting, or dancing. Nevertheless, there is some rock art in India showing cultivation. For example, in Fig. 2.1 we see depictions of a person ploughing with a pair of bullocks.

Terracotta models of various designs of bullock carts have been found in the ruins at Harappan sites, dated from 3500–1900 BCE (Fig. 2.2). Most researchers think that they were made as toys, but they also provide records of what the actual bullock carts must have looked like. It is difficult to trace a progression in design of the carts, since a variety of designs have been found at one site. Even today, a variety of designs of bullock carts are being used for specific functions:

> The long continuity in cart designs of the Indus valley and the fact that many different types of bullock cart continue to be used even today in Pakistan and India indicate that the original styles of cart were quite effective and that the early designers were able to produce a form that came to be improved upon only with the introduction of ball-bearing axels and rubber tires. (Jonathan Kenoyer 2004)

The similarity to carts used nowadays (for example, see Fig. 2.3), also suggests that we may use the term 'traditional' to describe the general design of these bullock carts. However, there is a fair amount of variation in the design of the carts in use at any one time, and perhaps even at any one place.

Figure 2.4 shows a stone carving at the Buddhist site at Sanchi, in central India, dating from the first century CE. Besides the domesticated animals, it shows people

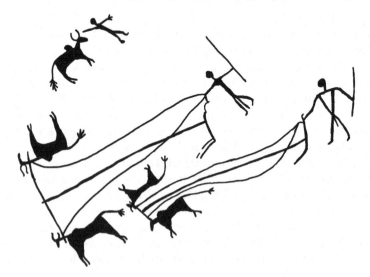

Fig. 2.1 Rock painting at Chaturbhujnath Nala from the Chalcolithic period. (From Erwin Neumayer 2013)

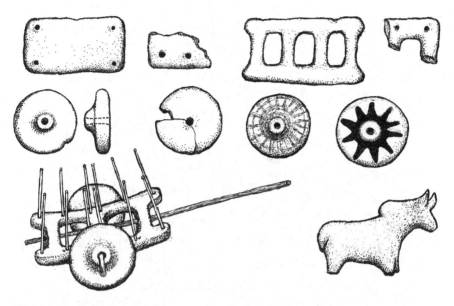

Fig. 2.2 Bullock cart models found in early Harrapan sites. (Redrawn from Jonathan Kenoyer 2004)

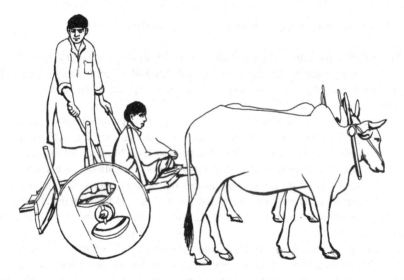

Fig. 2.3 Modern bullock cart in Sindh, Pakistan. (Drawn from a photograph found on http://www.travelmypakistan.com/portfolio-item/shahdadkot)

winnowing, grinding, and pounding grain (presumably rice). These stone carvings are illustrations from the Jataka tales, which sometimes mention the arduous drudgery and abject poverty of women and men of the lower classes doing agricultural labour (Uma Chakravarti 2012).

Fig. 2.4 A stone carving at Sanchi, showing people sifting and processing grain

We find a few paintings on paper that show paddy cultivation from as early as the sixteenth century, such as that shown in Fig. 2.5, from Malabar. In this picture, sickles are being used by the harvesters, wearing what appear to be reed coverings, perhaps as protection from the sun.

Paddy cultivation is occasionally shown in the Company school of miniature painting (Figs. 2.6 and 2.7). The British had commissioned local artists to paint such scenes from everyday life using styles influenced by both Mughal and European art. The British used these paintings to categorise various occupations, castes, and races as romanticised, exotic spectacles, and as records of flora, fauna, and 'native' culture and technology, which were useful for colonial rule and British development.

Figure 2.8 shows a pull-down chart on paddy cultivation that was printed in twentieth century Berlin. Such charts were used in German schools to teach children about the geography of foreign lands. Although it is entitled "Reisernte in Hinterindien" (Rice Cultivation in India), it seems that the artist had not been to India, and may have been painting from memory of paintings or photographs. The original pictures may have actually depicted scenes from various seasons in China or some other country. In one picture we see seven or eight different stages of paddy cultivation - fanciful, but an efficient use of wall space (reducing the cost of manufacturing multiple charts).

During the struggle for independence from British rule, and continuing into the struggle to create a more just and free India, some professional artists made artwork depicting ordinary people, including people working on farms (Fig. 2.9). These

Fig. 2.5 Harvesting in sixteenth century Malabar. (Redrawn from Fernand Braudel 1981, p. 107)

Fig. 2.6 A farmer using a pair of bullocks to plough a field, painted around 1860, near Delhi (A villager ploughing with two bullocks; a hilly landscape in the background, artist unknown, Museum number: 08111(IS), Victoria and Albert Museum, London)

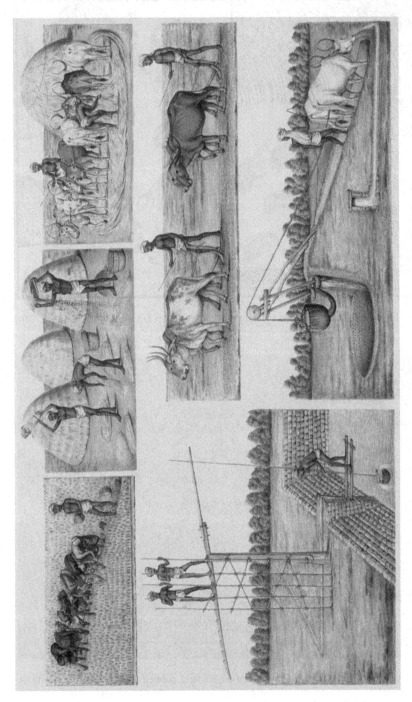

Fig. 2.7 Scenes of paddy cultivation from Tanjore, around 1830 (Six scenes of cultivation and irrigation, Museum number: IS.39:23–1987; Victoria and Albert Museum, London)

Fig. 2.8 Rice Cultivation in India: Images of foreign culture and crops. (Printed by Verlag Dr. te Neues, Berlin)

Fig. 2.9 Threshing, KK Hebbar, 1944

were made as historical records, as a way of educating urban people about the work and the problems of people in villages, as protest against the trend of making art that does not depict everyday life, and as a form of protest against the injustice to people who were engaged in cultivation. Chittaprosad even made ugly pictures of hungry people, in protest against the Bengal Famine (Fig. 2.10).

In some villages, people make paintings on the walls of their homes. Occasionally they may show scenes of paddy cultivation, such as this example from people of the Saura tribe in Odisha (Fig. 2.11).

Recently, such 'folk art' has been commodified and made on paper for sale. We also find some examples of commercial art by unknown artists depicting paddy cultivation that serves to romanticise or glorify village life, which multinational companies are selling on the internet. The example shown in Fig. 2.12 was being sold using the epithet: "Olden (Eco Friendly) & Best Ways". People in cities may be hanging these on their walls as a nostalgic reaction to the alienation that accompanies industrial development. And school children are sometimes asked to draw copies of such pictures in their art classes, even if they have never been to a paddy farm.

All of the above artworks appear to be depicting 'traditional' paddy cultivation. We will later refer back to them, comparing them to each other and to what we observe on the actual farm, in order to analyse the meaning of 'traditional' paddy cultivation and how it might be passed on from generation to generation.

2.1.2 Historical Accounts of Paddy Cultivation

Paddy cultivation was described in the Krishisukti of Kashyapa, probably written before 900 CE (as discussed and quoted by MS Randhawa (1980)). Kashyapa describes the entire process of paddy cultivation. We quote at length, as this will

Fig. 2.10 Copy (by Karen) of parts of a drawing by Chittaprosad of what he saw during the Bengal Famine of 1944

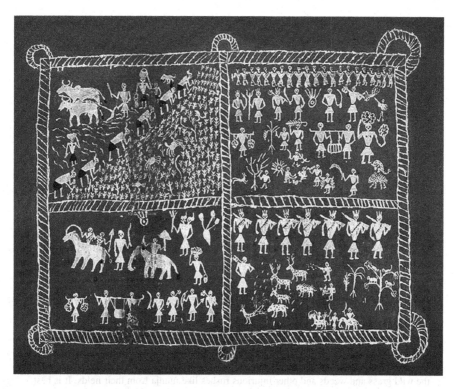

Fig. 2.11 A wall painting from Saura, Odisha (recent) http://www.bradshawfoundation.com/india/central_india/ethnology_india_rock_art.php

Fig. 2.12 A painting of a farm, being sold on the internet. https://www.clipart.email/download/6833248.html

also give an overview to readers who are not familiar with the process of paddy cultivation:

2.1.2.1   Krishisukti of Kashyapa

[**Irrigation:**] … wise husbandmen who have decided upon reaping a harvest of different kinds of rice like Kalama and Vrihi [also white Sali and red Sali are mentioned] should first irrigate their fields well by means of channels drawn from reservoirs, rivers or lakes and then till them with the help of oxen.

[**Transplanting seedlings from the nursery:**] Then the cultivator should order the uprooting of the wisps of paddy seedlings which have already been grown closely in a separate nursery, and then, tying together each wisp, he should, on an auspicious hour, have them transplanted by servants in rows evenly in the paddy field in which the clay has become soft by tilling and has been carefully dressed with the dung of cows or goats or with decayed vegetable matter. …

[**Growth of paddy:**] In this manner, the paddy seedlings transplanted in rows in a well-ploughed paddy field full of water, in a village, town, forest or woodland, irrigated by a canal and provided with several outlets for draining away the surplus water, are made to take root by the Creator (i.e. Nature) after the lapse of seven or ten days and then the new shoots sprout forth and make their blessed appearance. Then the land shines forth with that fascinating bloom which is found on the plumage of parrots or on the body of a damsel in the prime of her youth. …

There is an all-round growth of shoots and increase of splendour on the plants which feed on water daily and hold out a promise of rich harvest.

[**Weeding:**] Therefore the cultivators should systematically and assiduously weed out the wild grass and weeds and other injurious rushes like munja from their fields. It is best to destroy the wild grasses, rushes, weeds, etc., which affect the growth of grain and reduce the yield of crops, from their very roots. An experienced person should first fill the paddy fields with water and then gradually root out the weeds, etc., row by row. Or it is best to have the pest removed daily by the servants. When the weeds have been eliminated by the cultivators their fields shine forth. lustrous and luxuriant. …

[**Ripening:**] When the ears at the top of the rows of plants gradually grow solid and full of milky-juice (or sap) and appear to be somewhat bent, then they should be protected, especially from the parrots. …

Then by and by the juicy grains in the ear would become harder at the core and would· finally ripen into the rice. Till then regular irrigation of the fields is advisable and beneficial; otherwise there would be the loss of crop.

Therefore, the cultivators should continue to observe the development of sap frequently, and should regularly water their fields at the proper times for the sake of increasing the sap.

[**Protection against pests:**] It is extremely beneficial if the crop is protected from rats, locusts, parrots, and other pests.

The ripe paddy, which is so beneficial to the living beings, should be honoured by those who live on it, with circumambulations, and then preserved by them after having determined the period of life of the grain from its appearance, the stamina of its seed and the quality of the soil from which it is produced or from actual experience. …

[**Harvesting:**] When the ears of the paddy have gradually ripened according to their respective duration of time, then water should not be let into the fields. This is the advice of the agricultural experts and should be followed in the case of all paddy fields.

When the ears at top of the plants become ripe then their stalks bend their head very low to the ground. Seeing them bend so, the husbandman should himself or through his servants protect them in the field for a period of twenty days. Now, when the stalks have become ripe and assumed a golden hue, the cultivators should then reap them with sickles etc. They may have the harvest mown in one day or in several days successively with the help of their dependants and servants working in co-operation.

But reaping would be useful only if care is taken that the stalks are not damaged during the operations nor the harvest spoiled by rain or carried away by thieves.

[**Threshing**:] The cultivators should have the reaped plants with their ears stacked on threshing floors with the help of their servants and attendants. It is advisable to keep the harvest lying on the threshing-floors for three to five days. By threshing the ripe stalks from which the ears become loosened, the grains drop on the threshing-floor and sparkle like heaps of bright pearls. On the second round of threshing done by having the crop trodden over by bullocks and buffaloes; the remaining quantity of the grain also drops on the floor. After having the paddy threshed in this way on the threshing-floor, the cultivators should gather the assorted stuff that is thick and substantial.

[**Winnowing**:] The cultivators should carefully sift the superior grain from the inferior stuff by means of the winnowing-fans, etc., and gather the former into a heap. A wiseman should then dry them in the sun and get them cleaned of impurities.

[**Distribution**:] He should then, having apportioned one share to the gods, one to the king, one as a gift to a Brahmana learned in the Vedas, and one for the maintenance of the servants, keep the remaining stock of paddy in his own house.

[**Storage**:] He should carefully preserve the grain in suitable receptacles like kathinya or in well-baked clay pots or in vessels of strong glass, or in containers woven of ropes and plastered with mud, according to the custom of the locality. At some places, the farmer should dig a pit in the hard earth, provide it with descending steps and store the paddy into it, taking care that it is safe from the hazard of damp, thieves, parrots, rats and other noxious animals. (Krishisukti of Kashyapa, quoted by MS Randhawa 1980)

Note that this book was obviously not written for labourers themselves to read - they would have been illiterate. It was written for, and from the point of view of, those (zamindars) who were in charge of directing work by servants or slaves. The text does not just describe the 'objective facts' of cultivation - it includes many social aspects: religion, divisions of labour according to class, caste, and gender hierarchies, social problems such as thievery, and values of good and bad, caring, and beauty.

We will refer back to this text for comparison with what we observe in the process of paddy cultivation in Rudravali.

We will also refer to what Jyotirao Phule wrote in his book, Shetkaryaca Asud, written in 1881, focussing particularly on farming in Maharashtra (Phule 1881). Here again, social aspects are described, but this time from the point of view of the oppressed castes and classes, and from the point of view of women. We quote from his detailed observations, as translated by Gail Omvedt and Bharat Patankar:

2.1.2.2 Shetkaryaca Asud (Jyotirao Phule 1881)

It has become openly known that in these times of peace and plenty, four crores[1] of farmers don't get sufficient food to fill their stomachs even twice a day, and they don't see one day pass without experiencing the affliction of hunger.

---

[1] Four crores = 40 million

... [Farmers] have to make toilsome efforts even to meet the expenses for their cultivation. Then they go to Marwaris and take loans for meeting the land revenue charges. (Phule 1881)

It is significant that Phule describes cultivation through the work of women:

...the women of the farming households, ... after applying cowdung to their house, go with their husbands behind the drill plough, breaking up the clots of earth, digging up the weeds, putting the seedlings in the pits and stacking a bit of dirt up around them, treading the harvested plants, winnowing the grain, taking up the winnow and giving it to the man, taking on their head pans of heaps of ashes, dung or manure, bundles of grass or other chaff, and labouring as hired labourers for the full day in the hot season when there is less work on their own lands. (ibid.)

He describes the clothing, food, and shelter, this time from the perspective of a few different men, and farmers of different castes:

...the Shudra farmer dresses in a loincloth with a bag for tobacco tied around his waist, a rag of a turban on his head, and goes bare-bodied and barefooted, holding the handle of the plough throughout the blazing heat of the day, ..."

... His nourishment and meal in the afternoon and evening is a dry bhakri of jowar (ज्वारी, sorghum), nandni (another millet) or a dry chappati; and vegetables of carrots or roots; there may be a sauce of dried fish, and if this is not available then there is nothing to eat with the bhakri but chutney. Even chutney and bhakri may not be available on time! Since his home is next to the bull's shed, with three calves and she-buffaloes tied there in the muck, there is a squalid smell of urine on all four sides of the house. A torn cloth or worn-out black blanket is spread for sitting or lying down; there is a pond made filthy by all the buffaloes of the village sitting in the water; beside it a well for drinking water and his toilet; this is his rural mansion. If he should get cholera or a fever he would be extremely blessed to find good medicine and a knowledgeable doctor; besides this the sword of concern about finding money to pay the land revenue and other funds and taxes is hanging over his head. (ibid.)

Phule also describes the houses of slightly better-off farmers:

[One landowner] has a one-floor tiled-roofed house. In front and against the house, a shed has been made in which to tie the bullocks by throwing up a framework of thatch. His two or three decrepit bullocks are sitting chewing their cud. On one side two or three large corn-bins of 20-25 quintal^2 grain capacity have fallen empty. Outside on the veranda to the right is an old, eight-bullock cart. An unravelled woven basket lies fallen on that. On the left side of the house a big, four-sided earthen platform has been made with a tulsi plant on it, and next to that is another platform below which earthen water pots are set. Two or three earthen pots filled with water and a steel pot have been placed on that. Next to the water pots a three-sided frame has been built with small walls and a small bathroom has been made inside by putting tiles cross-wise. The water that flows from it collects in a small pond outside; it is swarming with insects... (ibid.)

Phule goes on to describe living conditions in detail, and also how they compare to the lifestyle of English and Indian (mainly Brahman) government employees: they live in government housing with lamps, furniture, bedding, bathrooms, water, and have mutton and superior quality grains, "and good quality liquor like English port, unadulterated oil, ghee, milk, sugar, tea, salt, chillies, spices, a knife and fork, ..."

---

^{2}A quintal is now 100 kg, but at the time this was written is probably meant an imperial hundredweight (50.8 kg).

2.2   Traditional Cultivation in Rudravali

Paddy cultivation on the farm in Rudravali is done by the extended family, with relatives and other villagers also contributing occasionally. At the height of summer, at the end of April, the nursery plot is prepared by rabbing (collecting and burning organic waste on it), using crude 'home-made' implements (Fig. 2.13).

Gobar is used to fertilise the field, as mentioned by Phule. Ploughing is done in the beginning of June, with a wooden plough and iron share, pulled by a pair of bullocks (Fig. 2.14). Last year's stubble is in the field, which is very dry since the last rain was more than six months ago. Note the similarity to what was shown in paintings made one or two hundred years ago (Figs. 2.6, 2.7, 2.8, 2.11, and 2.12) - and even the rock painting from much earlier (Fig. 2.1). The description and division of labour between men and women is similar to what Phule described.

A wooden plank on a split bamboo pole is attached to the bullocks to level the field (Fig. 2.15).

The rice seeds are broadcast by hand (Fig. 2.16).

Clods of earth are broken and the seeds are covered by striking the clods with sticks (Fig. 2.17).

The seeds sprout only when the rains arrive - hopefully within a few days of sowing. Then the main field is ploughed and levelled (Fig. 2.18). Pratik has been replaced by a large stone. Note the similar design of the cape shown in Fig. 2.5.

At the same time, the seedlings are transplanted onto the main field by hand (Fig. 2.19). Note the similarity to transplantation shown in Figs. 2.7, 2.8, 2.11, and 2.12. The process is just as was described by Kasyapa.

Fig. 2.13 Smita lighting the rab

Fig. 2.14 Namdev ploughing a nursery plot in the centre of the field. Namdev is holding the plough, a rope to each bullock, as well as a stick, the same as is shown in the prehistoric stone art (Fig. 2.1)

Fig. 2.15 Pratik sitting on the leveller as Bhimraj guides the bullocks

Fig. 2.16   Namdev sowing the seeds by broadcasting them across the nursery plot (4 June 2015)

Fig. 2.17   Bhimraj and Smita breaking clods of earth to cover the seeds

Fig. 2.18 Levelling the main field before transplanting

Fig. 2.19 The family transplanting the seedlings (26 June 2015)

Fig. 2.20 Smita weeding the paddy field (2 September 2015)

As the paddy grows, weeding is done by hand (Fig. 2.20). Water is supplied only by rain. Water for drinking and household use is carried on the head from a well a kilometre away (Fig. 2.21).

The crop is reaped by hand with sickles (Fig. 2.22). The sickles seem to be of the same design as those used in the 1600's (Fig. 2.5). And just as in those days, the reaped plants are laid out in the field for further drying. They are tied in bundles with rope which is hand-made from paddy stalks. This does not seem very different from the harvesting that was shown in artwork in the past (Figs. 2.5 and 2.9).

The threshing floor is plastered with mud and gobar and the threshing is done by beating the paddy with branches (Fig. 2.23), among other methods.

A bullock cart is used for transport (Fig. 2.24). Notice the similarities to the bullock carts from the Harappan Civilization (Fig. 2.2).

Winnowing is done with a bamboo winnowing basket (soop), as shown in Fig. 2.25. The design of the soop is the same as was shown in the nineteenth century (Fig. 2.7), and even in the very early stone carvings at Sanchi (Fig. 2.4). After milling, the family finally cooks and eats the rice produced for and by themselves through this difficult process.

Fig. 2.21 Smita
collecting water

Fig. 2.22 Reaping the paddy (16 October 2015)

Fig. 2.23 Grains of rice are separated by beating with branches (4 December 2015)

Fig. 2.24 The family's bullock cart

The beauty of the paddy, the satisfaction of having produced it, and the sustenance it gives when it fills their stomachs is what makes it valuable to the family. It is a physical thing for people with a physical need for food.

At least superficially, this appears to be very similar to what was described by Kashyapa many centuries ago. The main differences are in rabbing, irrigation, and threshing. Kashyapa does not describe any process of rabbing. The family's fields are too far away from the river for irrigation, and they have to depend solely on rain, which is plentiful for the kharif crop. Threshing is not done by bullocks or buffalos (as in Fig. 2.7) - instead an electric or foot-pedalled thresher is used.

Fig. 2.25 Prabhavati sifting rice before cooking

We do not notice any major differences in paddy cultivation between what we observed in 2015 and what Phule observed in 1881. The houses that Phule described are not as clean or well-kept as what we saw, but the basic structure is similar. The major difference is that the house now has electricity. But it does not have any source of water closer than the well 1 km away.

The clothing has changed since the descriptions of Phule and what is shown in the older artwork we showed above. The family now wears some ready-made, manufactured clothing. Especially, the men and boys wear ready-made trousers, shirts, and rubber chappals. But the dhoti and towels are similar. A similar sari is still worn by the women, although now it may be made of terrycot rather than cotton, and may be manufactured rather than produced at home or by local artisans.

We can refer to artworks and historical texts as well as archaeological and earth science studies to see that even the environment of the cultivated area we are studying seems to have remained much the same over the past several hundred years.

Thus, what we see on the farm at least superficially appears to be an example of 'traditional' paddy cultivation that is based on a traditional 'body of knowledge' that is unchanged over hundreds, if not thousands of years. It seems to be a way of life in balance with nature.

> ... you have to do just the same the next year, and the same all over again the year after that. 'Nature only works.' [quoted from Hammett 1950]

2.3 Passing on 'Tradition'

How is cultivation passed on from generation to generation? Initially, we thought the older generation teaches the 'traditional knowledge' and the younger generation learns it.

For example, just before sowing the nursery plot, Namdev instructed Bhimraj on how to do the ploughing. He wanted Bhimraj to do it the way he (Namdev) thought was correct, so he kept directing him, "त्याच्या पायापासनं फिरव" (Turn their [the bullocks') feet around!). Namdev also told Bhimraj not to press the plough hard (too deep) into the soil.

While doing the harvesting in the field, Ankita asked Smita and Bhimraj how they learn:

01:41	Ankita:	हे तुम्ही कसं करायला शकिलात काम? (How did you learn to do this work?)
01:44	Bhimraj:	हे आसती-आसती येतं एकमेकांचं बघून. (it comes slowly, slowly, seeing each other.)
01:45	Smita:	[over Bhimraj]: कसं शकिलात म्हणजे काय? मोठी माणसं घरामध्ये करायचीच, ना? (Learned how means what? At home the elders just used to do it, nah?)
01:51	Ankita:	कोणाच्या? (Whose?)
01:52	Smita:	मोठी माणसं घरामध्ये करतातच, ना? तर आम्ही पण शकिलो त्यांचं पाहून. (At home the elders just do it, nah? Like that, looking at them, even we learned.)
01:54	Ankita:	इकडची की तिकडची? माहेरची? (Over here or there? At the maika [mother's place]?)
01:57	Smita:	तथि माहेरी पण! (Even at the maika!)
01:59	Bhimraj:	महेरी पण हेच! (At the maika!)
02;00	Smita:	माहेरी पण आम्ही हेच करायचो. घरची माणसं - (The same is done at home, the people at home -)

Maybe Bhimraj and Smita are saying that they just pick up cultivation from having grown up in a farming family, which sounds simple. But if it happens very slowly, maybe it is not so simple. Maybe it does not even make sense to use the word 'शिकणे' (to learn) in the context of agriculture, because learning agriculture is so different from the usual meaning of learning that happens in school. In school, students 'learn' the 'body of knowledge' by hearing, reading, copying in writing, remembering, and recalling. How could this sort of 'learning' be done in a paddy field?

Paddy cultivation seems not to change over time or from village to village. Smita and Bhimraj appeared to be talking vaguely, without distinguishing between the methods of the two different families in which they grew up. The ways of knowing how to do paddy cultivation were also mentioned somewhat more specifically in a conversation we had about putting layers for rabbing:

00:23	Bhimraj:	[gesturing towards the still-smouldering rab with the tool in his hand] आणि ते राबासाठी आणि धान-बीज टाकलं ना अपुन, (And, in/from this rabbing, the paddy seeds you put,)
00:25	Kalpana:	हुम्मम्. (Hmm,)
00:25	Bhimraj:	का ते जोसाने एकदम येते वर. (—suddenly burst up [with 'josh': Heat, passion, fervour].)
00:28	Kalpana:	हुम्मम्. (Hmm,)
00:28	Bhimraj:	असं. (Like that.)
00:32	Kalpana:	तुम्हाला कसं कळलं की, म्हणजे, हा एक layer टाकल्यावर दुसरा layer, दुसरा layer- (How do you know that you have to put one layer and then the next layer, the next layer -)
00:36	Bhimraj:	हाँ, (Yeah,)
00:36	Kalpana:	म्हणजे, कश्याच्या नंतर काय टाकायचं हे - कसं काय ठरवलं? (—like what to put after what - how do you decide that?)
00:39	Bhimraj:	मग ते - पाहिल्यापासनंच ते - आ आ - हे. (It - ah - ahhh - came from beginning - then.)
00:41	Kalpana:	अच्छा, (Ok,)
00:42	Bhimraj:	पाहिल्यापासनंच ते आमच्याआमच्या इकडं ... आजोबांच्या ...[inaudible mumbling]... (From the beginning [first, past, oldest times] here in our place ... In our grandfather's ...)
00:45	Kalpana:	अच्छा. (Ok.)
00:46	Bhimraj:	जुनी माणसं पहिल्यापासनं अश्या, अश्या पद्धतीने, लावतेत [...] (From the beginning, in such, in such a way, the ancestors do it [...])

Here again, it seems that traditions are passed along from generation to generation, unchanged.

Their manner of speaking also says something about how Smita and Bhimraj understand the process. Perhaps to those of us who are scientists, Bhimraj's description of the effects of the rab does not sound at all like the way scientists speak or work. We might say the language and mannerisms are not 'pure' or highly refined - it is the direct, rustic language of peasants: both blunt and poetic. We may not expect to hear a scientist using an emotional word like, "जोसाने" ('josh': heat, passion, fervour).

Also, rather than saying 'I' or 'we' put the seeds, Bhimraj uses the impersonal 'you' ('आपण'), which we take as an indication of a traditional tendency to be less individualistic and more concerned with 'collective agency' rather than 'personal agency'.

Neither does this sound like a textbook or like a classroom discussion. Kalpana is not acting like typical teachers in classrooms, who usually ask questions for which they already know the answers. She asks because she wants to find out the answers to her questions. In that sense, her questions are authentic. But neither are any of the people in these conversations acting like typical students when they silently listen to the teacher tell the answers or give a lecture. Kalpana is participating in the traditional 'haankar' style of conversation in which Bhimraj's sentences are punctuated by Kalpana's short interjections of acknowledgement, agreement, or understanding, with the corresponding eye contact back and forth between the two.

The work of cultivation and maintaining the house and farm seems to require the entire family's attention. Talk also centres around the seemingly trivial, unending series of daily tasks: what to do next, bringing tools, carrying things, taking care of the animals, cooking, eating, cleaning, etc.

While working, everyone seems to be concerned with the immediate, everyday, local environment rather than with general matters of the wider world. Everyday learning and teaching within the family seems to be pragmatic - concerned with these practical things, rather than abstracting, theorising, or generalising.

At first glance, it also appears that everyone acts according to a traditional hierarchy of authority in which the father-in-law decides and tells the husband what to do and the husband and father-in-law tell the wife what to do.

For certain tasks there appears to be a traditional division of labour. Ploughing is done by men. Transplanting, harvesting and threshing are done by both men and women. Weeding is done mainly by women. Applying artificial fertiliser is done by men. The father is the one who broadcasts the seeds. The heaviest sacks of grain are carried by men. Most of the cooking and house cleaning is done by women.

We also noticed various rituals associated with cultivation. A few mango leaves are attached to the plough on the first day of ploughing. An onion is in the basket of seeds which are being sown (Fig. 2.16). A coconut is placed on the threshing floor before threshing is begun. There are also images of gods displayed in a small shrine

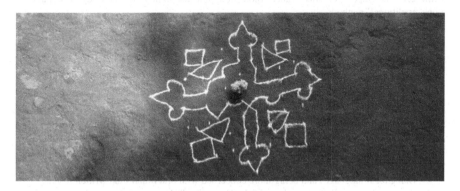

Fig. 2.26 Rangoli made on the ground at the front of the house

at home, there is a sacred tulsi plant in front of the house, a rangoli is sometimes made at the doorstep (Fig. 2.26), and various Hindu festivals are celebrated.

2.4   Summary of Superficial Appearances

In summary, on this farm we superficially see evidence of the characteristics listed in Table 2.1. All of these characteristics may be considered to be indicative of doing 'traditional' farming, and learning and teaching 'traditional knowledge' (also called 'indigenous knowledge'). The table includes references to researchers in various countries who have claimed that these characteristics are traditional and that they differentiate 'traditional knowledge' from 'modern western science'.

Thus, apparently, the family is not doing 'science'. It seems that rather than experimenting and developing new methods of cultivation, they are just passing along the old ways.

But maybe we need to investigate this question more thoroughly, concentrating not just on *what* the methods of cultivation are, but on *how* practices are acquired and passed on. A closer look is required. We also do not see any reason to assume that different 'indigenous' peoples in different communities and different parts of the world have the same kinds of knowledge or the same ways of learning or passing on 'traditional knowledge'.

Researchers in other countries have pointed out that it is not really clear how farmers learn and how they know what to do. We need to question whether or not they use methods of investigation that are different from or similar to the methods that professional western scientists use. For example, Susanne Kummer and others

Table 2.1 The Superficial Appearance. Indications that the family we are studying is passing along 'traditional' methods

(1)	Stating that they learn (or 'know') from tradition - and that the traditions are relatively unchanging and harmonious - in balance with nature (Glen Aikenhead and Masakata Ogawa 2007)
(2)	Using primitive materials and methods (e.g. home-made rather than manufactured tools) (Ray Barnhardt 2007)
(3)	Being vague or approximate rather than precise
(4)	Being pragmatic rather than abstract or theoretical - being specific rather than making generalisations (MN Maddock 1981 and Gloria Snively and John Corsiglia 2001)
(5)	Acting locally and being concerned with the local area (Chimaroke Nwaoha et al. 2015)
(6)	Having values such as respect for elders and authority figures (Ray Barnhardt 2007)
(7)	Using local languages and being more concerned with oral rather than written communication (Gloria Snively and John Corsiglia 2001)
(8)	Being less individualist and more concerned with 'collective agency' (Cassie Quigley 2009)
(9)	Learning by following rituals and engaging in spiritual ceremonies (Glen Aikenhead and Masakata Ogawa 2007);
(10)	Seeming not to have knowledge of modern agricultural methods

have pointed out that "there has been little attempt to study the nature, characteristics, and factors associated with the experimental processes of farmers in a systematic, comprehensive way" (Susanne Kummer et al. 2008, p. 1).

Through closer observation and analysis, we hope to get a better understanding of what is meant by 'tradition' and 'science'. In particular, in subsequent chapters, we will look at their work and the process the family is doing, not just the 'body of knowledge' or results of the process.

References

Aikenhead, Glen S., and Masakata Ogawa. 2007. Indigenous knowledge and science revisited. *Cultural Studies of Science Education* 2 (3): 539–620.

Barnhardt, Ray. 2007. Creating a place for indigenous knowledge in education: The Alaska native knowledge network. In *Place-Based Education in the Global Age: Local Diversity*, ed. Greg Smith and David Gruenewald, 113–153. Hillsdale: Lawrence Erlbaum Associates.

Braudel, Fernand. 1981. *Civilization and Capitalism, 15th–18th Century*. 1st U.S ed. New York: Harper & Row.

Chakravarti, Uma. 2012. Women, men and beasts: The Jataka as popular tradition. In *Everyday Histories: Beyond the Kings and Brahmanas of 'ancient' India*, ed. Everyday Lives, 198–221. New Delhi: Tulika Books.

Fischer, Ernst. 1971. *The Necessity of Art: A Marxist Approach*. Harmondsworth: Penguin Books.

Hammett, Frederick S. 1950. Agricultural and botanic knowledge of ancient India. *Osiris* 9: 211–226.

Kenoyer, Jonathan Mark. 2004. Wheeled vehicles of the Indus Valley civilization of Pakistan and India. In *Wheel and Wagon – Origins of an Innovation*, ed. M. Fansa and S. Burmeister, 87–106. Mainz am Rhein: Verlagg Philipp von Zabern.

Kummer, S., R. Ninio, F. Leitgeb, and C. R. Vogl 2008. How do farmers research and learn? The example of organic farmers' experiments and innovations: A research concept. 16th IFOAM Organic World Congress, Modena, Italy, June 16–20, 2008, 789–792.

Maddock, M.N. 1981. Science education: An anthropological viewpoint. *Studies in Science Education* 8 (1): 1–26.

Neumayer, Erwin. 2013. *Prehistoric Rock Art of India*. New Delhi: Oxford University Press.

Nwaoha, Chimaroke C., A.N. Calista Zugwu, and Angela A. Osele. 2015. Infusion of indigenous knowledge system approach into Nigeria music education pedagogy: Historical review. *African Review of Arts Social Sciences & Education* 5 (1): 156–173.

Phule, Jyotirao. 1881. *Shetkaryaca Asud*. 2017 Marathi edition, Translated from Marathi to English by Gail Omvedt and Bharat Patankar. Pune: Mehta.

Quigley, Cassie. 2009. Globalization and science education: The implications for indigenous knowledge systems. *International Education Studies* 2 (1): 76–88.

Randhawa, M.S. 1980. *A History of Agriculture in India. Volume I Beginning to 12th Century*. New Delhi: Indian Council of Agricultural Research.

Snively, Gloria, and John Corsiglia. 2001. Discovering indigenous science: Implications for science education. *Science Education* 85 (1): 6–34.

Chapter 3
Dialectical Conflicts

In the previous chapter we saw that superficially, in the field, the family seems to be doing, learning, and teaching 'traditional knowledge', and does not appear to be doing 'science'. But let's take a closer look to see what is actually going on.

3.1 Dialectical Conflicts Between the Old and the New

Although the farming techniques seem at first to be quite 'traditional', when we look more closely, we see some contradictions, and a mix of the old and the new.

We might think that irrigation is an indication of modern technology. But quite advanced systems of canals and irrigation were being used since early times in India (see for example, Fig. 2.7). Although it is doubtful whether irrigation was or was not used in the Harappan Civilisation (Eleanor Kingwell-Banham 2019), it was used as early as 1000 BCE in southern parts of the Indian subcontinent and in the period from 500 BCE to 500 CE there is ample evidence of wells, runoff-fed reservoirs, canals, and masonry-reinforced earthen dams further north (Kathleen Morrison 2014). Irrigation continued to develop until the British arrived (Nirmal Sengupta 2018). In 1892, a British engineer wrote, '… in length, cross-sectional dimensions, discharging capacity, number and aggregate mileage, the Indian canals are the greatest in the world….'' (D. Kumar et al. 1991, 677).

The family's fields in Rudravali (Fig. 3.1) are not irrigated, and therefore they are used for paddy cultivation only during the monsoon. However, they also cultivate paddy as the rabi crop on a field on the other side of the village that belongs to the extended family. This field is irrigated by the river and the Left Canal from Dolwahal Dam, constructed in 1969, which is too far away to be used for their own fields. The house has no plumbing, and the nearest well is about 1 km away, across the highway at the other side of the village.

© The Author(s), under exclusive license to Springer Nature Switzerland AG 2021
K. Haydock et al., *Learning and Sustaining Agricultural Practices*, International Explorations in Outdoor and Environmental Education 7, https://doi.org/10.1007/978-3-030-64065-1_3

Fig. 3.1 Rudravali, showing the house and fields in April 2016, after the nursery plots (dark rectangles) had been rabbed. The rabi crops are on irrigated land near the canal and river. The distance from the house to the wells is about 1 km

The Konkan railway line (completed in 1990) passes in between the canal and the river, although Rudravali (2011 population 825) has no station. Even the nearest station, 2 km to the south in Indapur, is not labelled on google maps (which caters to private automobile owners).[1] The main highway connecting Mumbai and Goa is right in front of the house – separating it from the rest of the village. A small part of the family's land was taken away five years ago for its construction, and widening of the highway is still in progress. The noise of a considerable amount of traffic can be heard day and night: people travelling by two-wheeler, automobile, bus, truck, tempo, bicycle, bullock cart, and by foot. There is no barrier between the house and the road, but even the chickens learn not to venture too far in that direction. The family does not have any vehicles other than a bullock cart and bicycles.

The house has electricity, lights, and fans, although not usually for 24 hours a day. Although there is no fridge, there is a radio and TV (but not always in working condition). The family did not have a phone when we first started visiting them in 2015, although in 2018 they had one, but without internet access. The family also owns an electric thresher as well as a foot-operated thresher. Cooking is done by chulha, using wood or cowdung fuel, as well as with a stove connected to an LPG gas cylinder (delivered from Indapur).

---

[1] Railways have been side-lined in favour of highways, and there are only two passenger trains that halt daily each way in Indapur/Talashet (population 4554), which is one of the larger towns in Mangaon Taluka (population 159,613). Mangao Taluka has 187 villages and towns and two railway stations.

Some containers for use on the farm and in the house are made of cane (some plastered with cowdung) and some are made of plastic. Plastic sheets and plastic bags are used for various purposes. While some tools have wooden or steel handles, others have plastic handles. There is a small amount of plastic waste here and there.

In addition to using rope handmade from rice stalks, some store-bought rope made of jute, nylon, or cloth is also used. There is very little written material in the house besides some important documents and the children's textbooks.

Manufactured fertilisers are used in small amounts (e.g. urea is applied to the main field during transplantation). Manufactured pesticides are not used in the paddy field, but they are used to kill rodents and preserve seeds at home. Herbicides are not used.

The house is made of plastered brick, with a tile roof mounted on wooden trusses, in the tradition of village houses made over the last more than 100 years in this area. There are two living/sleeping rooms, a storage room, and a large kitchen, as well as sheds for the pair of bullocks, the two buffalos, and for storage. There is also a large attic for storage.

This is not just a balance or mixture of old and new things, or a gradual, orderly process in which the new replaces the old – it is a process involving dialectical conflict. For example, there is a conflict since tractors exist, but cannot be used because the family cannot afford to pay for them. There is a conflict since electric lines pass by but electricity may not be available due to power shortages or due to lack of money to purchase electricity or electrical appliances. There is a conflict since trains pass by but do not stop, and since there are many people passing by, but they are confined to separated metal boxes and do not communicate with each other or with the people on the sides of the road. There is a conflict since a modern road (and development itself) enhances transportation and communication and also divides the village and separates people from their source of water.

All these conflicts are dialectical processes since they are characterised by inherent, interdependent opposition that gives rise to continuous change, which may suddenly become radical, qualitative change. For example, an abrupt change comes about when the family acquires a phone and is suddenly able to instantly communicate with distant friends and family members.

3.2   Following Tradition or Doing Science?

Whether or not what we do can be described as the process of doing 'science', depends on our definition of science. Over the course of this study, the farmers and the researchers have discussed and argued, agreed and disagreed, about what is science. As our research proceeded, we identified a few characteristics which may be indicative of doing science – or of not doing science. Following are some examples that we found in the process of paddy cultivation.

3.2.1 Understanding the Reasons for Doing

Bhimraj first says that the rabbing (Fig. 2.13) is done according to the old ways (see 00:39–00:46 above). But later in the same conversation, he and Namdev discuss more about the order of the rab layers with the researchers:

01:06	Karen:	So, what is he saying? How did they first learn? Did he say how did he first learn how to do it?
01:13	Kalpana:	From his grandfather.
01:14	Karen:	But did he-… ask him why are these layers? Why not the second layer then the third – then the first layer? Why not in a different order? Why not? What would happen if different layers -
01:22	Kalpana:	Because he said-… he's just following his grandfather only …
01:25	Karen:	Yeah but, just ask what would happen if he did it in a different order.
01:27	Bhimraj:	[speaking on top of Karen] Old, old हो (Old, be old)
01:29	Kalpana:	तुम्ही असं उलटं-सुलट नाही टाकत? हाच layer त्याच्यानंतर हा, त्याच्यानंतर हा? (Why not upside-down? This layer is afterwards, is afterwards?)
01:31	Bhimraj:	हां, हां, (Yes, yes,)
01:33	Ankita:	असं काय, आता वर पाला टाकला ना? (The reverse, put it on top, no?)
01:35	Bhimraj:	हां. (Yes,)
01:36	Ankita:	तो खाली टाकला तर? (If it was underneath, then?)
01:37	Bhimraj:	अह, नाय मग ते जमिनीला हे होत नाय ना. (Well, then this will not be happening to the ground.)
01:39	Namdev:	जमिन भाजत नाय अन गवत जासूत होतो. (The ground does not get roasted, the grass increases.)
	…	
01:51	Ankita:	Grass अह – गवत जासूत पाहिजे की नकोय? ("Grass" is – more grass is needed?)
01:54	Bhimraj:	नकोय. (Not needed.)
01:54	Ankita:	नकोय. (Not needed.)
01:55	Bhimraj:	शेतीमध्ये म्हणजे आपण धान टाकलंय ना, तो धानच जासूत पाहिजे. आपला- बीज! त्यासाठी अन्य, त्यात – दुसरं काही नाही पाहिजे. असंय. (In the field, you put the rice, no? So you want more rice only. Your- seeds! For that, anything else – in that we don't want anything else.)
02:04	Karen:	पर ऐसे क्यू है? (But why is it like this?)
02:06	Bhimraj:	वो आयेगा, तो बाधा आयेगी ना अपनेको? एये याने – (That happens, then an obstruction happens, no? Meaning -)

From this episode, we can see how confusion and contradiction gives rise to questioning – and learning – by the researchers, Kalpana, Karen and Ankita. They began with authentic questions, which arose because they (especially Karen and Ankita) did not understand how and why rabbing was done. Initially, Karen guessed that rabbing was just a matter of burning stubble or other waste in order to fertilise the plot (Fig. 2.13). But when she saw the rab being constructed, she was surprised and confused because it seemed to be a very difficult process with so many (five!) layers of different materials in a particular order (gobar, dry branches, dry grass, paddy straw, and lastly, dry leaves). To make it even more confusing, on top of all this, the farmers sprinkled some soil and water, before lighting the rab! That didn't make any sense to some of the researchers. Why sprinkle with water if you want things to burn! They did not understand how this order of the layers could be important, since they would all be mixed up in the burning. Maybe something that is more flammable should be below something that is less flammable – but then, the researchers did not think that gobar could be more flammable than branches, or that branches could be more flammable than leaves – and water on top?! They were confused, and therefore more question arose in their minds. Maybe there was also some subconscious condescension on their part – did the researchers really think the family could understand what they were doing and why they were doing it? Maybe they subconsciously thought the family was unscientific or was doing it wrong. And the rab looked more complex (or maybe more 'scientific'?) than they expected from something that people called 'traditional'. In the above conversation, Karen wanted to clear up these questions, and she thought that maybe they would tell more if they asked a more detailed question, like, "Why not in a different order?"

Bhimraj first gives a general reason for why the layers should be put in the order that they were put: that the soil will not be prepared properly. Then Namdev makes it more specific by explaining that it is because the ground will not get as roasted. He adds that the grass increases. He is talking about how the rab is needed to heat the soil and kill the seeds of grass weeds it contains. But Ankita thinks he is talking about the layer of grass that was added for rabbing, and assumes that more grass in the rab layer would be better. Karen is not understanding the Marathi, and continues asking about why the layers are put in this particular order. Bhimraj sees that Ankita is confused, but interprets her question to mean, "Why don't you want weed grass?" He explains that there is a conflict between paddy and weeds: if a lot of grass grows, the paddy growth will be obstructed. Thus, he has explained how cultivators are confronted by a conflict between growing paddy and weeds. In order to solve this conflict, they have developed the process of rabbing. There are also a number of other dialectical conflicts here. The grass is not wanted because it inhibits the growing paddy. But the grass is also useful in order to be used for rabbing in order to prevent the growth of grass.

Miscommunication and confusion are being confronted in the conversation when Ankita asks her question at 01:51. If she had not asked, Bhimraj would not have realised that she is confused. This is different from what typically happens in a classroom, as we will discuss in Sect. 4.1.1. When a teacher is lecturing, students are usually not encouraged (or allowed) to ask questions, and therefore the teacher may not realise when students are misunderstanding. and dialectical conflicts do not openly arise. This is partly because the classroom format consists of one teacher and a large number of students. Ankita is able to ask a question because the number of people in the discussion is very small, and she is in a relative position of power. This enables her to ask a question which otherwise might be embarrassing. Perhaps also because the teacher (Bhimraj) is not in a more powerful position, he does not ridicule her. Without being condescending, he takes her confusion seriously and explains the reason for grass being not needed, even though he might think it is obvious.

At the time, Karen did not understand Ankita's confusion, because she did not understand the Marathi. Afterwards, when analysing this conversation, Karen (being in a position of relative power over Ankita) realises that maybe it's just as well that she did not understand her confusion, because, assuming she would not have had the same confusion, she might have spontaneously said something belittling to Ankita, such as, "How silly! No, he's talking about the weeds, not the grass layer! Nobody wants weeds!" She would probably not have been as polite as Bhimraj – partly because of the different power relations between them.

During the discussion, Karen did not understand Bhimraj's answers. But instead of awkwardly asking Kalpana to explain right away (and instead of admitting her ignorance!) she thought that she could take a stab in the dark, and encourage Bhimraj to talk in more detail about whatever he had said, by asking (at 02:04), "But why is it like this?" So, as a result Bhimraj probably thought that even Karen does not understand why weeds are not wanted. She was trying to act smart, but ended up showing extra ignorance.

So we see how typical hierarchal social divisions that impede learning and teaching are to some extent stood on their heads, when (1) a person who is lower in the hierarchy becomes the teacher, and (2) language barriers reduce the power of a more powerful person.

Not only did Bhimraj, Smita, and Namdev explain how rabbing is done, they also explained why. They combined practical observations of physical reality with abstractions and theories. For example, Bhimraj explained why water and soil have to be sprinkled on top of the rab before it is lit: "हां, वो उपर..वो क्यू, वो झाड का पत्ता डाला था ना, वो हवा आताय ना, तो उड जाताय, तो इसकेलीये वो पानी डालेगा ना, उसके उपर पानी, और वो मट्टी, तो उसमे कैसा, वो ऐसा बैठ जाताय...हवा आयेगा ना, तो उठ नाही सकताय उपर." (Yes, on top, when you put the leaves of trees on top, no? So the wind comes, so they blow away. So that's why when you put water, no? on it – and that soil, so what happens to it? It stays down. The wind comes, it can't fly away.) Here again, there was a dialectical conflict: fire is needed to burn the rab, but the fire may get out of control and burn trees and houses. In order to solve this conflict, they confine the fire by adding water on top. They do not just add water because that is what their ancestors did: they understand the reasons and they have observed what happens if they do not do it.

In the course of several conversations and interactions on different occasions, the family ended up mentioning a number of reasons why rabbing is done. These are listed in Table 3.1:

Table 3.1 The beneficial effects of rabbing, according to the family

(1)	It gives strength to the seedlings, producing a better crop.
(2)	It heats the soil underneath.
(3)	The heat kills grass seeds in the soil, so that there are fewer weeds.
(4)	It makes the soil light and dry, which makes deeper ploughing easier, produces better seed sprouting, and makes it easier to pull up the plants for transplanting.
(5)	It provides nutrients to the seedlings.
(6)	Burning breaks down the fibres of the rabbing material.

This indicates that the family does rabbing not just because it is 'traditional', but because they understand that it has many specific benefits. This sounds scientific – and it is. When professional scientists have studied the process of rabbing, they have found similar benefits (Piyushkumar Patel 2005). But the family had not just memorised the reasons for rabbing that some authority had told them. They discerned these reasons themselves, and at times also questioned and disagreed with each other about the reasons, as will discuss later (Sect. 4.2.3).

As the family was showing and telling the researchers the way things are and the way things are done, they would tend to take the initiative to also explain why. For example, during transplantation Smita explained that the आवण (seedlings) were taller than what was ideal, but the transplanting could not have been done sooner because there was not enough rain and the young plants may break if they are pulled out of soil that is not wet enough. They also explained that if too many (say, 10) seedlings are transplanted together in one place then each plant will not get an adequate share of the cowdung fertiliser and they will not grow well. Smita explained that ploughing should not be too deep because then the clods of earth will be too big and that will increase the clod-breaking work, which is necessary because plants will not grow properly with large clods. During harvest, Bhimraj explained why the reaped paddy is sometimes dried in the field and sometimes at home: 'वो गिरि जाता है, ऐसा नीचे गिरिगा तो वो ख़राब हो जाएगा! … ज़्यादा बड़ा रहता हैं ना, वो सुखाने के लिए जगह घर में नही रहती. तभी वो खेत में सुखा देते हैं. और उसमे कैसा हैं, नजर रखना पड़ता हैं, कि पिंछी-बनिछी आके उठाके लेके जाते हैं.' (If [grain] falls down [if it dries in the field], it will be spoiled. … But if there is too much [paddy], there is no place to keep it for drying in the house. So then it's dried in the field. And how it is in it, it has to be kept an eye on, [or] birds come and take it.) Also, Smita explained that threshing is not done immediately after harvesting because the reaping is so exhausting that they have to rest from such hard labour for a few days.

When ploughing the field for sowing, we noticed that the stubble of the last paddy crop had two sets of roots (Fig. 3.2). Bhimraj explained that this was because one set of roots grows before transplanting, and another set (the upper set) emerges from the first node after transplanting. Both sets continue to function, the deeper set keeping the plant alive when the top of the soil dries out.

Once the researchers had asked what science is, the farmers had said that science explains 'the why'. They said that they are not scientists because they do not understand 'the why'.

Fig. 3.2 From last year's
stubble, a rice plant had
two sets of roots

But there are many examples, such as the above, where they actually did seem to
be investigating and understanding 'the why'. These contradictions lead us to doubt
the meaning of 'traditional' and ask whether the farmers are actually doing science.

3.2.2 Hypothesising

In the same discussion on why the layers for rabbing are put in this particular order,
the researchers asked whether they are sometimes put in a different order, maybe by
mistake:

03:21	Bhimraj:	[laughing] नाय, नाय! सध्या अजूनपर्यंत आमच्या आजोबांनी, म्हणजे पहिल्या old माणसाने सांगितलं ना? (No, no! Until now our grandfather, means our old men have said, nah?)
03:25	Karen:	नहीं, पर फिरि कैसे मालूम – ऐसे – ठीक तरीका है? (No, but then how do you know – this way is a good – method?)
03:29	Kalpana:	हेच बरोबर आहे हे कसं काय तुम्हाला समजलं की आजोबांना – वैगरे – असं? (How do you and (your) grandfather know [understand] it is right?
	...	
03:38	Bhimraj:	त्यांनी ते. पहिला अंदाज़ घेतला असेल, त्याचा परमाणे होतं. (He – first he guessed, according to that-)

Bhimraj was guessing that in the beginning his ancestors may have guessed how
to do things. This sounds similar to the way science is done by hypothesising
answers to questions. The word that he used, अंदाज़, means either guess or

hypothesise. There is an indistinct boundary between guessing and hypothesising, but in this case we think it was hypothesising because the design of the rab must have been based on various types of accumulated experience and observation of plants and gobar decomposing and burning, rather than by just making a very random guess.

We also noticed many other instances of hypothesising while working. For example, they hypothesised how much of the harvested paddy each person would be able to carry on their heads. Sometimes adjustments were required after finding that a load was too heavy. Sometimes this hypothesis was retested after rehypothesising that more weight could be carried. They hypothesised how long it would take the paddy to dry in the field, and whether or not it would rain. They hypothesised how much they would get for selling the crop. In some cases, it is hard to say whether these examples involve hypothesising or guessing – it just depends on the degree of explanation and understanding of the reasons.

In another example, at one point, Smita gave the researchers some flowering paddy plants to take home, so that they could draw, examine and photograph them under a microscope. But she said that probably the flowers would be disturbed and the anthers would fall off on the journey back to Mumbai. We think this is a hypothesis, not just a guess, because it is based on her experience and understanding of the nature of paddy flowering. The anthers are the most prominent sign of flowering, but they are very tiny and fragile (see Fig. 6.3). They appear in the morning, only last a few hours, and dry up or blow away by afternoon. Her hypothesis was correct – partially. But actually, we did succeed in photographing flowers, because as the old flowers withered away, new flowering was also occurring, even though the plants were uprooted.

One of the most important instances of hypothesising occurs each year when a decision has to be made as to whether to use the paddy that was produced as seeds for the next season, or to go buy new paddy seeds in the market (see below, Sect. 3.2.3). This entails making a hypothesis about how good the crop will be. One cannot just make a wild guess. Making the hypothesis will depend on comparing past experience with the same type of seeds and with other seeds – both first-hand experience and the experiences of others. First-hand observations are always valued more than second-hand reports or advertisements. Farmers know from experience that there tends to be a reduction in the quality of hybrids and high-yielding varieties if they keep using the new crop as seeds for the next year. But there are also other factors: most importantly the cost of the seeds, the family's present financial situation, the expected yield, and the expected price the next year's crop might fetch in the market. All of these interdependent factors make the hypothesising very complex indeed. This dispels any illusion that the kind of hypothesising that a professional scientist does is more complex or involves deeper levels of understanding.

3.2.3 Comparing, Categorising, and Testing

While threshing was going on, there was a discussion about the rice that had been produced, in which the following exchange occurred about the yield of crops produced from rice seeds that were grown in the previous years:

03:34   Karen: तो अगर – अगर ऐसे करते हैं एक साल, दूसरा साल, इसके बच्चो आया – मतलब बीच का – बच्चो – हैं? [laughter] और एक साल के बाद इसके बच्चो, फिर इसके बच्चो – (So if – if it's like this one year, the next year, their children come – meaning, the children of the seeds? [laughter] And after another year their children -)

03:46   Bhimraj: हां. वैसे ही चलता हैं. (Yes, it goes like that.)

03:47   Karen: तो इतने साल के बाद कुछ फर्क हैं? (So is there any difference after so many years?)

03:51   Bhimraj: नहीं, फर्क पड़ता हैं कभी-कभी, लेकिन वो कभी-कभी ऐसा हैं की ऐसा अच्छा नहीं दखि रहा हैं ये सब, अपना ये खेती अच्छा नहीं पका, इस साल मैं अपना ये खेती अच्छा नहीं पका, तो बाज़ार से दूसरा बैग लेके आनेका seeds. (No, [but] sometimes it's like this that it doesn't look good, all of it. your own crop isn't well done. So then we go and buy another bag – seeds – from the market.)

Karen is asking whether they notice differences between generations of the paddy. Bhimraj is explaining that there is always variation in the paddy crop. (This is a dialectical conflict.) Each year they compare the paddy to what was produced in previous years from the same line of seeds. This comparison is used to make a hypothesis as to how the next crop will grow. Based on this hypothesis they decide whether to use this rice as seeds for the next crop or purchase new seeds from the market.

This is the same sort of process of trial and error that is done in science. The various factors that are considered when deciding whether to buy new seeds (the cost of the seeds, the family's financial situation, the expected yield, and the expected price) are dialectical conflicts: inherent opposing forces that give rise to questioning, hypothesising, comparing, testing, and making changes in the process of cultivation. One might argue that this sounds more like gambling than doing science. But even gambling can be based on scientific hypothesising.

Other examples of making comparisons occurred while harvesting. We had a discussion about how many plants are held in one hand while cutting with the sickle. This question arose because the researchers who were being taught needed to know how many plants to grab. Smita compared the crop of the field we were in to the rabi crop they grow in the other, irrigated field. She explained that they have one big field where they can hold only a single plant, as the soil of that plot is more fertile. It produces more tillers per seedling as compared to this field.

Not only is a comparison being made, and a difference noticed, but the reasons for the difference are also mentioned. They mentioned two differences: the types of seeds and the soil qualities. In cultivation, it is difficult to make comparisons since there are always a number of variables that distinguish one crop from another: seeds, soil quality, fertilisation, weather, amount of water, pests, diseases, etc. Controlling variables is difficult. It has been claimed that one difference between modern science and traditional ways, is that in the latter not many variables are controlled. We could see that on this small farm, with limited resources, it would indeed be difficult to individually test the effects of different variables in different plots. This forces the family to use a slower, less reductive, more holistic approach in which they compare differences in several variables over different years.

Another example of comparing was when transplanting, Smita told Ankita to go to the centre of the plot rather than pulling up the seedlings from the edge because the soil was too firm there. They had compared the texture of the soil at different places and how it made pulling up seedlings at some places more difficult.

Comparing is an important aspect of doing science, and comparisons such as these are not trivial. Transplanting is one of the most labour-intensive parts of paddy cultivation. Anything that makes it easier to complete the work faster will lessen the workload, and decrease labour costs. Also, the plants can easily be harmed if too much force is required to pull them out of the ground for transplantation. Some newer varieties of paddy have been bred to produce more tillers per plant and more grains per plant. But these varieties usually require more fertile soil and more water. Various qualities need to be compared in order to figure out which varieties to plant at which place and at which time, and which methods to use.

These examples also illustrate the conflicts and difficulties in making decisions. Difficulties arise because in real life things are complex and there are many interdependent factors that need to be taken into consideration. The nature of nature is that it is full of conflicts. In addition, the social and the natural exist in dialectical relation to each other. For example, it may be known that a certain practice (e.g. rabbing) will produce a better crop, but if it is too costly or requires too much time, it cannot be done.

3.2.4 Doing Things Differently

Things do not always go according to expectations. Actually, in cultivation it may not be an exaggeration to say that things never go according to expectations. Plans sometimes have to be changed in order to solve unexpected problems.

3.2.4.1 Different Rabs

In rabbing, sometimes some material does not burn completely, so the unburned material is piled up and relit:

11:30   Gurinder: ये ऐसे भी था ना, कि आप आग लागते है और वो बठिया था ना,
                     सारा layer, वो डाला, आग लागाया पर वो जल नाही राहा था.
                     (It [could have been] like this, no? That you had lit the
                     fire, and laid down all the layers that were put, fire was lit
                     but it was not burning.)
11:37   Bhimraj: हां. कभी-कभी कच्चा रेहता है ना. ये अभी- (Yes. Sometimes
                     it's 'kachha' [unfinished, green, raw]. This now-)
11:42   Gurinder: सुखा नही है, अच्छा? (It's not dry, ok?)
11:43   Bhimraj: हां, अच्छा सुख गया नही, अच्छा पा- हो गया नही, तो उसके
                     बाद मै क्या करनेका, वो अभी ये कर दिया ना? तो फिर उसको
                     इकठ्ठा करने का एक जगह. जितना जला नही, ना? उतना
                     गोल-गोल ऐसा चुनता, वो – झाड़ू से ले. ये कर सकताय आपुन
                     और उसको ज्वाल करके फिर- (Yes. it wasn't good and dry.
                     Not p- well done. So what we do after that is, now that it
                     was done, no? Then collect it in one place, however much
                     is not burned, no? That much is picked out like this and
                     rounded up with a broom. You can do that and then light it
                     again then -)

As Bhimraj goes on to explain, the rab is actually not always done exactly the same way:

13:23   Gurinder: सुखा, सुखा सरिफ़ सुखा होना चाहिये तो जलेगा? कुछ भी जला
                     देंगे? (Dry, Dry, you only want it dry for burning? Anything
                     will burn?)
13:26   Bhimraj: हां. कुछ भी. अभी अपुन ने लकड़ी-बकिड़ी डाला हैं, वो भी नही
                     डालना – कभी भी लकड़ी नहीं डालेगा, ना? तब भी चलेगा. वो –
                     ये रहता हैं, ना? वो पेड़ का पत्ता रहता हैं ना, वो- लेकिन सुखा
                     मनता है. वो भी डालेगा तभी भी चलेगा, वो – असर हो जायेगी.
                     ऐसे. (Yes, anything. Now you put pieces of wood.
                     Sometimes wood is not put, no? Then also it goes. It –
                     what is here, no? These leaves of trees are here, no? – but
                     you want it dry. Whenever you put them, it also goes. That
                     will be the effect. Like that.)

He also says, that sometimes there may not be enough gobar. In that case you may have to add extra urea mixture later on.

Ten minutes before this, Ankita had asked what would happen if, in the rab, the wood is put first, before the gobar:

02:32	Bhimraj:	नाय नाय, मग ते कसंय, हे बरं – (No, no, then how it could be-)
02:34	Namdev:	[speaking over Bhimraj] नाही नाही ते टाकलं तरी जळणार तो, तसा काय सवाल नाय, पण तो राब कमी पडतो. (No, no, but it actually doesn't matter, if it is put, it would still burn, but the rabbing won't be good enough.)
02:38	Bhimraj:	राब कमी पडतो. राब मग कमी होणार तो… (The rabbing would be less – then rab would be less.)
02:39	Namdev:	[Gesturing towards the rab]: आणि असा हे केल्यानंतर भरपूर राब. आणि तियाने काय, जमिनी जळून जाते खाली. (And it's like this, when you do like that the rab is more. And what happens, the land underneath is roasted more.)

There seems to be a difference of opinion between the father and son as to how carefully one should adhere to the usual method of doing the rab. Namdev does not seem to be as careful to adhere to one method. But they both give reasons why their method should be followed – not just because it is the way it has always been done. And there is some variation in the way the rab is done, not necessarily by choice, but because some material may be lacking.

3.2.4.2 Growing Different Varieties of Paddy

One of the biggest changes between the present and the past generation is the type of paddy that is grown. In the past there were many more different varieties of paddy grown in each village and in each part of the district. Most of it was grown for the farmers' own sustenance – rather than to be sold. As we will discuss later (Sect. 6.2.1), these varieties were what are called 'landraces': they are developed by farmers. In the latter half of the twentieth century, varieties of rice that were introduced that had been bred by companies and government institutions such as Indian Agricultural Research Institute (PUSA) and the International Rice Research Institute (in the Philippines). These are locally referred to as 'HYV' (High Yielding Varieties). Now, the family grows three types of paddy: Komal and IR as kharif crops, and Jaya as the rabi crop. Komal is a landrace, and is the main variety that the family eats (e.g. boiled and eaten with veg or non-veg dishes, or ground into flour). IR is an HYV and is only grown as a cash crop. Jaya is also a HYV, which is sold, but the family also eats some of it after it is made into poha. Although the yield is more for the HYV rice, it is not eaten by the family because they prefer the taste, aroma, and cooking quality of Komal. It also has a more stable yield. The family decided to grow HYV rice because they get more profit from it when they sell it. Thus, you might say that although they have 'learned' some modern ways, many old ways persist. This leads us to ask why some old ways persist and why some new ways emerge. In order to understand the reasons, we have to consider historical and economic development, which we will discuss in Chaps. 6 and 7.

3.2.4.3 Making Baskets

One kind of change between the generations that we noticed is the use of manufactured items (commodities) to replace things that were home-made. For example, Kalpana saw Namdev making cane baskets (Fig. 3.3). He had collected the bamboo, brought it home, selected, cut, and shaved it into strips of particular widths, thicknesses and lengths. He arranged the spokes and wove the basket with long, thin strips, keeping the pieces in place with his feet as he went around. He could decide the size and shape of each basket that they required, and make the basket accordingly. It is clear that making these baskets requires quite a bit of skill. Namdev had not taught this skill to his son or grandsons. Bhimraj did not have the time to make baskets, since whenever he could take time away from his tasks around the farm, he tried to find a job doing construction work. Namdev made baskets according to his own decision, taking his time, and taking breaks as desired. He worked sitting on the floor of the veranda, while his wife Prabhavati, was sitting on the side, sorting grain and talking to Kalpana's mother. Young children, including Kalpana's daughter, were running around. Each basket is slightly different from the others, due to its size, shape, weaving pattern, and imperfections that may be due to the material or the process of construction.

In addition to using these baskets, we saw that the family also uses plastic baskets that they purchase in the market. They are manufactured using technology developed in more 'developed' countries, using plastic injection moulding machines imported from China (Fig. 3.4). The machines are automated, only requiring an operator to pour in plastic pellets, press a few buttons, and receive each basket as the machine ejects it. A machine may produce 100 plastic baskets in one hour. Some labour may also be required to remove plastic sprues and inspect and pack each basket. Petroleum-based raw materials are obtained from abroad. Plastic baskets are

Fig. 3.3 Namdev making a basket from bamboo strips

Fig. 3.4 A plastic injection moulding machine

fairly inexpensive, because they are mass produced, most of the production is automated, and the few workers who are required do not receive the true value of their labour. The factory is usually in a building that is separated from the home, family members, and daily life of the workers. Workers are made to conform to timings and production goals that are decided by their managers. Therefore, there is some disruption of the workers' lives with their families and friends. Workers are exposed to a number of hazardous irritants, poisons, and carcinogens. This can be minimised with expensive filters and fume extraction systems, which are rarely used in India, because they reduce the profit margin for the owners. An untrained worker (often a teenager) learns how to operate the machine in a few minutes. Once learned, the work is mechanical and repetitive, not involving any questioning or investigating on the part of the workers. Repairing the machine would require more skill, although there are just a few steps to follow to locate the faulty part and replace it. The plastic baskets the workers make do not belong to the workers, and they have no say about what happens to the baskets after they are made. The owners decide who to hire and fire, and they are free to close down the factory or relocate at any time in order to increase their profits. Usually the owners of such small scale industries are not forced to give workers safety equipment, minimum wages, pensions, or other benefits.

When Namdev constructs baskets, he does every step himself, and every step requires integrated planning, decision-making, creativity, experience, and manual skill. In comparison, the process of manufacturing plastic baskets is highly differentiated into specialised types of labour. One set of workers construct oil rigs and extract crude oil, other workers transport the crude oil, others do specialised jobs in oil refineries, others trade, transport, design machinery, manufacture machinery, and others operate the machine. As industrialisation develops, it becomes more and more impossible to even list the workers who are involved in producing a basket (Who produces the steel? Who produces the electricity? Who designs the equipment? Who teaches the designers?). Most of the work requires little training, and the training that is required is highly specific. Only a very small number of highly educated engineers are required to design hundreds of machines that will produce many tonnes of baskets. Even these engineers need not be very creative or very highly skilled, because they do not independently design a machine – engineers develop designs by building upon each other's research and design over time – and because each machine consists of a number of components designed and produced at different places.

Clearly, a craftsperson who performs all the tasks in making a bamboo basket is much more highly skilled than a worker who performs only one task in manufacturing a plastic basket. And even less skill is required to operate a machine in large-scale industry. The work of weaving a basket by hand is much more complicated than the work of pouring plastic pellets, pushing buttons, and removing a plastic basket from a machine. Hand tools require much more skill than using machines for the same task. As industrialisation develops, tools become more specialised and machines replace hand tools, there is more division of labour, and the labour each person does becomes more uniform. Production becomes more efficient.

However, the cane baskets that Namdev makes have no price or exchange value because he makes them for his own use. He makes them from bamboo that he gathers himself without having to pay anything. The cane baskets have use value, not exchange value, since they are not commodities, made to be sold. Cane baskets have some qualities that are very different from plastic baskets – e.g. they are not air-tight or water-tight. This could be an advantage or a disadvantage, depending upon exactly what they are used for. For example, paddy will rot faster in a plastic container than in a cane one. Although they cannot be repaired or recycled, plastic baskets last much longer than cane baskets. This is first an advantage, but later a disadvantage, as it harms the environment after disposal. In using a plastic basket, Namdev no longer has the freedom to decide the exact size and shape, since only one size is available in the market. But he can choose between a number of bright colours, providing a chimera of 'freedom' and modernity. Whatever their use value or disadvantages, plastic baskets exist because they give the companies a good profit, which is the bottom line in capitalism.

Another major difference between making bamboo and plastic baskets is that the plastic baskets are made by people who are getting paid wages – they are selling their labour power. Namdev does not get paid for his labour. His labour power is not commodified.

3.2.5   Questioning, Investigating, and Experimenting

While working, we noticed that the family often engages in discussions in which they ask each other questions. We will discuss a few examples of different kinds of questioning that we observed.

3.2.5.1   Confirmation Questions

In Sect. 2.3, we mentioned that when Smita explained how they learned cultivation, she said "मोठी माणसं घरामध्ये करतातच, ना?" (At home the elders just used to do it, nah?). Adding ", ना? " (, nah?) to the end of a statement is a very common way of speaking, and can be seen in many of the conversations that we are reporting. We call this a confirmation question, although it may not really be much of a question.

Rather, the ', nah?' may be added in order to check for agreement, and/or soften the statement so that the hearer is more likely to agree. It may indicate a dialectical sort of logic, in which someone is combining a statement and its anti-statement (antonym / opposite) in one. Sometimes there is no discernible pause or inflection and the "nah." is just appended to the end of a statement. In that case the question may be less of a question, but it still sounds less assertive than if there is no "nah". We see these statement/questions as being dialectical: inherently neither statements nor questions, but arising through some amount of conflict, and thus driving the conversation along.

This use of the tag ', nah', which is common in Hindi and Indian English as well, has been referred to by Tamara Valentine as a linguistic device that creates active involvement and cooperation between people in a conversation. She writes, 'Involvement is created in everyday oral discourse by the participants' use of devices such as first- and second person pronouns, egocentric phrases, evidential particles, questions, direct quotations, emphatic particles, speaker-monitoring cues, the informal style or expressions and the emphasis on people and their relationships' (Tamara Valentine 1988, 144). We found evidence of all of these in our discussions.

3.2.5.2   Implicit Questioning

We think that the family was continuously asking themselves and each other implicit questions about the work they were doing. And the researchers were also asking themselves implicit questions while they were trying to learn how to do things.

For example, during rabbing Bhimraj asked Smita to bring him the baila (see Sect. 3.4.1). Just before asking, he had gone off to bring a pile of branches from another field. But when he saw the pile, there was a conflict: he realised that the pile was large and the branches had thorns, making it difficult to carry. We surmise that because of this conflict, maybe the question arose in his mind: 'How will I carry the branches?' Then he thought that the baila would be useful, but he did not have it (another conflict), so maybe he asked himself, 'Where is the baila?' and 'How do I get it?' Then he probably saw that it was closer to where Smita was, so he asked her to bring it. All of this was part of the process of learning and figuring out how to do something.

Some of these sorts of questions are very routine and may seem insignificant, but we think they are still important as indications of how questioning arises due to the conscious (or subconscious?) recognition of conflict. It also helps us understand how interdependent questioning occurs, with one question leading to another, or one question evolving into a different sort of question.

Another example occurred when Karen was trying to reap the paddy. She picked up a sickle by its handle with her right hand, since she is right-handed. By briefly observing the others, she had seen that they were holding a bunch of paddy plants steady in one hand while cutting it with a sickle held in the other hand. But she had not paid much attention to exactly how the paddy or the sickle were held. She just realised that the general idea was to hold the paddy steady with one hand and cut with the other. So she just went over to the edge of the standing paddy and reached out for a handful with her left hand, and tried to hack it with the sickle. She had some difficulty. There was a conflict. She was feeling uncomfortable, and the plants did not easily get cut. She was not confidently reaping like the others. She asked herself the implicit questions: 'How big a handful should I cut at once?' 'With how much force should I apply the sickle?' 'How far down should the paddy be cut?' 'What am I doing wrong?' 'Would it be more comfortable if my left hand was upside down, with my thumb pointing down rather than up?' All of these questions occurred in her mind more or less at the same time, although none of them were voiced. As a result of the questioning, she tried changing her hand position and the way she was cutting. She found that it was even more awkward with the left hand upside down. And someone saw her and suggested that she should hold it the other way. She felt ridiculous, and wondered if anyone was laughing at her. But this was how she was learning – from trial and error – how to reap the paddy.

In all of these small examples of implicit questioning, someone is figuring out how to do something, but the person is not alone. Bhimraj was working together with Smita. Karen was working with a number of experienced reapers.

It's interesting that even though they are working in groups, questioning may be implicit rather than explicit. The other members of the group may become involved as a result of questioning that they do not hear. There may be implicit group questions. They may be all observing what is happening and this may bring them to ask similar implicit or explicit questions. Maybe Smita also realised that Bhimraj would need the baila because she also saw the same difficulty of how to pick up the branches. Maybe she would have brought the baila even without being asked. Probably the other reapers saw Karen struggling, and they asked themselves, 'What is she doing wrong?' The result was some coordination and cooperation between people in order to learn and/or solve problems.

It is important to note that in all of these examples, questioning is not separate from investigating in order to find answers to the questions that arise.

3.2.5.3 Sprouting Questions

When Bhimraj was tossing bundles of paddy down from the stack of harvested paddy in order to begin threshing, they noticed that some of the paddy on this stack was a little wet. They had been having problems with bundles getting wet and starting to sprout (Fig. 3.5).

Smita raised the question: "तो भाग पंखयावर टाकून वाळून घया?" (Then put that part under the fan for drying?) The question is: should something be done about the damp bundles; and if so, what should be done; and will the proposed solution succeed; – or what will happen if nothing is done? It is an investigatable question, in the sense that it can be part of a process in which there is an interconnected network of aspects (not necessarily in one particular order): questioning, observing, generating and collecting some original data, analysing the data, reasoning, hypothesising and making some progress towards answering the question on the basis of available first-hand evidence.

Fig. 3.5 Some of the paddy at the bottom of the stack kept for threshing had started to sprout

In this case, after discussing, they decided that the problem had to be solved, and an investigation was done. The damp bunches of stalks were set aside for drying. Smita's question led to other questions of where to do the drying, whether to use a fan or the sun, when to do it, etc. Depending on observed results, the method could be continued, modified or abandoned. It is a process of trial and error, but it is also based on past experience and an understanding that water can be evaporated by heat and/or wind. At that time of year, there was not much wind, but plenty of sunshine, no rain, and the temperature could be expected to be fairly high during the day. If they had wanted to dry something during the monsoon, wind would have been used. Fans may not have been used because of the availability and cost of electricity.

Questioning is basic to science. Doing investigations in order to search for answers and solutions to practical problems is also part of doing science. Thus, our interpretation is that such questioning suggests that the family is doing science, at least to some extent.

But, did the family think they were doing science? Did the family think that questioning was a central part of their work in cultivation?

3.2.5.4 Do you Have any Questions?

While doing the harvesting (Fig. 2.22), Karen asked Bhimraj and Smita if they had any questions about their own work:

05:30	Karen:	[To Ankita] So you could ask also if they have any – ah particular questions about anything they are doing right now. Any questions come to their mind.
05:42	Ankita:	[Laughs nervously] Madam वचिारतात की – अ – तुम्हाला डोक्यामध्ये l – doubt काही आहेत का...म्हणजे, आमच्यावषियी नाही, तुम्ही करता आहात त्याच्यावषियी काही परश्न पडलेत का, कामावषियी?- (Madam is asking that if you have any questions in your mind, any doubts. Means, not about us, but about what you are doing, about that. About the work you are doing.)
05:55	Bhimraj:	नाही. (No.)
05:55	Ankita:	हे कधी होणार, हे कधी-काय कधी करणार- (When will this happen, when this is to be done-)

05:59 Bhimraj: नाही. हे तर प्रश्न पडतातच. (No. These questions
are there.)

06:00 Smita: आता आम्ही काय बोलतोय, पाऊस पडलं तर हे धान्य पडेल,
नाही काय? मग ते आम्हाला असं वाटतंय कि पाऊस पडू नये
पंधरा दविस. जेणेकरून आमचं. धान्य घरी जाऊ द्या! असं वाटतं!
आम्ही प्रयत्न करतोयच पटापट नेण्याचा, पण आम्हाला असं
वाटतंय की पाऊस पडू नये पंधरा दविस. जेणेकरून – आमचं धान्य
सुकाच्या सुका घरी जाऊन राहील. मग झोडता येतं ना घरी.
(Now what we are saying is, if there is rainfall, this crop
will be damaged. Isn't it? Then we think that it should not
rain for the next 15 days so that our crop will be taken to
our place! That's what we think. We are trying to move it
fast, but we don't think that it will rain for the next
15 days. For that reason, our dhaan [reaped paddy] should
be as dry as possible, and we can take it and keep it at
home. Then we can thresh it in the house.)

Teachers often find that there are no interesting responses when they ask the
students in their classes, 'Do you have any questions?' This is partly because questions naturally occur in response to conflicts between people or between people and
their observations or actions (Gurinder Singh et al. 2019). If students are not allowed
to talk or voice their questions whenever they occur, and if they do not have stuff to
observe and manipulate, they may not have any questions when asked. Also, when
asked to ask, they are often more concerned about figuring out what sort of question
the teacher wants them to ask than asking an authentic question (one for which they
do not think they know the answer – and need to know).

So we were not surprised to hear Bhimraj first say that he did not have any questions. But then he agrees with Ankita that there are actually many practical questions that arise while working, like deciding when to do something. Then, Smita
gave an interesting reply. The question that occurred to Smita was the most important question at the moment: Will it rain within the next 15 days? She explained that
if it rains before they have finished harvesting, the crop may be spoiled.

After hearing this, Karen wanted to ask her what determines the answer: what
determines whether or not it will rain? Actually, Karen thought that it is not really
possible to predict exactly when it will rain or to understand why it rains or does not
rain on a particular day. But she asked in order to find out what Smita thinks. Karen
was wondering whether Smita thinks that rain is determined by the gods or by one's
'karma' or 'kismet', whether she has some superstitious belief, or else whether she
might be quite sure that it will not rain due to some observations or past experience.
So, Karen asked Ankita to ask the question:

07:30 Ankita: पाऊस पडेल की नाही हे तुम्हाला कसं कळेल? (How would
you know if it will rain or not?)

[Ankita, Smita, Bhimraj laugh]

07:37	Bhimraj:	पाऊस सांगता येत नाही बलिकुल. पाऊसाचं बलिकुल नाही सांगता येत. (Rain cannot be told [predicted] at all. Rain really can't be told [predicted].)
07:38	Ankita:	नाही कळत असेल तर सांगा नाही कळत. (If [you] don't know, don't tell.)
07:39	Smita:	नाही कळत, ना! तसं काही कळत नाही. पण मग आम्ही काय करतो माहितिये? अ- जर आम्हाला पटकन न्यायचं असेल तर मग आम्ही माणसं घेतो. जेवढी माणसं घेतो आणि पिटापट आवरतो. आम्ही दोन तीन लोकं एवढं पटपट करू शकत नाही- (Don't know, nah! Don't know anything like that. But then what do we do? You know, if we have to do it faster, then we hire people. Take as much as required and finish it quickly. We two-three people…)

Karen did not get any of the answers she was expecting. Bhimraj and Smita said that it cannot be known [predicted] whether it will rain or not. In other words, they realised that (at this time of year) this is a question without a definite answer.

Every year they have to try to predict when it will not rain so that they can do the harvest. But sometimes they observe that it does rain even when they predict it won't. From their empirical evidence, they realise that it is really not possible to predict the rain.

Karen thought that in order to do the harvesting, one has to be very sure of one's prediction that it will not rain for a few days. But actually, they did make a prediction that it will not rain and decide to harvest based on this prediction, even while realising that the prediction is not very accurate and rain is unpredictable. It is a dialectical conundrum.

Also, Smita explained that they may have to hire people to help do the harvesting faster in order to avoid rain damage. The stark reality of the conflict between money and time has to be faced.

3.2.5.5   How to Set up the Kanaga and Fill it with Paddy

A problem arose when it came time to figure out how to set up a new kind of kanaga (कणगी – large canister for storing grain) in the storage room in the house. Namdev had just made the kanaga, with the help of Bhimraj and a few others, by forming a thin steel sheet into a cylinder (about 1.5 m tall and 80 cm in diameter). They had never made this type of kanaga before. Every house in the village has a kanaga of a similar size, and traditionally they are made of woven cane plastered with cowdung.

This new steel kanaga had no top or bottom. They were planning to cover the filled kanaga with a plastic sheet, and then seal the bottom and top with mud to protect the grain from rodents and preserve it until it could be brought to the mill when it was needed 2 or 3 months later (because unmilled paddy is easier to preserve than the milled form).

You might think that a bottom would not be necessary if it was set upon the floor: the floor would be the bottom. But actually, if you try to pour grain into a cylinder that sits on the floor, the cylinder may rise up and grain will spill out the bottom – or the entire cylinder may fall over. Some interesting 'physics' is actually involved here.

Therefore, they had to figure out how to seal the bottom of the kanaga in order to fill it with paddy. Namdev brought some old cartons, and began dismantling them to get some pieces of cardboard to put underneath the steel cylinder. But Bhimraj told him not to tear up the cardboard, but to leave it in one large, unfolded piece. Then Bhimraj brought over a large plastic sheet. They began to lay it down on top of the cardboard, but then Bhimraj said that they should put the cylinder on top of the cardboard first. Then, the plastic sheet could go inside, lining the bottom and lower sides of the cylinder.

But there was a problem. The grain could keep the plastic in place once it was added, but in the beginning there was no way to make sure that the grain would not be accidentally poured into the gap between the plastic and the sides of the cylinder (and thence out of the bottom of the cylinder).

The entire discussion between Namdev, Bhimraj, and Smita that occurred regarding this problem is given in Sect. 3.4.2.

A solution was given by Smita. She said that a person should go inside the kanaga. The grain would be handed to this person by the others who carry it in on head-loads from the large pile they will make outside after threshing and winnowing. (It is not uncommon for someone – maybe a child – to be lowered inside a kanaga in order to empty it.)

Bhimraj said that the person to go inside would be Namdev. But now there was another problem, that was asked explicitly by Namdev. He was very worried, and he said that this was a big problem: How would [the person] get out [of the kanaga] at the end?

Bhimraj and Smita explain that "Just as the rice would keep going down, down – at the same time, the person will keep coming up-".

Namdev may have thought that a person inside will not rise as the grain is added: the person may instead become buried in the grain inside the kanaga. Or the kanaga may topple over. On the other hand, the coupling holding the kanaga together may break if the grain is just thrown into it from above, rather than being gently poured by a person standing inside. Actually, it is not so easy to tell what would happen unless one has some previous experience with such a kanaga, which none of them had. This was the first time they were making a steel kanaga. Previously they had made them from reeds and mud/gobar plaster.

Would a person rise as the grain falls down? If you put a spoon in a jar and fill the jar with rice, the spoon does not rise as the rice is poured in. Even if you shake the jar from side to side, the spoon does not rise. But if you shake the jar up and down, the spoon rises! So what will happen to a person as rice is poured into a kanaga? Apparently Namdev did not want to be the guinea pig in this experiment.

3.2.5.6   Investigatable Questions

This episode involving the kanaga provides a number of examples of investigatable questions, which were not only asked, but also investigated:

- How can the bottom of the kanaga be sealed?
- Can the bottom be made of cardboard?
- Will grain spill out the bottom of the kanaga if it has no bottom?
- Would it be better to put the plastic sheet inside or outside of the kanaga?
- How can grain be filled into the kanaga? Can it be done without a person going inside?
- If a person is inside, standing on the grain as it is being filled, will the person rise with the grain, or be buried in the grain?
- If a person goes inside the kanaga to fill it, how will they get out?

These are design and engineering problems, but they are also physics questions. The experimentation that occurred arose out of the need to solve practical problems.

3.2.5.7   Experimenting with Rabbing

When we first went for the rabbing, we did not notice that there was much if any experimentation occurring on the farm. However, we later had an interesting discussion. About experimenting with rabbing.

We wondered whether the family had ever experimented with trying out different ways to do the rabbing in order to find out which ways are better. When we asked them if they ever did rabbing differently, they said they always did it the same way. But later, the following conversation occurred:

27:17	Kalpana:	न लावता कधी केलंय का तुम्ही? (Have you done it [sowing paddy] without [rabbing]?)
27:18	Bhimraj:	हां. सब. हो स- (Yes, all. It -)
27:19	Gurinder:	कब कयिा वह आपने? (When did you do it?)
27:20	Bhimraj:	हां, हां, होता- (Yes, yes, it happens-)
27:21	Karen:	कब कयिा? (Did it when?)
27:22	Bhimraj:	वह- अगले time – ऐने, कभी कभी ऐसे मम – time नही मलिगा कभी, ज्यादा काम रहेगा। कुछ ये रहेगा. (It- next 'time'- and, once in a while, sometimes [we] don't have – like, a lot of work is pending.)
27:28	Gurinder:	हां, हां. (Yes. Yes.)
27:29	Bhimraj:	तो, ऐसा एक दो खेत है ऐसा रेह्ताय. (So, it's done like this in 1–2 fields?)
27:31	Gurinder:	अच्छा. (Ok.)
27:32	Bhimraj:	तो तभी भी ये करना पडताय. ऐसा ...[...? करे का बढ़ता?] ... (Then we have to do this sometimes [...] like this.)

27:34	Gurinder:	तो उसका जो- जैसे लगाके आपने – (So it's whatever-whatever you put -)
27:37	Bhimraj:	तो उसको वैसा है? वो – खत रेहता है ना, वह केमी- आह – [to Kalpana:] खताला काय बोलतात? त्याला काय बोलतात?? (So, how is it? That fert- That chem-aahh-[To Kalpana] What do you call khat as? What do you call it??)
27:42	Kalpana:	[and Ankita at the same time] Fertilisers.
27:43	Bhimraj:	हां, फर्टलसिा – वह रेहताहैडालते, ना? वह ज्यादा परमाण से use करना पड़ता है. (Yes. Fertilisers. That is there, no? That we have to use in more amounts.
27:47	Gurinder:	ज्यादा लगता, उसको? (More [of it] is required?)
27:48	Bhimraj:	हां. उसको ज्यादा लगताया। (Yes. More of it is needed.)
27:49	Karen:	और पैसे भी ज्यादा? (And money [the cost] is also more?)
27:50	Bhimraj:	हां. ये – पर पैसे का मेहनत कम करने – पैसा कम करने के लिए – तो यह करना पड़ता। (Yes. This – but the money – to reduce the hard work. – For reducing the hard work, so this is necessary.)

Thus, paddy was cultivated in two different ways: with and without rabbing. This was not purposely done in order to find out which method is better. It happened by necessity, because there was a conflict: there was too much other work and therefore there was a shortage time. Rabbing requires a lot of labour and time. To solve the conflict, rabbing was not done in one or two fields. Because paddy was therefore cultivated with and without rabbing, the family necessarily compared the results and experienced first-hand the differences with and without rabbing. They found that without rabbing, the beneficial effects listed in Table 3.1 may not occur. Then they have to use more artificial fertiliser (urea), which is more costly. Ordinarily, the family cannot afford to apply so much fertiliser. They are aware of environmental problems due to overuse of artificial fertilisers, but the reason they use very little of them is that they are too expensive. They never said that they thought rabbing should be continued in order to preserve tradition.

In doing science, sometimes scientists purposely do experiments in which they do something two different ways in order to compare the results. The comparison may tell them something about how and why things happen. If they can keep all variables the same except for one, then it is likely that different results may be due to the different variable. In the case of cultivation, growing the same crop in different ways in two different fields may tell which method is better. Of course, cultivation is a very complex process with many interdependent variables (macro and micro differences in soil, sunlight, temperature, water, pests, etc.), so these kinds of experiments are not easy. It is not really possible to just change one variable at a time, or to rule out 'extraneous' effects.

Doing cultivation with and without rabbing, and observing and comparing the results, is an empirical science experiment. It may not be a very rigorous experiment, because there may be too many variables that were not controlled. The fields with and without rabbing are necessarily in different locations, even if they are next to each other, and so they may be different in other respects (drainage, water, sunlight?). Different results could be due to many factors besides whether or not rabbing was done. Nevertheless, some tentative, probabilistic conclusions may be indicated.

Since Bhimraj had said that they had compensated for not rabbing by using more artificial fertilisers, we wondered which method would be preferred if money was not a problem, so we asked him:

28:23	Ankita:	जर तुमच्याकडे पैसे आहेत तर तुम्ही हे कराल की खत जास्त टाकाल? (If you have money, then would you do this or put more fertiliser?)
28:27	Bhimraj:	मग खत प्रमाणात टाकणार. (Then more fertiliser would be put.)
28:29	Ankita:	जास्त टाकाल? (Put more?)
28:30	Bhimraj:	हां...मग काय करणार?! असंय. (Yeah, then what else can be done?! That's how it is.)

If they had more money, Bhimraj says they would use fertilisers rather than rabbing. But he says it with resignation, knowing that there are many beneficial effects of rabbing in comparison to using chemical fertilisers. He goes on to explain that applying gobar and compost (in the rab) is actually better than applying chemical fertilisers:

| 29:13 | Bhimraj: | जमीनीला थंडावा येतो आणि जमीन एकदम अशी सुपिक होते. म्हणजे हलकी प्रमाणात होते. म्हणजे जी कडक जमीन असते ना, म्हाजे खतं वाळून टाकून-1. खतामुळे जमीन कडक असते, पण ते जर तुम्ही शेण टाकलात ना, तर ती जमीन हलकीच्या-हलकी राहते. असंय. (The land cools down and it becomes fertile. Means it lightens up. Means, that hardens land, that means the one (in) which we have put fertilisers. Because of the fertilisers, the land gets hardened, but if you put shenkhat [compost with gobar] instead, that land remains lightened. This is how it is.) |

In order to convince us, when we were in the field, he bent down and picked up a handful of soil after it had been rabbed, and gave it to us to feel for ourselves. We were surprised at how light and airy it was.

The technology of rabbing and fertilisation with gobar is different in many ways from the technology of using artificial fertilisers. The design, production, and distribution of artificial fertilisers is not in the hands of the family – it is done by institutions, companies, and government agencies, and is controlled through the economic system. The compositions and quantities of artificial fertilisers to be applied are usually decided mainly by protocols recommended by external authorities. Of course, the family may not adhere to the protocols, depending on what they can afford, and on their own experimentation. However, with rabbing and using gobar that is produced on the farm, the family has more direct control over the process, and the process is slower, less uniform, and more variable, depending also on how much time and energy the family has. The family is more intimately involved with the process of producing and maintaining fertile soil – manipulating materials and soil with their own hands. When urea was used, Bhimraj simply scattered it over the field, in one quick step that did not require much hard work. Comparing these two types of technology, we can see how technology is interdependent with mental conceptions, understandings, social relations, and relations to nature. The kinds of questions the family asks and the kinds of investigation and experimentation they do in each case depend on which technology they are using. In doing rabbing, they ask how much of each type of material to put in each layer, where to find the material, which order to put the layers in, what the effect will be, who does what, and when they should do it. In applying urea, they ask how much to use, how much it costs and how much money they have.

Another experiment was done regarding the soil and water sprinkled on top of the rab to prevent wind from spreading the fire. Smita and Bhimraj mentioned that they knew of one time when some neighbours of theirs had not put soil and water on top of the rab. They put stones on top of their rab instead. This may have been either a deliberate or an inadvertent experiment.

Putting stones rather than water and soil may be based on what educationists used to call a misconception, and now (having been influenced by postmodern relativism) call an 'alternative conception'. The neighbours may have reasoned that water and soil should not be used to keep the leaves from flying away because the water will seep down and it will douse the flames, whereas stones are heavier and will keep the leaves down without dousing the flames. One might argue that both ways of thinking and both methods are equally valid and should be allowed, or at least tried. But what happened is that when the stones were used instead of water, the wind blew, and the fire spread, causing harm to neighbouring fields and trees and threatening others' homes and animals. The danger of experimenting in this case is clear.

3.2.5.8 How Do Questions Arise?

Note that the questioning that we observed on the farm did not occur out of the clear blue sky, nor out of one's 'inner-being', while an isolated individual was sitting back and thinking. It occurred when people were busy working. Questions arose out of a number of different types of conflicts:

(a) Conflicting observations by one person
(the paddy looks ripe – the paddy looks unripe) – Is it ready to harvest?
(b) Conflicting observations of different people
(the stack of paddy is dry – some of the stack is wet) – Is the stack dry?
(c) Conflicts between different beliefs* of one person

(rabbing is too much work – rabbing is cheaper and more beneficial) –
Should we rab?
(d)   Conflicts between different people's beliefs*
       (we have put enough branches in this layer – we have not put enough
       branches in this layer) – Should we put more branches?
(e)   Conflicts between beliefs* and observations
       (the stack is dry – some of the paddy at the bottom of the stack is
       sprouting) – Is the stack too wet?
[* Note that by 'belief' we mean justified or unjustified belief.]

The conflicts and the questions do not necessarily occur sequentially or separately from each other. Questions may arise out of conflicts while at the same time
conflicts arise out of questions. It may not be possible to tell which came first. The
conflicts are inherent to the process of questioning. We call the process of questioning as dialectical because it consists of inherent contradictions.

Conflicts may or may not be explicitly noticed or remarked upon.

Sometimes such conflicts may not give rise to questioning. There can be many
reasons why questioning is inhibited. The conflicts may not be recognised, or
thought to be very significant. It may be (or it may be considered to be) more important to try to avoid conflict or hide conflict. Questioning and conflicts that give rise
to questioning may be suppressed due to hierarchies of power and agency. It may be
thought that there is a lack of time or resources needed to answer a question, so the
question is not even raised. But without questioning, and without the conflicts that
give rise to questioning, how could science be done? We will discuss this in Chap. 7.

3.2.6   *Approximating vs Precision*

As the transplanting of the seedlings began (Fig. 2.19), Kalpana announced that she
would give it a try:

07:20   Kalpana: थोडंसं try करते. चांगलं आलं तर लागेल. ([I] will give it a try.
                     It will be good if it comes out right.)
07:24   Ankita: किती घेता काड्या? (How many seedlings do you take?)
07:25   Bhimraj: ३–४ काड्या (3–4 seedlings)
07:26   Sandeep: ३–४ जातात. (3–4 goes)
07:27   Ankita: हमूम (hmm.)
07:28   Bhimraj: हे- ४ काडी. हमूम (This- Four stalks. Hmm.)
07:31   Smita: अंदाजाने (Guessing/approximately)
07:35   Bhimraj: चार काडी, पण ते अदाजानी बरोबर येत्यात. (Four stalks, but
                     it's approximately all the same.)
09:25   Bhimraj says [teaching Ankita] that the exact number of plants to put
         each time in transplanting is not important. More is also ok.

[At 09:12 Sandeep watches over how Kalpana is doing the process.]

09:21	Bhimraj:	ते बघत नाही बसायचं किती चारच काड्या आहेत...(Should not sit watching at it, if we have four stalks or not.)[You should not fixate upon if it is four stalks or not.]
09:23	Sandeep:	जास्त-जास्त झाल्या तरी चालेल! (More-more is also alright.)
09:25	Bhimraj:	हा, ते येईल तसं लावायचं. असं. (Yes, should plant as [you] can)
09:30	Sandeep:	जास्त झालं तरी चालल. (More is also alright.)
09:36	Bhimraj:	कारण ते तसं जरी लावला तरी चखिलाला लागतो. असंय. (Because even if it is put that way, it is still in the mud. That's how this is.)
09:38	Kalpana:	हम्मम. (Hmmm.)

Similarly, when trying to make rope, Karen carefully counted the number of stalks to use in the beginning. Family members saw her and explained that the exact number is not important.

The farmers emphasised the importance of approximation more than the researchers did. As learners, we sometimes requested precision. But the farmers were imprecise about certain measurements, and about certain dates (e.g. their own ages).

It may appear that counting and trying to measure precisely are characteristic of being more scientific. But actually, approximation is widely practiced in doing science as it is necessary in measuring, hypothesising, doing qualitative analysis, and comparing theoretical predictions (under 'ideal' conditions) to observations of physical reality. In doing science, precision is actually necessarily limited and dependent on the method of measurement or calculation, as well on as the practical necessity. Being used to book-learning, the researchers at times tried to ask for information and precise instructions before trying to do something. They did not realise where and when precision is or is not necessary.

Another area in which a lack of precision is obvious is in the final goal of paddy cultivation: cooking and eating. You might think that the family eats traditional food, prepared in the same way, generation after generation. But members of the family do not learn cooking by listening to the older generation or an experienced cook tell how to cook or how much of each ingredient to add. Cooking is learned by doing, sometimes alongside an experienced cook, but not as part of a formal process of teaching and learning. When cooking, people do not precisely measure ingredients. Neither do they refer to any written recipe. As a result, there is quite a bit of variation in the way different cooks prepare the same dish, and even between how one cooks prepares the same dish at different times. Sometimes one ingredient or another is not available. Sometimes mistakes are made. Cooking is actually a process of trial and error in which experimentation is an inherent part of the process. Hypotheses are made in order to decide which materials and methods to use. The results are determined through observation of taste, smell, looks, texture, temperature, etc. Food preferences are very dynamic. Taste is a matter of individual and social preference, and also depends on mood, context, health, and other factors. Thus, learning and doing cooking is one of the most important processes in which people need to do science.

3.2.7 Directly Observing

As we will discuss later (Sect. 4.2.3), we noticed that Bhimraj and Smita tended to show us things directly rather than just telling us about them. Direct observation was essential for them. Here is a list of some examples of close observation that we noticed:

- Picking up soil from underneath the rab after the fire went out to feel how warm it is. Handling the soil to observe its lightness.
- Observing that a paddy plant grows a new set of roots from a node above the original roots after it is transplanted (Fig. 3.2).
- Identifying the different types of paddy by closely observing the shape, size, colour, and other characteristics of the grains.
- Identifying different kinds of weeds and removing the ones that had to be removed, eating ones that were edible, and discarding the others.
- Going to the field to search for flowering plants even after the flowering season has long passed, finding some, and bringing them back to show the researchers.
- Taking some grass plants and dissecting them to see the flowering.
- Feeling the developing grains to find out if they are milky, soft, or hardening.
- Handling stalks and panicles of paddy to test how strong they are.

In each of these examples, the observation was done purposely, in connection with the investigation of some questions. Observing was done continuously and simultaneously with questioning and possibly also included hypothesising, comparing, testing, identifying, and/or categorising. Observing was part of an inductive process of investigating and reasoning. Accordingly, the premises – the specific observations – give strong support, but not absolute proof for general conclusions. In other words, the reasoning is concerned with probabilities rather than absolutes.

For example, we simultaneously observe the development of the grains by feeling their hardness, as we ask whether the paddy is ripe. We observe because we do not know if it is hard. Our questioning (our not knowing) gives rise to our observing (in order to know). But it is not clear how hard is hard, some grains are harder than others, the percentage of hard grains varies on each plant, different plants have different percentages of hard grains, etc. Thus, the observing has an inherently contradictory nature: we observe that the grains are hard / not hard. Due to this contradictory nature, the observing gives rise to questioning, while the questioning gives rise to observing. Without the questioning, there would be no scientific observation. There is a dialectics between knowing and not knowing whether the paddy is ripe, which gives rise to and sustains the interdependent process of questioning and closely observing. Note that the way of determining whether the grain is ripe and ready for harvest is much the same as what was described by Kashyapa 2000 years ago (Sect. 2.1.2.1).

As the seedlings were growing, the researchers kept asking when transplantation would start, so that they could be sure to be there. At first the family gave them a rough idea (hypothesis) of the date, based on observations from previous years. They kept observing the growth of the plants and observing the weather and their estimated day of transplantation kept being advanced. The final decision was made only a day or two in advance. This is an example of how in farming, everything can change quite quickly, and unexpected things keep happening. It would be problematic if farmers just expect what is hypothesised, without realising that there is actually an inherent opposition between the expected and the unexpected. The conflict between the expected and the unexpected is what drives them to keep observing in order to resolve the conflict and find out what will happen. Ideally, farmers should be ready to observe both the unexpected and the expected, and also be ready to react to both possibilities.

The family had to know how their crop was doing throughout the season: how tall the plants were, whether they were flowering, whether the grains were getting ripe, whether it had the correct amount of water, whether there were too many weeds, etc. They had to keep track of the plants in order to know what to do and when to do it. They did not find out about such things by remembering what happened last time, or by listening to some authority guiding them. Working on a farm requires direct observation of what is happening.

Thus, the reason for observing was not only that the farmers just happened to observe – although there was that also. When looking for flowers, we may find pests. We have to have opportunities to observe, as well as reasons and needs to observe. Actually, there is a dialectic between chance and necessity that gives rise to close observation.

The farmers showed that they cared about the plants, animals, and land. The pair of bullocks were kept at the side of the family house, and all family members, including the young boys, interacted with them intimately throughout the day, calling them by name, talking to them, feeding them, giving them water, washing them, and looking after them in many other ways. This caring was not distinct from their need for the bullocks to produce gobar and to pull the plough and the bullock cart. The family's caring about ploughing, rabbing, fertilising, transporting, and cooking was interdependent with the family's caring about the bullocks. There was a very direct interdependency between each person's daily life, social interactions, and their observations, interactions, and relations to plants, animals, and the rest of nature.

In case there was lodging of paddy (when the mature plants fall over because the stems are too weak or tall), or in case the cattle got into the field and began eating the crop, their emotional reaction was obvious. It was due to their caring for the plants, while at the same time they needed the plants for their own livelihood. The farmers cared about and were genuinely interested and curious about the plants, animals, and the rest of the environment, and their lives depended upon it. Therefore they both wanted and needed to observe. They kept observing the paddy as it was

growing because they cared about it and found the growth of the plants to be wonderful and beautiful. But they also observed it because their lives depended on it.
They had to keep checking to see if weeds were taking over, if there were problems
with pests or diseases, if cattle, birds, or any other animals were getting into the field
and eating or destroying it, etc. If there should be any problem, the sooner they saw
it and resolved it, the better. Thus, there is a dialectic between needing and wanting
(caring, feeling for the farm). This dialectic gives rise to, and is required for close
observation.

These processes of observing were not just a matter of glancing or superficially
looking at something. They involved taking time, having patience, and going to
extra trouble to find something or check something. They involved not just looking
but also handling and manipulating things.

We discussed how the family told us about the benefits of rabbing (Sect. 3.2.1).
How did they come to understand it themselves? They observed the effects of rabbing on their own crops. They felt the temperature of the ground underneath. They
felt the lightness of the soil after rabbing, in comparison to the hardness of the soil
in areas that had not been rabbed. They observed a decrease in the number of weeds.
They never studied it in school. They were never told by some authority. Their
understanding is based on their own direct observation and experience, and not just
on some theory. They also asked us to observe the effects for ourselves.

Direct observation by the family and the researchers was possible because we
were all there on the farm in the midst of what we were observing. This allowed us
to observe with a number of senses: seeing paddy flowers, hearing the sounds of
plants being held against a thresher; smelling the approaching rain before sowing
and the ripe paddy while harvesting; tasting the final product; and feeling the temperature of the soil. We should also mention the more emotional kinds of feelings,
such as feeling tired after working in the field, and feeling disillusioned for not getting a better price. These are also senses through which we made direct observations.

Farmers have to care about the bullocks, but if they care so much that all they can
see is how tired the animals look when they are pulling the plough, they may not be
able to continue ploughing. If a family cares so much for the paddy that they cannot
bear to look at it after a pest attack, to see how much it has been destroyed, it will
be problematic. They may observe less (or more) damage that there really is.
Something like this happened in Karen's family: they had a fruit tree that they could
not cut for a long time after it became diseased, partly because they were not able to
observe how badly damaged it was, because they had such love for the tree. Farmers
need to care about the plants, animals, and the rest of the environment, but at the
same time, they need to have some fortitude, indifference, or lack of fear. Otherwise
they will not be able to observe, or their observations may be blinded by their emotions. This is related to the dialectic between objectivity and subjectivity. While it is
neither possible nor desirable to be entirely objective, we need to recognise our
subjectivity, and sometimes we need to try to change our subjective beliefs if we
find that they are having negative consequences. We may need to try to step away
from our personal point of view in order to make observations that do not contradict
physical reality. Our observations are actually the result of a struggle between the

opposing tendencies of objectivity and subjectivity (and between indifference and caring).

When examining the standing crop to see if it was ready to harvest, the family looked closely at individual plants, checking individual grains to see if they were full and hard. But they could not harvest plants individually. Their decision of when to harvest had to be based on the over-all crop. They had to make generalisations, and make general observations of the colour, how bent-over the panicles were, how wet or dry the soil was, etc. Similarly, when selecting paddy to use as seeds for planting, Bhimraj explained that they select the seeds by observing and selecting the entire plant. Karen had thought that they would select seeds without regard for which plant they came from. But actually, cultivators do not just want certain types of grains, but also want stalks to be strong enough, many panicles, many seeds on each panicle, many tillers, etc. These are just two examples among many of how the family tended to focus simultaneously on the particular and the general, or the specific and the holistic. They were able to do this because they were on the farm and so could directly observe it. Conflict between the particular and the general gives rise to closer observation.

The family's observations were often more qualitative than quantitative. There may be conflicts between qualitative and quantitative observations (measurements). Qualitative observations may seem to be too vague or general. Quantitative observations may be too specific or reductionist. For example, measuring a crop by counting the number of bundles at the time of reaping may sound convincing, but be deceptive: there may be a great number of plants with fewer or smaller panicles. The yield of paddy will be different if it is measured by volume or weight, and before or after milling. The weight will depend on the water content, which will decrease as it dries.

We have already mentioned how the farmers often used approximation (Sect. 3.2.6). There is a conflict between being exact and being approximate, and we discussed how this conflict arose when the researchers were trying to learn how to transplant or make rope. Farmers made observations that may have been more or less approximate or precise. The dialectic between precision and approximation, also gave rise to more observation.

We mentioned above that Bhimraj had observed that the paddy plants have two sets of roots, and that he knew the reason for this (Fig. 3.2). The upper set grows from the first node on the stem, after transplantation. When we later looked this up on the internet, we found that the upper roots are technically called adventitious roots. In what appears to be a very authoritative source, the International Rice Research Institute (IRRI), we read, "After 10-20 days of growth, all roots of the rice plant are adventitious roots" (International Rice Research Institute 2019). Clearly, the farmers have more experience than the Institute: mature rice plants do have two sets of roots. Do both sets continue to function and grow after transplantation? How do the two sets of roots respond in case of water stress? There seems to be a deficit of research reports on such questions (Dongxiang Gu et al. 2017), although farmers may have some understandings based on their own observations. The actual authority is the plant itself. We need to observe actual plants in order to raise questions and do investigations, not just read and theorise. It occurred to us that if farmers had access to research reports, and if farmers wrote research reports, and if science and farming was done by the same people, science of and for all would be enhanced.

We previously discussed comparing and categorising (Sect. 3.2.3). Observation is needed in order to compare, categorise, identify, and name what is observed. For example, when winnowing, one has to categorise and distinguish between empty (undeveloped) husks and full (rice) grains, so that the former can be winnowed out. A decision has to be made as to when the grains are sufficiently winnowed. This process requires questioning and the recognition that one may have mis-identified, and that reclassification may be necessary. People need to have the agency to name, or rename, their observations. Hired workers may not have this agency. There is also a danger in naming. The act of naming something may actually inhibit further observation and questioning, if it is taken to be a conclusive, definitive fact. It may seem that the purpose lies in naming rather than in investigating and trying to understand. Thus, there is a dialectic in identifying, naming, and classifying: things are not necessarily what they appear to be. This dialectic drives us towards further observation.

3.2.8   Keeping Records: Oracy vs Literacy

Although both Smita and Bhimraj had completed school up to Class X, they did not use much reading and even less writing in their work of paddy cultivation.

We found some indications of record keeping (e.g. past yields, and expenditures). A record for selling rice that was written by Kalpana's grandfather in the same village is shown in Fig. 3.6. 148 kg of rice was sold for Rs 296, which is Rs 2/kg. For comparison, the retail price of Kolam rice in Mumbai in 2016 was about Rs 40/kg, and at that time the retail rate for 1 litre of full cream milk in Mumbai was about Rs 40. Most people would say that this record does not look very scientific (although we know professional scientists that write similar cryptic notes of vital data on tiny chits of paper!) Actually, he was the doctor in the village, and he was of the few who kept written records in books.

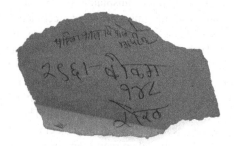

Fig. 3.6 A receipt of "First rice sold 13/05/8" indicating the amount Rs 296, for 148 kg of Kolam rice

However, records were often unwritten, unlike in professional science. Unwritten records were sometimes more qualitative than quantitative. For example, rather than keeping a written record, farmers might remember that they got a good price for a crop in a previous year, but not exactly how much they got for it.

Kalpana explained that since people like Bhimraj earn very little and spend whatever they earn right away, they probably do not find any need to keep written records of their accounts. They can't fix the prices of their products. Whatever the purchaser gives, they have to take it. Keeping records would just add to their tension about how little money they got.

The family did have a bank account, and there were written records from the bank. There were also other official documents, but not too many that were made by the farmers and directly related to farming. Also, except for the official land records, which included a map, we did not see many graphic records. The family had a few photographs of each other, but we did not see any photographs of the field, the crop, or the animals (this was in 2015, before the prevalence of phones with cameras). Observation that is more scientific may involve more record keeping, and more communication of records.

3.3 The Role of the Gods

The researchers saw that the family practices a number of rituals in connection with cultivation, and they wondered what the role of religion is, and whether there is a conflict between rituals and science in cultivation. The role of rituals in cultivation was discussed as we all stood in the field watching the rab burn down:

29:54	Karen:	एक और प्रश्न. ये हो सकता है, – ये, अह, अह, ये crop- ये फसल, अच्छी तरह से आ रहे- (One more question. Is it possible, that, uh, uh, this crop - this crop will come out good-)
30:04	Bhimraj:	हम्म. (Hmm,) [Both Smita and Bhimraj are looking at Karen as she is talking, Smita smiling, amused and Bhimraj looking very serious, with both hands behind his back.]
30:05	Karen:	- भगवान् के वजह से. (– because of bhagwan [god]?)
30:06	Bhimraj:	हम, (Hmm,)
30:07	Karen:	ऐसे है? (Is it like that?)
30:08	Bhimraj:	नही. मेरा- जितना आपुन मेहनत कर रहे है, वह उतना उसको उप्पर से (No. My- However much hard work you are doing, it's based on that much.) [Bhimraj shakes his head vigorously in negation, and begins smiling as he completes his sentence]
30:08		[Smita laughs and raises her hand, turning her palm upward as if to say, 'what nonsense!']
30:10	Karen:	[laughing] क्यों? (Why?)

30:11   Bhimraj: [smiling] आपुन ज- जब आपुन मेहनत दी- [bringing both hands in a gesture towards the ground in front] उसकी लिए इसके उप्पर इसके पास म- गया नहीं, [gesturing towards his left with both hands] तो ऐसा कैसे आएगी? [palms turning upwards as he asks] – ऐसे है, ना? [turning from Karen to look at the others] (You, wh- when you give hard work- for that- based on, for that – If not, how can it come? It's like that, na?)

30:18   Smita: Hmpha [a nodding laugh with a shake of her shoulders, –and Bhimraj glances at her]

30:19   Karen   तो फरि, अह्- भगवन भी जरुरी है? (So then, ah, bhagwan is also necessary?)

30:22   Bhimraj: नहीं, नहीं. नहीं, नहीं, नहीं. नहीं, ऐसे कुछ नहीं है. [shaking head in the negative and smiling] (No, no. No, no, no. It's nothing like that.)

30:26   Karen: [laughing] अच्छा. (Ok.)

30:27   Bhimraj: आपने तो सब- (Even all your-)

30:28   Ankita: [interrupting Bhimraj] कोंबड्या-बमि्बड्या कापता? (Do you sacrifice chickens?)

30:30   Bhimraj: अं? (Huh?)

30:31   Kalpana: कोंबडा कधी कापता? (Do you cut [sacrifice] chicken sometimes?)

30:31   Bhimraj: [smiling and shaking head in the negative] नाही, नाही! (No. No!)

30:31   Ankita: नक्की काय! (Certainly not!)

30:32   Bhimraj   [smiling and shaking head in the negative] नाही. तसं काही नाही. (No. Nothing like that.)

30:32   Ankita: नाही. ठीक आहे. (No. Ok.)

30:34   Karen: So-

30:36   Ankita: आणि पीक झाल्यावर पण नाही कापत? (And, not even after the harvest is done?)

30:37   Bhimraj: [smiling and shaking head in the negative] नाही! ते तसलं काही नाही. (No! Nothing of that sort.)

30:40   Smita: [smiling and shaking head in the negative] काहीच नाही! (Nothing!)

Karen was surprised by the answers to her question. Thinking that they are 'traditional', she assumed that at least they would say that gods are also necessary, in addition to hard work. But Bhimraj and Smita are very adamant that the paddy yield does not depend on any god, but rather on their own hard work. They may be laughing because it seems absurd to think that god could determine the outcome, or they may be laughing at Karen for asking such a question.

They also said very vehemently that they do not even perform any chicken sacrifice rituals – as some people in the area may do when faced by a grave problem.

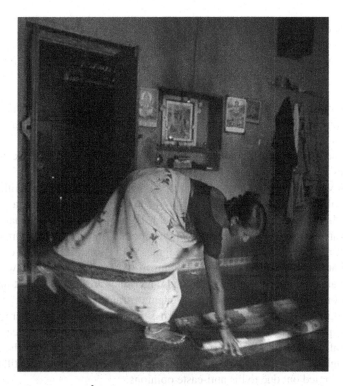

Fig. 3.7 Smita in front of the देव्हारा inside the house

But just after this, as everyone was walking into the house, Karen noticed inside there is a देव्हारा (a small alcove in the wall for worshipping, with some idols and pictures of gods, as shown in Fig. 3.7), and she asked about it:

01:51	Karen:	[looking and gesturing towards the देव्हारा (small altar) inside the house] वो- अह, अच्छा, भगवान तो है यहा पर? (That- uh, okay, so bhagwan is here?)
01:53	Namdev:	[laughing] हैं. वो मंदिर हैं. (It is. It's a mandir.)
01:54	Bhimraj:	नही नही ये, ये [... indistinct word] हैं- (No, no – that, that is only [...].)
01:55	Karen:	थोडासा मदत दे सकते हैं, नहीं? [laughing] (It can help a little bit, no?)
01:57	Bhimraj:	नही नही, ऐसा हैं, ऐसा कुछ नहीं. [Karen is still laughing] (No no, It's like this. This is nothing.)
01:58	Gurinder:	भगवान? (Bhagwan?)
02:00	Bhimraj:	वो अपना अपना अपना पूर-पूर प्रणाम प्रमाण कर लेंगे दे, आपुन. ऐसा. (This is for you to do r- r- respect [obeisance], for your own- your own- your own self. Like that.)

02:03   Karen;   अच्छा! इसलिए. (Okay! [with a patronising tone]
                     That's why.)
02:04   Bhimraj: हां. वो temple- आह- घर सामने रहता हैं. (Yes. That 'temple'-
                     ah- it is kept inside the front of a house.)

Karen was implicitly comparing what they had just said about gods and rituals with this evidence that they do practice rituals. Bhimraj explained that rituals are for themselves (for people, not for the crop). The implication is that these rituals are what people do, but it does not mean that the people doing rituals think it is what determines the crop yield.

Abhijit was not present during either of these recordings. When he initially saw the video of the first episode, he felt glad to hear that Bhimraj and Smita were discounting the role of rituals in the crop yield. Later, upon on seeing that they do practice various rituals, he felt somewhat disappointed. These feelings are because he himself rejects Hinduism – because he agrees with BR Ambedkar that caste must be annihilated and caste is integral to Hinduism (Ambedkar 1944). Drawing upon his personal experiences, he thinks that gods and rituals are seen as the ultimate way out when one suffers a huge personal problem or tragedy and when science seems to fail. Abhijit says he was reluctant to discuss this episode with the other researchers, and especially with the family, because he comes from a lower caste, and is therefore afraid that he may have a different set of beliefs and opinions than others. Even though he feels that his views on gods have evolved over time, he still feels he might be singled out due to his anti-caste opinions.

Abhijit was disturbed at hearing Karen's constant laughter throughout the second episode. This was because he felt that she was trying to make fun of their earlier statement that they do not believe that gods help in cultivation. She wanted to show them they were having a contradiction. He pointed out that the laughter may indicate condescension or scornful derision. He noticed that Karen was the only one we heard laughing.

After thinking about it, Karen agreed. She also started to wonder whether she may have been acting like a white, foreign missionary (or an anti-missionary? – or a scientist?) by pointing out the family's 'contradictions' to the family, and trying to 'teach the natives'. And in fact she is white and foreign, although she has been living and working in India for almost 35 years. Her extended, nervous laughter and somewhat patronising tone reinforces this interpretation, and it may be that she was actually laughing at the family or at their customs. On refection, Karen now thinks that probably she was subconsciously trying to teach them a lesson in rationality – albeit while thinking she was being a learner rather than a teacher. She was acting as if she thought she was an impartial/objective observer. She now sees her own racism and arrogance in this. This is what she has learned from this episode.

Abhijit disagrees that Karen was being racist, but still he does not think she should have laughed like that. He now feels indifferent about whether the family's beliefs and rituals are 'correct' or not, even though he realises that they must be different from his own. He suggested that we should analyse the family members' views on rituals and religion without bringing in our own points of view and without letting our own feelings interfere. We discussed whether this is preferable, necessary, or even possible. It is clear that the researchers have various different points of view on religion. If we try to analyse the episode by explicitly considering all our various points of view, the analysis will be lengthy and complex, and we probably would not come to any agreement or single understanding of the episode.

An alternative, or additional explanation is that Karen's laughter may have been an attempt to reduce the seriousness of her accusation, relax the tension and decrease her own nervousness. The nervousness may have been the result of realising that she had said something that maybe she should not have said, because talking about religion may be hurtful or derisive, and we wanted to avoid confrontation.

There are some things that were not talked about in any of the conversations between the family and the researchers. For example, we all avoided asking or mentioning our castes. Initially, we purposely did not ask each other's surnames, which may give some indication of caste. Up until these conversations, we avoided talking about religion, and even here we never explicitly mentioned which religions we belonged to. However, Karen's white skin signified Christianity, Gurinder's turban signified Sikhism, and the others could be assumed to be Hindus. People, especially those who are not from dominant backgrounds, may be afraid and hesitant to talk very openly about their own caste and religion because these are very sensitive issues. At the time of writing, there continue to be major instances of communal violence, rioting, and even genocide in different parts of the country, as well as numerous cases of persecution and injustice due to religious conflict. Even if we want to be open-minded and tolerant of the beliefs of others, it is not easy.

The beliefs of Smita and Bhimraj regarding rituals reminds us of what Jyotirao Phule had written almost 150 years ago. He had described the numerous festivals and rituals that farmers participate in throughout the year, and how each time they are made to give donations to the Brahmans who perform the rituals (Jyotirao Phule 1881). He does not mince words in describing the perfidity of the Brahmin priests. But interestingly, he does not mention any donations or rituals that are done for the purpose of getting better harvests. All the rituals he mentions are done ostensibly for the well-being of the people: for births, deaths, weddings, eclipses, pilgrimages, in addition to numerous religious festivals for various gods and goddesses. It might be that it is not uncommon to find that people who do not believe that the rituals actually have much direct effect on the outcomes of their cultivation.

We will analyse the meaning of this episode in a larger context – in relation to the conflict between religion and science – in Sect. 7.2.1.2.

3.4   Power Relations and Problem Solving

3.4.1   Who Tells Who What to Do

In Chap. 2, we mentioned that the work of cultivation and maintaining the house and farm involves a lot of discussion around daily tasks and seemingly trivial matters. This simultaneously involves a lot of learning and teaching of how to do things. We can try to understand how teaching is occurring by seeing who is telling who what to do, who is asking what is to be done, and how decisions are made and communicated.

For example, as Bhimraj, Namdev, and Smita were walking in single file along a path to get more branches for rabbing, Bhimraj realised that a tool called bailla was needed to carry the branches and he called out:

00:09	Bhimraj:	बैला – बैला आण, बैला! (Baila! bring the baila, baila!)
00:13	Smita:	कुठनं आणू? (Bring it from where?) [turning back to look for it]
00:15	Bhimraj:	[pointing] तथिनं आण, बैला. (Bring it from that side. Baila.) [and Smita goes to get it]

Even without calling her by name, it is obvious that he was asking his wife rather than his father. And Smita did what her husband requested, without hesitation. Later on, Namdev asked Smita to bring a soop full of soil and a bucket of water to sprinkle on the rab. Namdev told Bhimraj to spread the soil and he would sprinkle the water himself. In context, these requests seemed neither artificially polite nor domineering. They seemed to be straightforward and practical.

These were the directives that we first heard when we went to the farm. We began with some preconceptions. We had thought that things would be done according to a hierarchal order that might be expected in any traditional patriarchal society: grandfather > father > wife.

The position of the children would be expected to be at the bottom. And we occasionally heard the adults asking the boys to do things. For example, when starting to make the layers of the rab, Bhimraj asked his older son to get a basket:

00:52   Bhimraj: पुरतकि टोपली घेऊन ये जा. (Pratik, bring the basket.)
[Bhimraj continues working and discussing the rab with Gurinder.]
02:56   Pratik: [faintly, from a distance] भरलेली का रकिामी पप्पा? (Full or empty, Papa?)
[The work continues.]
03:02   Pratik: [interrupting Gurinder talking to Bhimraj] पप्पा, भरली का रकिामी? (Papa, filled or empty?)
03:05   Bhimraj: [turning towards Pratik, who has approached from the back] अ? (Huh?)
03:05   Pratik: भरली का रकिामी? (Filled or empty?)

03:06 Bhimraj: [softly] अर्रे रि-रि-रिकिमी-रकिमी! (Arré, em- em- empty. Empty!)

03:08 Pratik: [mumbling to himself] रकिमी. (Empty.) [walking away to get the basket]

Bhimraj was not overbearing when he made this request. And he sounded more amused (or maybe slightly frustrated) rather than irritated when he had to explain that the basket should be empty – something he must have thought would be obvious. It seems that Pratik was ready to do what his father asked – but without acting scared or being meekly obedient. Maybe it was not just that he accepted his role in the hierarchy, but that he also understood that a basket was needed, and that family members help each other. We may have gotten this impression because we did not see him constantly being asked to do things, or any sign of him being annoyed or not wanting to do something he was asked to do. He seemed to be free to come and go as he pleased. The boys were the first ones to happily run out to the field at the beginning of the rabbing, and they continued running around and playing while the others were working.

Later, on the same day of rabbing, Namdev paused his back-breaking work in the sun and asked his younger grandson (Pranay) to bring him a drink of water:

10:38 Namdev: बाब्या! कुठे गेला बाब्या?! (Baabya! Where did Baabya go?!) [as he continues working]

10:56 Namdev: ए बाब्याला सांग पाण्याचा तांब्या घेऊन ये रे! (Hey tell Baabya to bring a taambya of water.)

11:00 Namdev: जरा कटिलीत पानी आण. (Get a ketli [container] of water.)

11:13 Bhimraj: ए छोटू, पाणी घेऊन ये तांब्याभर! पाणी! पाणी! पाणी घेऊन ये प्यायला! (Hey Chotu, go bring a taambya of water! Water! Water! Go bring water for drinking!)

11:25 Namdev: ए बाला बस कर चल! हे घे – हे घे चल. (Hey Baala [child], enough now! Take this – take this c'mon!)

12:09 Namdev: [shouting] ए बास झाला चल! ये! ए बाल्या! पाणी आण जा! (Hey! Enough now! C'mon! Hey Baalya! Go bring Water!)

12:29 Namdev: अर्रे बास कर रे राजा! (Enough now, my dear [prince]!)

Pranay was younger (only 6 at the time), and he was preoccupied with his own games, although he was usually playing somewhere nearby where the family was working. He had to be asked several times to get some water. Here again, Namdev and Bhimraj may have been somewhat frustrated, but they did not appear to get angry, or to say anything derogatory to Pranay. Rather, they used coaxing and called him lovingly by various pet-names.

So paddy cultivation might appear to be proceeding according to the expected traditional power relations. Through these small requests for help in fetching, carrying, gathering, and spreading material, people also teach and learn how to do the work that is being done. If the work is more traditional, then we might expect there would be one authority (the grandfather) at the top of the command, who teaches what he knows is to be done (as he has learned from his teachers).

However, we also noticed many instances in which this hierarchy was contradicted. For example, while rabbing, Bhimraj told his father what to do:

04:00   Bhimraj: [To Namdev] गवताची मोळी बांधा! गवताची मोळी बांधा! (Tie
                               the bundle of the grass! Tie the bundle of the grass!)

The tone of voice was not at all subservient. Bhimraj was quite insistent.
Smita sometimes told Bhimraj what to do. For example:

01:42   Smita: साफ़ करुन पण देखवाल न लगेच? साफ़ करुन पण? (We can show
                               them the cleaning [threshing] now, the cleaning?)
01:44   Bhimraj: [nods yes]
01:46   Smita: चला बस जाला। या अगोदर, इथुन काध्ता येतलि, ना? मग एक
                               काध्या। (Done enough. Come down. From here we can take
                               one, nah. Then take one [bunch of paddy].)

Here we see that Smita is directing Bhimraj, but her tone was not nagging. She seemed neither overly authoritarian nor submissive.

3.4.2 Arguing About the Kanaga

The problem mentioned above about how to set up and fill the kanaga illustrates how the family works together to experiment and investigate questions arising from conflicts that are both social and physical.

The argument concerning this plan was the most emotional sort of disagreement that we witnessed on the farm. It was filmed by Karen, who could not understand much of what was going on, but did not say anything because she just wanted to record the others interacting with each other, and she was afraid that her own questions may interfere with their discussion. Even so, it is probable that her filming did affect the discussion to some extent, although no one looked directly into the camera. Although it was a rather mild argument, it was emotional.

The problem and its resolution were defined and carried out as a group effort of Namdev, Smita, and Bhimraj. It began when they were figuring out where to set up the kanaga:

		[All are in the storage room. Smita is sweeping. Bhimraj is dragging out a plastic sheet. Namdev is standing with one hand on the kanaga. Dnyaneshwar is standing to the side.]
00:01	Namdev:	[looking at Smita] घे तिकडे घे. काढ पाय. हा [… an inaudible word] घे जरा. खालन घाण काढ ती. (Take it until there. Remove [your] foot. Yes. Take it there a bit. Just remove the dirt from underneath.
00:03	Smita:	[to Dnyaneshwar] भाऊ जरा ते पाय-पुसनं द्या ना. (Brother, can you just give that foot mat [doormat], na?) [Dnyaneshwar immediately does several things which cannot be seen very well on the camera, including moving a bicycle aside.]
00:04	Namdev:	[looking at Smita] ?
00:05	Bhimraj:	[with a somewhat raised voice, pointing to a place on the floor] इथे – इथे – इथे ठेव ना. (Here. Here. Keep it [the kanaga] here, na?)
00:05	Namdev:	इथे ठेव ना. ना? (Keep it here, na. Na?)
00:06	Smita:	इथे ठेवायचाय? (It has to be kept here?)
00:09	Bhimraj:	हां. इथे ठेव ना. (Yes. Keep it here.)
00:11	Namdev:	[raising his voice as he speaks] इथेच ठेवायचंय. इथे घे ना. इथे – इथे – इथे घे ना. जरा खालची घाण काढू दे. (Have to keep it here only. Take it here. Here – Here – Here – Take it here. Let [her] remove the dirt from below.)
00:13	Smita:	[softly] मग. चांगला आहे ना तिथि. (Of course! Its clean over there.) [Smita pauses for a few seconds while Bhimraj and Namdev move the bin to the side that was already swept. Then she sweeps where the bin was, as Namdev and Bhimraj watch her.]
00:20	Namdev:	[Looking at Bhimraj] हा तर मोठा प्रश्न आहे.. (Now this is a big question:)
00:22	Bhimraj:	[turning his head to look at Namdev] काय? (What?)
00:23	Namdev:	खाली कसा करायचा? (-how to get down?)
00:24	Bhimraj:	[raising his head in a questioning gesture] खाली? म्हणजे? (Down? Means?)
00:28	Namdev:	हा, जो- (Yes, which-)
00:29	Smita:	[interrupting Namdev, stops sweeping, and turns towards Namdev and Bhimraj] माणूस थांबायला लागतो. आतमध्ये. आतमध्ये एक माणूस थांबायला लागतो. (A person has to be there. Inside. A person needs to be waiting inside [of the kanaga].) [As she begins speaking, Bhimraj glances at her, but then turns to look at the floor. As she continues, Namdev turns to look at her.]

00:31	Bhimraj:	[gesturing as if spreading something on the floor, raising his voice slightly] हे पुठ्ठे पहिलि आंथरून घेतले-तुम्ही घेणार म्हणताय ना? (These cardboard sheets you will – you want to put, is that what you are saying?)
00:34	Namdev:	[looking at Bhimraj, with a slightly raised voice] ते पुठ्ठे बाहेरच ना – बाहेरच राहतील ना! पुठ्ठे बाहेर राहतील. (Those sheets [will be] outside only – will be outside only! Sheets will be outside.)
00:38	Bhimraj:	[in a somewhat raised voice] अहो, पण नंतर आपण – तुम्ही- आतमध्ये जाणार आहात, ना? नंतर भात टाकते वेळी. (Oh [addressing Namdev], but then we – you – are going to go inside, na? -at the time of putting rice.)
00:42	Smita:	हां. एक माणूस आतमध्ये राहील. (Yes. One man will be inside.)
00:43	Bhimraj:	हा मग! (Of course!)
00:44	Smita:	तेव्हा तर एक माणूस देईल ना वरून. (Only then can another one hand it from above.)
00:45	Bhimraj:	आपुन काय लगेच भात टाकणार आहोत? अजून time आहे भात टाकायला! (We're not going to put rice right away, are we? There is still time to put rice!) [as he sets the folded plastic sheet on the rim of the kanaga]
00:46	[[Namdev brings an opened cardboard carton from nearby, and tosses it down on the swept floor next to the kanaga.]
00:47	Smita:	अहो पण समजून तर [laughing nervously] दाखवा ना यांना, लगेच राग कुठे दाखवताय?! (Hey, but you can make him understand this. Why are you being angry?)
00:52	Bhimraj:	मग तेच काय! (Yeah, of course!)
00:57	[Namdev tears off a piece of the cardboard from one side and places it on the other side]
01:00	Bhimraj:	अखंड राहू द्या ना तो! (Let it be intact [in one piece]!)
01:01	Namdev:	कशाला अखंड?! ... [inaudible dialog] (Why intact?! ...) [Namdev does not continue tearing the cardboard sheet.]
01:04	Smita:	[To Bhimraj] ओ राहू द्या! (Oh [hey Bhimraj], let it be!)
01:07	Bhimraj:	अह (Huh)
01:08	Namdev:	टाक त्याच्यावर ते प्लास्टिक टाक! (Put, Put the plastic on it!)
01:08	[Smita makes a 'tch' sound]
01:09	Smita:	[softly and hurriedly] टाका! टाका! – टाका! टाका!! (Put! Put! – Put! Put!)
01:10	Namdev:	[raising his voice as he speaks, and gesturing towards the plastic sheet] तू प्लास्टिक टाक! हाथर! आता हे साफ करताना – आतमध्ये शरिताना – हे करून टाकू – भरून टाकू. (You put the plastic! Spread it! Now, clean it – while going inside – we can do this – fill it up.)

01:11	[Bhimraj removes the plastic sheet from the rim of the kanaga, and lays it on top of the cardboard, then turns it over so that the blue side, rather than the white side, is on top (may be cleaner), with Namdev helping.]
01:22	Smita:	हां जरच करा! एक आतमध्ये माणूस थांबाय लागेल. (Yes. Do it then! One person has to wait inside.)
01:26	[Bhimraj and Namdev together start moving the kanaga, placing it on top of the plastic sheet]
01:27	Namdev:	[muttering] हे बोलणं काय? (What is this way to talk?)
01:28	Smita:	मग तो माणूस बरोबर बोलेल. (Then the person will talk correctly.) [as she finishes the sweeping]
01:40	Bhimraj:	बास? हाच जर आता आतमधी पडला असता ना आत्मधून- (Enough? Now had this thing been inside then-)
01:46	Namdev:	हम्म? (Hmm?)
01:47	Bhimraj:	-तर तो कडेला असा गोल – ह्याच्या – राहिला असता. (-then it would have been – this rounded on the side of this – of this)
01:48	Namdev:	वरती परत कसा निघिणार शेवटपर्यंत?! (How would [one]come out at the end?!)
01:51	Smita:	[Looking at and walking towards Bhimraj and Namdev] काय?! माणूस निघितो बरोबर! (What?! A person comes out alright!)
01:52	Bhimraj:	[Bhimraj and Namdev are standing very close to each other, looking at each other as they talk] आहो हां त्याच्या – भात जसा पुरा- – जसा-जसा भात खाली-खाली हे होत जाणार, तसं वरती येणार माणूस- (Hey this – his – just as the rice completely – just as the rice would keep going down, down – at the same time, the person will keep coming up-)
01:56	Smita:	वर येणार! ... (-will come up! ...) [something more she says here which is inaudible as she is walking away]
01:58	Bhimraj:	माणूस तथिनं वरतीच येणार! (The person will come right up!) [he raises his hand to his head, palm up, as a gesture of frustration]
02:02	Namdev:	[moving his head slightly downwards] मग मी जातो आतमधी! (Then I will go inside!)
02:03	Bhimraj:	[looking at Namdev, with an angry expression on his face] आता नाय! अजून time आहे! (Not now! There is still time!) [As he starts walking off, Namdev catches the eye of Dnyaneshwar and smiles at him. Dnyaneshwar smiles back. Bhimraj rolls his eyes towards Dnyaneshwar, and they both smile.]

02:05	Namdev:	[smiling and laughing, looking at Dnyaneshwar] हाततच्चियातर! (Oh dear god!) [Dnyaneshwar smiles, looking from Bhimraj to Namdev]
02:06	Bhimraj:	[laughing] मग! (See!)
02:06	Namdev:	अजून time आहे! (There is still time!)
02:07	Bhimraj:	आहो, नाही तर काय! (Hey, isn't that what I have been saying!)
02:08	Dnyaneshwar:	मी येऊ म्हशीला घेऊन? (Shall I bring the buffaloes?)
02:09	Bhimraj:	जा! जा! जा! जा! जा! (Go! Go! Go! Go! Go!)
02:09	Smita:	जा. (Go.)
02:11	Bhimraj:	चल थोडी- (Come some-)

One might think that setting up and filling the kanaga would be directed by Namdev, since he is the elder and could be assumed to be the authority figure. Namdev did first direct Smita in sweeping the floor. But later on it became apparent that he is not the main taskmaster or teacher, nor is Smita the silent, obedient daughter-in-law, nor does his son always accept Namdev's authority. Smita and Bhimraj are actually teaching Namdev, who seems to be more of a learner. But we saw that the roles kept getting interchanged, with cooperative teaching and learning from each other and through collective experimentation. This is despite the fact that all this did occur in a patriarchal context. But this traditional patriarchy was different from what we expected.

Namdev and Bhimraj disagree about exactly where the kanaga should be placed, and at 00:05 Bhimraj uses a slightly raised voice to tell his father where to put the kanaga. But then, without any verbal communication, Bhimraj and Namdev work together at 0:15 to move the kanaga aside for the sweeping. (Although it was not heavy, it was flimsy and awkward and had sharp edges, so it required two people to move it.)

At this point, 00:20, Namdev is the one to state the main problem of how to fill the kanaga, which he identifies as being a big question – and it is a genuine question for which he does not know the answer. We have discussed above how we interpret this as being an investigatable question (Sect. 3.2.5).

Bhimraj is not sure he understands the question, and asks for clarification. But interestingly, Smita immediately understands the question, and (at 00:29) she interrupts the discussion between Namdev and Bhimraj to assertively state the solution to the problem: that a person needs to go inside the kanaga (Fig. 3.8).

Bhimraj was the first one to say that Namdev would be the one to go inside the kanaga. Maybe he decided this as he was speaking, clarifying that 'we' means 'you': "अहो पण नंतर आपण – तुम्ही- आतमध्ये जाणार आहात, ना?" (Oh [addressing Namdev], but then we – you – are going to go inside, no?).

Namdev seemed indignant at the suggestion that he should go inside. He may have had past experience of going inside a kanaga for loading or unloading, and it can be tiring as well as claustrophobic. In this case, it was not clear if it was even

Fig. 3.8 Smita says that a person needs to go inside the kanaga

safe. As we discussed, it is not clear whether he would get buried or rise up with the grain as it is filled.

Both Bhimraj and Namdev seem frustrated with each other for disagreeing and not understanding. Bhimraj softens his stance by saying at 0:45 that it is not yet time to go inside the kanaga, and Smita tries to diffuse the situation and stop them from arguing. But Smita and Bhimraj are clearly supporting each other, both in opposition to Namdev.

Namdev spreads out a sheet of cardboard (from a dissembled carton) on the floor. Bhimraj and Namdev also disagree about whether the cardboard sheet should remain intact, but Namdev more or less acquiesces to Bhimraj.

In the middle of their conflicts, and without any verbal communication about it, Bhimraj and Namdev work together (at 1:11) to spread the plastic sheet over the cardboard, to turn it over so that the cleaner side is facing up, and (at 1:26) to place the kanaga on the cardboard, even while Namdev (at 1:27) mutters that this is no way to talk: "हे बोलणं काय?" This indicates that despite their disagreements, they must be understanding that the plastic needs to be spread and the kanaga needs to be moved, and that it takes two to do these things, so they cooperate. Neither Bhimraj nor Namdev hold all the power over the other.

Then (at 01:48), Namdev explicitly states his worry: "How would [one] come out at the end?!)" Bhimraj and Smita explain that "Just as the rice would keep going down, down – at the same time, the person will keep coming up-". Bhimraj is clearly angry and frustrated (Fig. 3.9). Finally Namdev agrees to go inside the kanaga, and the situation is diffused when everyone laughs and smiles at each other (Fig. 3.10), and they all go out to do the threshing.

This episode is interesting in that it shows how a conflict arises: probably because the elder father is being told by his son to do something that he does not want to do, or is afraid of doing. Those patriarchal relations do exist and create conflicting emotions. Bhimraj becomes annoyed, maybe because his father is not agreeing and not

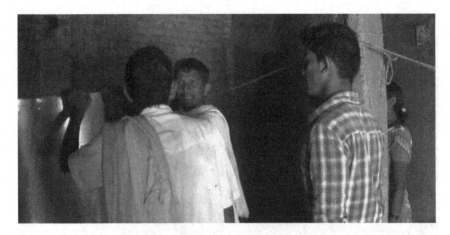

Fig. 3.9 Bhimraj and Namdev are angry

Fig. 3.10 Namdev, Bhimraj, and Dnyaneshwar laughing afterwards

understanding. Apparently Bhimraj does not think there is any risk of getting buried or of the whole thing toppling over.

But underlying the conflict, it seems that everyone also wants to avoid conflict and get the work done, and in the end they do not mind laughing the whole thing off. Maybe they do not take themselves or their arguments so seriously that they do not see the humour of the situation, or maybe they are a little embarrassed. Even in the midst of the conflict, Bhimraj and Namdev work together cooperatively to move the kanaga, spread the plastic sheet, etc. Getting the work done seems to take priority over conforming to the power relations.

After threshing and winnowing, it was time to fill the kanaga. Everyone loaded grain into baskets and carried them on their heads into the storage room (which is rather dark). This is what happened:

01:45		[Dnyaneshwar starts unloading the basket in the bin.]
01:45	Bhimraj:	थांब! थांब! टाकू नको. दादांना येऊ दे. (Wait. Wait. Don't put. Let father come.)
01:52	Dnyaneshwar:	कशाला दादांना? (Why sir?)
01:53	Smita:	जरा आतमध्ये ठेऊ दे त्यांना. (Just have him inside.)
01:54	Bhimraj:	थांब कसा काय बोलतात ते बघू दे, कुठे ठेवायचंय. (Wait to see how to do it, to keep it where we want it.)
01:56	Uncle:	[holding a child] पहिलि दुसरं कणगा होता तो दाखव. (First show the other kanaga.)
01:57	Smita:	का नुसता टाकून ठेवायचाय विचारा? (Or why not put it first?)
01:59	Kalpana:	काय? (What?)
02:00	Bhimraj:	नाही. टाकून तथ्यून direct आता- (No. Dump it directly from there.)
02:02	Kalpana:	मामांना विचारा? (Ask mama?)
02:07	Smita:	ते जातायेत का विचारा अगोदर, आतमध्ये? (First ask [him] about going inside.)
02:12	Bhimraj:	[calling Namdev]: दादा! (Dada!)
02:13	Uncle:	आताशी टाकून तरी होईल मग हे नको करायला… (We can put it right now so that we don't have to do it later…)
02:15	Bhimraj:	अर्रे, कफ्लगि सुटली म्हणजे त्याची मग? (Arré, what if the coupling breaks?)
02:18	Smita:	हां ते असं आतमध्ये जाया लागतंय एक माणूस. (Yeah, one person needs to go inside.)
02:20	Bhimraj:	कफ्लगि जर side ने तुटली तर …[inaudible] (If the coupling breaks from the side …)
02:23	Uncle:	मग दुसरं चांगला असतो- कनुग्याचा. ह्याचा.. बाजूने…टाकायचा नाहीस.. (Then the second one is better – of the kanaga…[you] should have put it from this side…)
02:26	Bhimraj:	हे – हे – आहे ना, ह्यात जर तू ओतलास, आणि हितिन सगळा भात निघाला तर? (This- this is there no, if you pour in this, and if all the rice comes out, then?
02:29	Uncle:	आला! [Namdev enters] ([I] came!)
02:30	Namdev:	चल, मला आतमध्ये – मला आता- (Come on, let me inside – Let me now-)
02:31	Smita:	थांबा! तो कणगा उचला! अहो तो कणगा उचला! आन ते बरोबर टाकतील. (Wait! Pick up that kanaga! [To Bhimraj] O! Pick up that kanaga! And he will be put in it the right way.)
02:33	Uncle:	[Inaudible word] आणायची. (Should have bought [Inaudible word])

02:34	Smita:	नाही! कशाला?! खालून जातील बघा आरामात! (No! Why?! [He] will go from below, easily/comfortably!)
02:36	Bhimraj:	बरं, आणखी तिकडे पाहिजि का बास? (Okay, do you want it more on that side or is it alright?)
02:38	Namdev:	बास! काय करायचं? काय करायचं?! (Enough! What do we have to do?! What do we have to do?!)
02:39	Smita:	[To Bhimraj]: नको राहू द्या. बरोबर इथंच आहे. ओ नका नका. (No let it be. It is alright here. [To Bhimraj] O No. No.)
02:42	Namdev:	आणखी कुठे जायचं?! कुठे न्यायचं नाही काय करायचा नाही जा-जा- फक्त घाण जायला पाहिजि. (Where more do [we] have to go? [We] don't have to take it anywhere, have nothing to do with it. J-j-j- Just the dirt needs to go.)
02:47	Namdev:	[To Dnyaneshwar] अरे! आता खाली ठेव ना खाली! (Arré! Keep it down, now!)
02:49	Bhimraj:	[To Dnyaneshwar] ए! कशाला भरतोस?! तिकडं कशाला भरतोस?! (Hey! Why filling it up?! Why filling it up that side?!)
02:50	Dnyaneshwar:	अरे दादाला आतमधी शरियला. (For Dada to get inside.)
02:54	Smita:	[To Bhimraj] ओ तुम्ही उचला ना! (Hey! You pick it up, no!)
02:57	Dnyaneshwar:	[To Namdev] ते राहू द्या! (Let that be!)
02:59	Namdev:	राहू द्या नाही, आता आतमध्ये टाकायचं. जरा जरा जरा. बस. (No. Not let it be. Have to put it inside now. That's all.)
03:07	Smita:	येताना वर, बरोबर, आपोआप येतील. (While coming up, [he] will come automatically.)
03:09	Bhimraj:	हां ठेवा! करा उभा!! (Yeah, keep it! Make it stand!!)

[As Bhimraj & Dnyaneshwar tilt it up, Namdev crawls inside the bottomless kanaga.] (Fig. 3.11)

03:18	Bhimraj:	बरंय? आता प्लास्टिक गोल करा. (Alright? Now rounden the plastic.)
03:30	Smita:	जाम उंच आहे ना कणगा?! (It's too tall, nah? – the kanaga?)
03;32	Uncle:	ब-हे-पाट बारीक आहे. (The width is a bit narrow.)
03:33	Smita:	हां. पाट बारीक आहे. (Yes. The width is a bit narrow.)
03:34	Bhimraj:	[To Namdev] झाला सगळा? केलात? सगळा गोल करा. (Done with everything? Round it all out.) [To Dnyaneshwar] Mobile दे तुझा. (Give your mobile.)
03:40	Bhimraj:	[Shouting to someone far away] ए battery दे रे तुझी! Torch दे तुझी – mobileची! (Hey, give your battery! Give your torch – of the mobile phone!)
03:51	Namdev:	काय रे काय करू आता?! (Now what has to be done?!)
03:52	Bhimraj:	सगळा गोल करा! हा प्लास्टिक आहे ना, सगळ्या बाजूने गोल करा. (Round it all out. Round this plastic from all sides!)
04:07	Bhimraj:	[To Namdev] हां. ठीक आहे. (okay.)

04:15	Bhimraj:	[To Namdev] बस? झालं आता बरोबर packing? (Alright? Finished up with the packing properly?)
04:20	Namdev:	हा. भात बाहेर नको यायला. (Yes. Rice cannot come out.)
04:24	Uncle:	[Holding a Child] हां. टाका. टाका भात टाका. (Yes. Put. Put the rice.)
04:27	Namdev:	असती-असती द्या. असती-असती द्या. (Give it slowly-slowly. Give it slowly-slowly.)
04:32	Namdev:	स्टूल द्या. स्टूल द्या. (Give a stool. Give a stool.)
04:33	Smita:	स्टूल नका. असंच घ्या. तुम्ही बरोबर वर याल. (No need for a stool. Take it this way only. You will come out right.)
06:45	Namdev:	अरे असा आडवा-वाकडा नाही ना नघित? (It's not tilting, is it?)
06:48	Bhimraj:	नाही. नाही. (No, no.)

Dnyaneshwar happened to be the first one in the line of people carrying grain into the storage room. Even though he had previously heard the decision that Namdev was supposed to first get inside the kanaga, he started dumping the grain in the kanaga without anyone inside it. It's unlikely that he had forgotten that argument. Probably he did not have sufficient agency to ask Namdev to go inside, or even to ask anyone else whether someone should be inside. Bhimraj had asked him to start filling the kanaga, so he started filling the kanaga.

It is interesting that although it had seemed like the big problem of how to fill the kanaga had been solved at the time of setting up the kanaga, it had only been solved in theory. When it came time to actually do the filling, the problem again cropped up, and became complicated with additional problems. Even the question of the exact place of the kanaga on the floor was asked again by Bhimraj, this time, seemingly, in deference to Namdev. Unlike earlier, now everyone seemed to be more agreeable and less annoyed, and Namdev did go inside the kanaga (Fig. 3.11, 3.12).

Fig. 3.11 Namdev going into the kanaga

Fig. 3.12 Dnyaneshwar handing a load of threshed paddy to Namdev, who is standing inside the kanaga to fill it

But new questions also arose: Would the coupling that held the seam of the kanaga together break as it was loaded? Would it be more apt to break if the kanaga was filled by throwing the grain from the top? How should Namdev get into the kanaga? Where is a light (to see if the plastic is arranged properly)? Is a stool needed? Is the kanaga wide enough for Namdev to get down and fix the plastic around the bottom and then stand up? This last question is related to the physics of counter-balancing to stand up when you are sitting – you need some space to lean forward or you cannot stand up.

All of this questioning was interconnected with the collaborative doing and investigating. There was quite a bit of trial and error as well as application of past experience. But bookish knowledge was absent. Relying on any kind of authority was very reduced. Various people contributed to the collaborative effort, all with the important and necessary aim of having to store the grain in the kanaga.

The researchers were surprised about the social relations and power hierarchies that they saw in these episodes involving setting up and filling the kanaga. There was something different from what they expected, and what they noticed when they first went to the farm. We will discuss this in Chap. 7.

3.4.3   Gender Relations

In the above episode in which there was an argument about the kanaga (Sect. 3.4.2), we might have expected Smita to be very submissive, but we see her running the show: (1) she gives the solution to the problem (a person must go in the kanaga); (2) she pacifies the argument between the men; (3) neither of the men ever argue with (or raise their voice against) her and neither does she argue with them; (4) none of the men criticise or complain about her; (5) her husband accepts and agrees to everything she says; (6) she does not get upset or frustrated, although she is concerned that the fight should not escalate.

Nevertheless, it is true that Smita is the one doing the sweeping, which may be considered to be the more lowly kind of work. And she does not complain.

So this could be seen as a dialectical conflict: Smita is at the same time a victim of gender-based oppression and she is leading the decision-making, problem-solving, and inter-personal relations in order to accomplish the cooperative work the family is doing.

Looking beyond the superficial appearance of a division of labour according to gender and class, we see much more complex power relations. In another example, when carrying 70 kg gunnysacks of grain to a tempo for transport to the mill in Indapur, although Smita is present, she does not help carry the gunnysacks. Each gunnysack is carried by two men. However, she does participate, as shown in the following episode. Smita and Arjun are watching Bhimraj pull a gunnysack of grain off a pile that had been stored for two months in the storage room:

01:40	Smita:	[looking at a tear in the gunnysack] फोडलेत काय??! मगापासून बोलली तुम्हाला बघा म्हणून फोडलेत काय. (Is it torn? Since a long time [I] have been telling you to see if they are torn.)
01:43	Kalpana:	उंदराने? (Rats?)
01:44	Smita:	हो. ते बघा! ती दुसरी आणू गोणी? ओ सांगा? (Yes, Look at that! Should that other gunnysack be brought? Oh [addressing Bhimraj], tell?)
01:48	Bhimraj:	गोणी नको. Packing packing घेऊन ये. (Not the gunnysack. Packing-bring the packing!)
01:51	Smita:	[interrupting] किती भरायचेत? मग काय आणू तुम्हाला packing ला? Plastic पिशवी आणू का? (How much to fill? Then what should be brought for your packing? Should a plastic bag be brought?)
01:55	Bhimraj:	[To Arjun, who is putting his hand in the hole] थांब! तू फाडू नकोस!! थांब! थांब लावतोय बघ. Wait! Don't you tear it. Wait. (I) am putting, see!
02:00	Smita:	का [...]? प्लासटकि [...]? ([...]? Plastic [...]? [inaudible]).
02:01	Bhimraj:	ह? (What?)
02:01		[Arjun steps up on the pile to reach a carton from which he tears off a piece of cardboard and stuffs in in the hole. Bhimraj helps.]
02:03	Smita:	का पुठ्ठा टाकणार? प्लासटकि पिशवी आणू काय? (Or (will you) put cardboard? Should (I) bring plastic bag?)
02:04	Bhimraj:	हां आण. (Yes. Bring.)
02:06	Smita:	नाही तर मग दुसर्या गोणीमध्ये भरायचं. (Otherwise fill it in other bag.)
...		
00:00	Arjun:	तकिडून घे ना. (Take it from there, no?)
00:01	Smita:	नाही, त्याच्यातलाच घ्या ना. (No. Take one from there only.)

| 00:04 | Smita: | चांगले नाहीत ना गोण! राहू दया आपण हे बदलू ना! (The gunnysacks are not good, right? Let it be, we will change them!) |
| 00:09 | Smita: | [Asking Bhimraj] काय हो? (What do you think?) |

[Bhimraj removes a handful from the bag and shows it to Ankita and Smita.]

00:10	Bhimraj:	कुठंय हयाला? (Where is it in here?)
00:11	Smita:	हे बदलू. [? Indistinct word(s)] झाले ते. (Yes change. [indistinct])
00:12	Kalpana:	खाऊन टाकले ते? (Was it eaten?)
00:14	Smita:	तो घ्या मग. (So take it.)
00:15	Arjun:	कुठं घ्यायचाय? (Where do you want [it]?)
00:16	Smita:	त्या हया sideच्या. हया झाल्या पाच गोणी पूरण? (Those on this side. Are these five done? [removed from the pile, ready to take to the truck])
00:20	Arjun:	हम्म हम्म. हया गोणी? (Hmm, hmm. [yeah] These gunnysacks?)
00:21	Smita:	ओ? (Oh? [addressing Bhimraj])
00:22	Bhimraj:	[as he tries to lift the bag] हः धान्य खेच. तकिडून खेच इकडे. (Pull the grain. Pull it from that end and bring it to this side.)

[Arjun goes to lift the back of the bag.]

00:26	Smita:	हया झाल्या ना-त्या पाच झाल्या ना पूरण? (These are done. Those five are done.)
00:28	Bhimraj:	हो. त्या पाच झाल्या पूरण. (Yes. Those five are done.)
00:29	Smita:	मग आता हय तकिडच्या घ्या ना-हया सगळ्या फुटलेल्या आहेत हया- (Then take from this side. All of them are torn.)
00:32	Bhimraj:	हय दोन घेता येतील हया- (We could take these two.)
00:32	Arjun:	वड! वड!! हलि पण चांगली- फुटल ती! (Pull! Pull!! Do it carefully – It might tear.)
00:35	Smita:	फोडलेल्या आहेत ना मग त्या कशाला घेता? (They are all torn. Then why take them?)
00:37	Bhimraj:	[to Smita, grunting as he pulls the bag]: असू दे गं! (Let it be.)
00:39	Arjun:	तलि काढून घे. अजून एक काढून घे चल. (Take) that one out. Take out one more)
00:41	Smita:	मी तुम्हाला- तांदूळ आणलेल्या पशिव्या चांगल्या आहेत त्या? (I- your- The bags in which the rice was brought – are they good?)
00:42	Bhimraj:	चांगल्या आहेत. (They are good)
00:44	Smita:	मग देऊ? (Then I'll give?)
00:45	Bhimraj:	हो! (Yes.)
00:46	Smita:	मी आणते काढून. (I'll remove and bring.) [She goes to bring the plastic bags.]

00:49	Arjun:	[To Bhimraj] हिला पण पलटी मार बघ. (Turn it around, too.)
00:50	Bhimraj:	हम्म (Hmm.)
00:51	Arjun:	हम्म? (Hmm?)
00:52	Bhimraj:	केलं. (Done.)
00:53	Arjun:	तकि- तकिडे लाव पायल. (Put the pile th- there.)

In this group, we might expect that Bhimraj would be at the top of the hierarchy of power, followed by Arjun and Smita. But we see that Bhimraj, Smita, and Arjun work together cooperatively to load grain onto the tempo and to identify and solve the problem of the tear in the gunnysack. Smita's main role is in organising, clarifying, and directing, rather than doing the work. She verbalises the problem of the tear (at 1:40), and they all contribute to finding a solution. Smita tells both Arjun and Bhimraj what to do (at 0:01, 0:04, 0:11, 0:14, 0:16, 0:29, and rhetorically at 0:35). Bhimraj also tells both Smita and Arjun what to do (at 1:48, 1:55, 2:04, 00:32, 00:37, and 00:45). And Arjun tells Bhimraj what to do (at 2:01(?), 0:00 (although here he is over-ruled by Smita), 00:32, 00:39, 00:49, and 00:53). Arjun does not tell Smita what to do in this episode. Even though people are telling each other what to do, the tone of the entire episode is friendly and non-confrontational – except perhaps for the initial comment in which Smita may be chiding her husband for not checking the gunnysacks earlier.

What is actually happening in this episode is that the men are doing the men's work, – they are loading the very heavy sacks of grain onto the tempo. The men also do the repair of the sacks. Men will go in the tempo to bring the grain to the mill while the woman stays home. We should note that the process of milling rice that we saw is very different from what it was in the past, when the husks were removed by a very laborious process of pounding by hand with a mortar and pestle, as shown in Fig. 2.4. This has been done mainly by women and/or slaves. Mechanical milling of rice began in some places in the late 19th century, but hand pounding is still done in some parts of India.

The men will take care of the selling. The grain will be sold to more powerful men who will assign the price. The woman has virtually no voice in the money matters. The woman is not doing men's work. So there is no doubt that there is an underlying traditional patriarchal structure.

However, within this structure, Smita is certainly not acting very obediently or dismissively. She is assertively directing the others, but she is not the only one directing. Bhimraj is also directing Smita and Arjun. Even Arjun, who is the lowest in the hierarchy, directs Bhimraj. So actually, they are more or less directing each other, and working cooperatively. Smita's direction is sometimes in the form of asking suggestive or rhetorical questions rather than making direct requests or demands. The most surprising thing to us was that Smita is participating at all in this context where she is not allowed to do any of the physical labour. Thus, we see this as a dialectical conflict between the power relations.

A number of times, the researchers were surprised to see how involved Smita was in the detailed aspects of farming, even when she was not the one who was doing a particular part of the physical labour. Generally in India, women do most of the work in the fields, and this may have been the case on this farm as well – or at least the work was fairly evenly divided. When they hired labourers for help, most of them were women. Smita was always busy, and her work included all aspects of collecting material for the rab, and the rabbing itself. Namdev was the one to first light the rab, but Smita also lit parts of it. She worked together with the men and children to break the clods and cover the seeds after sowing. They all worked together for transplanting, harvesting, carrying the paddy home, threshing, winnowing, and storing. She did most of the weeding herself. She did a lot of the work to take care of the animals, and the milking.

Generally, (with rare exceptions) women do not sow the seeds, apply artificial fertiliser, or do the ploughing. But Smita was in the field as Namdev and Bhimraj were doing all these things, doing other, related work, like breaking clods of earth. During the ploughing, she was watching, and she was explaining all the details to us. We were surprised at the depth and detail of her understanding. She explained how and why the ploughing should not be too deep when the nursery plot was being prepared for sowing. She also told Bhimraj not to plough too deeply, as this would create large clods that would be difficult to break. She also explained the differences between the ploughing techniques of Namdev and Bhimraj.

So, there is a contradiction here. By convention, a woman does not do certain aspects of the work, but she may understand it, and help direct it as well.

References

Ambedkar, B.R. 1944. *Annihilation of Caste: With a Reply to Mahatma Gandhi*. 3rd. ed, Reprinted 2011. Nagpur: Samata Prakashan.

Gu, Dongxiang, et al. 2017. Quantitative classification of Rice (*Oryza Sativa L.*) root length and diameter using image analysis, ed. Ive De Smet. *PLOS One* 12 (1): e0169968.

International Rice Research Institute. 2019. *Acronyms and Glossary of Rice Related Terminology*. http://www.knowledgebank.irri.org/images/docs/acronyms-and-glossary-of-rice-related-terminology.pdf. Accessed 10 Sept 2019.

Kingwell-Banham, Eleanor. 2019. Dry, rainfed or irrigated? Reevaluating the role and development of rice agriculture in Iron age – Early historic South India using archaeobotanical approaches. *Archaeological and Anthropological Sciences* 11: 6485–6500.

Kumar, Dharma, Meghnad Desai, and Tapan Raychaudhuri, eds. 1991. *The Cambridge Economic History of India. Vol. 2: C. 1757 – c. 1970. Reprint*. Hyderabad: Orient Longman in association with Cambridge University Press.

Morrison, Kathleen D. 2014. Water in South India and Sri Lanka: Agriculture, irrigation, politics, and purity. In *History of Water and Civilization, Volume VII, Water and humanity*, ed. Y. Yasuda and V. Scarborough. New York: UNESCO.

Patel, Piyushkumar Maganbhai. 2005. Relative efficiency of herbicides and non-chemical means on weed dynamics and growth of chilli (*Capsicum Annuum L.*) seedlings in nursery'. M. Sc. Anand Agricultural University.

Phule, Jyotirao. 1881. Shetkaryaca Asud. 2017 Marathi edition translated from Marathi to English by Gail Omvedt and Bharat Patankar. Pune: Mehta.

Sengupta, Nirmal. 2018. *Traditional Knowledge in Modern India.* New York/Berlin/Heidelberg: Springer.

Singh, Gurinder, Rafikh Shaikh, and Karen Haydock. 2019. Understanding student questioning. *Cultural Studies of Science Education* 14 (3): 643–679.

Valentine, Tamara M. 1988. Developing discourse types in non-native english: Strategies of gender in Hindi and Indian English. *World Englishes* 7 (2): 143–158.

Chapter 4
Formal Education vs Learning Cultivation

4.1 Dialectical Conflicts in Mainstream Formal Education

We were not able to directly observe the schools in Rudravali. However, we have visited and worked in schools in areas that are similar to Rudravali. We also discussed with the family the school education they have experienced. We will discuss this in more detail in Chap. 8. They verified that the schools in Rudravali were similar to what we expected, based on our observations of other schools.

Throughout India, the school curriculum is segregated into distinct subjects, becoming more and more segregated as the students get older. Especially after Class V or VI, the topic of agriculture has to be fit into geography, biology, social science, history, or economics. Therefore, social aspects are separated from the 'facts' regarding where and how crops are produced. After Class IX, the syllabus in the science stream hardly includes any mention of social, political, or economic aspects of agriculture – and even the topic of cultivation is hardly mentioned.

The subject of agriculture (as well as the related subject of environmental education) is inherently interdisciplinary. Such subjects cannot be pigeon-holed into one particular discipline. They need to be considered in their complex interdependencies with natural and social sciences – and even with arts, languages, and mathematics. They are also both areas in which teaching objectives often go beyond just 'learning about' or understanding. At least according to the objectives stated in policy documents, they include efforts to change students' behaviours in order to solve environmental and agricultural problems. Therefore it seems to be all the more peculiar that the pedagogies concentrate on thinking and neglect doing or working with the hands.

© The Author(s), under exclusive license to Springer Nature Switzerland AG 2021
K. Haydock et al., *Learning and Sustaining Agricultural Practices*,
International Explorations in Outdoor and Environmental Education 7,
https://doi.org/10.1007/978-3-030-64065-1_4

4.1.1   School Pedagogy

Despite the recommendations of teachers, educationists, and policy documents, students do very little if any practical or hands-on work related to science or agriculture in most Indian schools before Class IX.

Even the practical work in Classes IX to XII is very limited, consisting mainly of teacher demonstrations rather than hands-on work by the students. If students do practicals themselves, they usually follow a 'cook-book' approach in which they do not have much chance to do anything new or different from the well-structured step-by-step procedure which is designed to give predetermined results (Singh and Khaparde 2013).

School work is mainly with the mind rather than the hands: teachers use a transmissive approach to try to transmit to the students a designated 'body of knowledge' in a predetermined order. At the primary and middle school level, much of the work students do in science consists of hearing, reading, remembering and writing words and their definitions, and descriptions of phenomena. Especially in the school days of Smita and Bhimraj, manual labour of any kind was hardly mentioned anywhere in the syllabus, even though the occupations of most families consisted of manual labour.

Bhimraj and Smita said that they did not do any science activities at school, except for once helping to plant trees on the campus. They said the teacher did not do any science activities or demonstrations before Class IX. Neither did they engage in group work or group discussions as part of the classroom practice. In the similar schools that we have visited, children do not do activities. Usually there are too many students in each classroom to make activities or group work feasible. Practicals begin in Classes IX and X, but they are mainly a matter of copying reports that the teacher writes on the board, perhaps with some teacher demonstrations.

In his PhD thesis, Gurinder Singh (2020, chap. 4) has reported on his observations of classroom practice in a few schools in Mumbai as well as in some rural schools in Punjab. He investigated the nature and dynamics of classroom discourse in order to understand the process of questioning by students and teachers. He found that the teachers did most of the talking. The student talk that did occur as part of the main classroom discourse, was either in chorus, in response to a request by the teacher, or was directed to the teacher rather than to another student. Each student utterance was also much shorter than the typical teacher utterance.

Student questioning was rare, and most of the questions they did ask were directed to the teacher and were either requests for permission or procedural questions about how to complete an assigned task. It was very rare for students to show disagreement with the teacher or the textbook. The roles of the teachers and students were rigid and predetermined. Teachers asked the vast majority of the questions, and their questions were inauthentic, in that they already knew what they thought the correct answers to their questions were before they asked. They asked mainly in order to find out if students remembered the answers that they were previously told, or that they read in their textbooks.

Gurinder claims that in classrooms, "students learn the implicit norms governing the social organisation of the classroom. They learn the norms of participation in classroom discourse. Students get 'schooled' for certain classroom behaviours and learn the classroom meanings of talking, asking, answering, etc." Other studies of classrooms in India also report a lack of student talk and questioning and the dominance of teacher talk and questioning (Kumar 1989 and Sarangapani 2003).

4.1.2 Learning About Plant Reproduction in School

The primary and middle school textbooks, which are the basis for the curricula, do contain a few topics that are related to agriculture. For example, plant reproduction is mentioned in the Class VII Maharashtra textbook (similar to the Marathi version Smita and Bhimraj used years ago). It has one paragraph on "Sexual reproduction in plants":

> In flowering plants, flowers are the organ of reproduction. The androecium and gynoecium in the flower are important for reproduction. The androecium is the male part and gynoecium is the female part. When the pollen grains from the androecium fall on the stigma of the gynoecium they begin to grow there. This is called pollination. As the pollen tube grows from the pollen grain, male gametes are formed in it. They unite with the female gamete in the ovary. This union is called fertilisation. Fertilization produces a single cell called the zygote. This leads to the formation of the seed and fruit. The seed takes root in the ground and a new plant grows. (Maharashtra State Bureau of Textbook Production and Curriculum Research 2008, 71)

It is difficult to imagine that students would understand much from this verbal description, and there is no accompanying picture. It is mainly just a list of difficult word definitions without any illustrations, and is almost gibberish. Even if the teacher had explained it with a labelled picture of a stereotypical flower on the board, how could the students even relate it to a rice flower (Fig. 4.1), which the teacher would certainly not have mentioned? Of what use is this sort of text to people who are sweating away in a paddy field? Perhaps investigating plant reproduction would be useful, but it is doubtful if this text helps students understand or investigate plant reproduction – or become interested in it. It must be difficult for students to memorise all this information, which they are required to do so that they can write it in the examinations.

The new Maharashtra Class VII textbooks (Maharashtra State Bureau of Textbook Production and Curriculum Research 2017) that Pratik and Pranay use have a labelled picture of a hibiscus flower (Fig. 4.2), and five paragraphs defining the names of its parts: mainly Calyx, Corolla, Androecium, Gynoecium. Most of us authors have forgotten or never heard of these terms – the last time most of us heard them was when we studied botany at the school or BSc level. In the new textbook, the above paragraph on reproduction has now become:

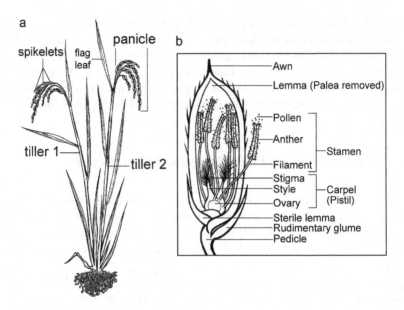

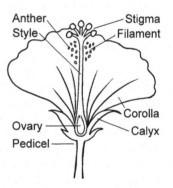

Fig. 4.1 (**a**) A mature rice plant having two tillers. (**b**) A rice spikelet, enlarged (actual length is about 1 cm – A grain of rice fits inside)

Fig. 4.2 Textbook picture of what is taken to be a stereotypical flower

After maturity, anthers burst and the pollen grains which are released fall on the stigma. This process is called pollination. Due to pollination, ovules (egg cells) in the ovary get fertilised. Fertilised ovules form the seeds and the ovary develops into a fruit.

Since the new text is accompanied by the picture of one type of flower, one could argue that it serves to help teach children the names of the parts of a flower and acquire a basic understanding of reproduction. This could be useful if the students were to also do some investigations of flowers. But still, we wonder whether they would do just as edifying investigations even without using such terminology. Even without knowing the standard terminology, we have seen that students make up their own terminology when they need to. For example, we observed students making up their own names for trees and colours of leaves they were discussing with each other (Singh et al. 2019).

Although their stated aims are to help children learn science, the textbooks are more concerned with giving names, difficult words, and word definitions – which students are supposed to memorise for examinations. Although questioning and investigating are inherent aspects of science (and doing paddy cultivation, as we saw) this sort of school science education discourages students from asking questions, doing investigations, or even having discussions with each other. Students learn that it is the teachers who ask the questions, and the students who answer the question – with the answers that the teachers already know. School science education is focussed on memorising a prescribed set of word definitions and 'known facts'.

There is a conflict here: it is doubtful that either textbook is of much use for learning anything about flowers or plant reproduction that is important or relevant to the students' own lives.

4.1.3 What Did the Farmers and Researchers Understand About Plant Reproduction?

We knew that Namdev had very little formal education (just up to Class V), but Bhimraj and Smita had both completed up to Class X. They must have been exposed to the above kinds of text regarding plant reproduction at school, in addition to whatever they picked up outside of school. So we wondered what they understood about the details of rice reproduction. Did they know about the male and female parts of rice flowers and fertilisation? We also wondered what they understood about the evolution of plants by natural and artificial selection. The topic of evolution hardly appears before the Class X syllabus.

All these topics arose in a discussion after harvesting, which began with Karen and Ankita asking Bhimraj about how and why the paddy crop may change by evolving through seed selection over the years. We wanted to find out how they selected seeds and how the paddy changed over generations as a result of this (see Sect. 6.1.1).

04:06   Karen: See just ask him this: if you keep choosing the best one each year and the crop is therefore better, then does it happen that after 5-10 years the crop is better or is it worse? Why does it get better or worse?

04:24   Ankita: म्हणजे तुम्ही- असं कधी-कधी झालं असेल, ना? तुम्ही - की तुम्ही खुप वर्षं चांगलं-चांगलं नविडता तर मग तुम्हाला मग असा फर्क दिसिला का की वर्षानुवर्ष ते अधिकिच चांगला उत्पन्न देतेय किंवा असा झालायं कमी झालायं उत्पन्न? (Like, you- this might have happened to you sometime, no? That you- that you selected the very best every year, then you might have seen a difference here, that year after year it's giving better yield - or it might have happened that the yield is reduced?)

04:39	Bhimraj:	जर उत्पन्न दुयायचयं जर कमी आलं, ना, आम्हाला असं वाटलं आता, आपली शेती बरोबर झाली नाही, कवि व्यवस्थति होत नाही, आपल्याला अंदाज असतं ना..... (So if the yield comes low, nah?, then we think that, your farming is not being done correctly, or it is not proper. That's what we hypothesise, nah?)
04:50	Ankita:	नाही, पण मैडम कशाबद्दल वचिरतायेत माहिविये का? तुम्ही असा सतत वापरलं असेल, ना? तर तुम्हाला फरक दसिलाय का? (No, but do you know what madam is asking about? That, you are using it regularly, no? Then have you seen any difference?)
04:57	Bhimraj:	हां, फरक दसितो! (Yes, There's a difference.)
04:58	Ankita:	मग कमी झालायं की- (So is it decreased or-)
04:59	Karen:	But अच्छा हैं या कम भी हैं? (But is it good or is it also less?)
05:02	Bhimraj:	ये कम होता हैं कभी कभी, और वो कडिी का प्रकार आता हैं, इसमें कैसा हैं, अलग अलग दाना mix हो जाता हैं दुसरा - (It gets less sometimes, and this kind of worm comes in it, inside it,. How is it, different grains get mixed. Other -)
05:09	Karen:	अच्छा इस लिए. (Ok, that's why.)
05:10	Bhimraj	दूसरा, ये खेती, ये चावल हैं ना? मैंने नकिोला लेकनि उधर रख दिया थोडा - वो - वो दखिाता था ना - उसमे कैसा हैं. ये चावल और दूसरा चावल वाला - ऐसा - mix रहता है. (Secondly, this field - this is the rice, nah, [that] I selected but I kept it a little bit over there - that - I could have shown you that - how it is in that - This rice and the second rice, like that, they get 'mixed'.)
05:26	Karen:	अच्छा mix मतलब ये pollination...he knows pollination?
05:29	Bhimraj:	कभी कभी black दाना रहता हैं... (Sometimes the grain is black)
05:34	Karen:	Can you ask him if mix means it gets pollinated by the other one?
05:38	Ankita:	Mix म्हणजे काय होतं? (What does 'mix' mean?)
05:41	Bhimraj:	[Repeats her question in a soft voice (to himself)] Mix म्हणजे काय होतं?
05:42	Karen:	No, ask in Marathi.
05:44	Ankita:	हां, हां. (Yes, yes.) He understands.
05:47	Karen	But, ask him to explain what is pollination.
05:53	Bhimraj:	इसमें जाड़ा आगएला हैं. (In this it has become fat.)
05:54	Ankita:	हुम्म्म. कसं आलं ते? कसं? (Hmmm. How did it come?)
05:55	Bhimraj:	ये ऐसा - ये, हो जाताय - कभी-कभी... (Like this - this, becomes, sometimes -)
05:59	Ankita:	म्हणजे, एकाच शेतात दुसरं? (Meaning, of this same field or another?)

06:01	Bhimraj:	नाही...नाही...भातामाध्ये जर थोडं तरी एखादं बीज...दोन-चार काडे बनून पडले ना...त्याचे एकाचे दोन होतात, दोनाचे चार होतात... अश्या पद्धतीने मग हे होतं. [No, no. Rice: if I have a little bit of seed, two or four bugs have also fallen in it. One of them becomes two, two becomes four. In this way it keeps going on.
06:11	Ankita:	अच्छा हे, ह्याच्यामुळे (Ok, it's because of this.)
06:14	Bhimraj:	आपण, हे आता, जमनिवर सगळं असं - (Now this, everything on the ground -)
06:17	Ankita:	पसरवून ठेवलंय, ना? (It's spread, nah?)
06:18	Bhimraj:	आणि त्याच्यामधी जर - (And if in that -)

Karen was wondering what he knew about the role of plant reproduction and pollination in the emergence of changing characteristics such as decreased yield. But, as we see here, the first reasons he gave for why the crop may be different from what is expected is that there may be some differences in what the farmers did – they may have done something wrong. He also mentioned that poor yields could be due to pest attacks: worms, or black fungus.

Then he also mentioned that a difference in the crop could be because the seeds got 'mixed'. When Karen heard this, she did not understand if he meant that different kinds of seeds might have gotten mixed up before sowing, or if he might be talking about cross-pollination. Karen had forgotten how to say 'pollination' in Hindi, so she asked Ankita to ask Bhimraj about pollination. But Ankita also did not know how to say 'pollination' in Marathi (having studied in an English-medium school). However, she asked him what he meant by the English word 'mix', that he had used. But (at 6:01) Bhimraj went on to explain how the bugs reproduce, not how the rice reproduces. There was a lot of miscommunication, which only got worse after Karen interrupted to ask about the role of the rice flowers:

06:19	Karen:	अच्छा, इसी में - इसी में फूल भी है, ना? (Okay, in this [pointing to the harvested stacks of paddy], in this there are also flowers, nah?)
06:21	Bhimraj:	हँ? (Huh?)
06:21	Karen	पहले फूल था - (First there was a flower -)
06:22	Bhimraj:	हाँ. (Yes.)
06:23	Karen:	वो - paddy मैं ना? खेत मै फूल था ना? ... हैं? (That.. in the paddy, nah? There was flower in the field, nah?... is?)
06:28	Ankita:	Tch! (makes sound meaning, No!)
06:29	Karen:	इस इसके, इसके.. इसके पौधों में, फूल भी है? (In this... in this - plants - there is also flowers?)
06:28	Ankita:	TCH! (makes sound meaning, an emphatic No!)
06:34	Karen:	No, let him say! [laughs]
06:36	[Bhimraj looks surprised. He has a half smile.]	
06:37	Ankita:	ह्याला फूल येतं का? भाताला? (Does this flower?! Rice!?)

06:39   Bhimraj: हा. येतं ना! आता हैं. (Yes. It does. It does.)
06:40   Ankita: हां?! [sounding taken aback - meek and surprised]
06:41   Karen: हैं!
06:42   Bhimraj: नहीं! नहीं!! अभी वो - (No! No! Now it -)
06:44   Karen: अभी तो नहीं हैं. (It's not now [flowering].)
06:45   Bhimraj: अभी धान पकड़ा, ना? फूलवाला - (Now the gathered rice, nah?
                     [reaching out and getting a plant] Flowering ones-)
06:47   Ankita: हां त्याच्या आधी फूल होतं ना? (Yes, there were flowers before
                     that?)
06:48   Bhimraj: [looking all around, slowly] -फूलवाला - आता शेती - (-ones
                     with flowers - coming in the field -)
06:51   Karen: जुलै मै. अभी तो नहीं हैं (In July. Not right now.)
06:52   Ankita: आता तर कुणाकडेच नसेल ना ? (Now it's nowhere?)
06:54   Bhimraj: ना, अभी [... unclear word]. पहलिं- पहि- प - फरि से नहीं. मैं याने,
                     इसका ये बीज जरा सा आता हैं धान, तभी इसका white, white,
                     white white बीज, आता है, इसके उपर. (No, [not] now. Before -
                     bef - b - not again. I mean, a little bit of this seed comes here
                     in the rice ... then white, white, white, white seeds come over
                     it.)
07:02   Karen: हां. हां, हां, हां, हां. मैंने देखा. (Yes. Yes, yes, yes, yes, I've seen
                     it.)
07:04   Bhimraj: हां, तो Kalpesh [Kalpana's brother] ने लिया होगा फ़ोन. (Yes, so
                     Kalpesh had taken a phone.)
07:07   Ankita: फोटो. (Photo)
... [talking about taking the photo] ...
07:18   Bhimraj: हां, September, September मैं, September महीने मैं. और उसके
                     बाद मैं ये बीज तयार होता हैं. (Yes. September, in September, in
                     the month of September. And after that the seed is ready.)
07:27   Ankita: तुम्हाला हे माहित्ये का, की हे, म्हणजे ही, हे काय आहे? प्रक्रिया
                     कशीये ती? म्हणजे कसं, फुल झालं, मग बिया येतात मग नंतर ते -
                     म्हणजे ती प्रक्रिया तुम्ही अशी explain करू शकाल का? (Do you
                     know what this is? How is the process? I mean, how did the
                     flowers come, and then the seeds come, and then they - So
                     like this, can you explain the process?)
07:39   Bhimraj: म्हणजे, जास्त भात हे. झाला कि नाही, तर असा, म्हणजे अशी.
                     ह्याजा-मध्न बीजांमध्न. जसं हे गवत आहे आपलं, हे गवत दाखवतो.
                     (Meaning, whether there is more rice or not, that's the way it
                     is. We have grass seeds. The grass shows.)
[Bhimraj walks to the front of the courtyard (आंगन), where some grass is growing. Others follow. He bends down and pulls up some grass plants, and shows
them to us.]
08:08   Karen: हां, घास में है. (Yes, it's in grass.)

08:10	Bhimraj:	ये देखो. ये ऐशी, ऐशी का - चावल का काडी हो जाताय. जभी वो बीज, वो धान का बीज ये हो जाताय, ना? अंदर? तो इसमे से ऐसा - ऐसा जाडा वाला ऐसा पराग पेहला ready हो जाता हैं. उसके बाद मैं, उसका, एक बीज - ऐसा वाला बीज - आता हैं. ये बीज हैं, चावल का, ऐसा बीज, साइड-साइड मैं, ऐसा आताय. उसके बाद मैं, ये पक्का हो जाताय. ऐसा. (Look at these [showing in the grass plant]. These like of - just like stalks of rice. Whenever those seeds, those seeds of paddy should be this, nah? Inside? So inside it's like - like this the pollen gets the fat one 'ready'. After that, its, one seed - that kind of seed - comes. It is these seeds, of rice, like these seeds, side by side, they come like this. After that, they get ripe. Like that.)

...

08:57	Bhimraj:	ये देखो. छोटा-छोटा तंतूतू तंतूतू नकिल रहा है और इसमे भी देखो - देखाते है, दखिाताय. वो ऐसे ही आता है. ये देखो. ये देखो..ये ऐसा ऐसा स्पर्श ये होतं है. उसके बाद मै ये ऐसा नकिलता हैं. (Look at this. Tiny little filaments and filaments come out. and also inside this, look, I'll show you, That comes like this. Look at this. Look at this.)
09:15	Smita:	असं असं नीट ठेवा, त्याला फोटो घेऊ द्या. (Keep it simple, let's take the photo.}
09:20	Karen;	Pollen भी हैं? Pollen भी हैं? (There is also pollen? There is pollen?)
09:22	Ankita:	[to Karen] But this is not rice.]
09:23	Karen:	What's the word for pollen? You know the word for pollen?
09:25	Smita:	[in the background to Bhimraj]: काय हो? (What is it?)
09:26	Ankita:	[meekly] हां, हां. Pollen grain. (Yes. Yes. Pollen grain.)
09:28	Karen:	It's a- there's a- What's that?
09:29	Bhimraj:	[To Smita] फूल वालं त्याना पाहिजि! (They want the flowering ones.)
09:30	Smita:	काय?! (What?!)
09:32	Bhimraj:	इनका. [To Smita:] नाही, आणु का मग शेतामध्ये जाऊन? (Their. No, why not go to the field?)
09:31	Karen:	अच्छा. Pollen को क- (Ok, so pollen is c-)
09:34	Ankita:	[Over Karen] She doesn't understand-
09:35	Smita:	[to Bhimraj] काय? काय? पाहिजि त्यांना? (What? What do they want?)
09:36	Bhimraj:	[to Smita] त्यांना हे असं फूल वालं ... हे पाहिजि [...? he continues:] हल्का - (They want flowering ones like this ... slightly -)
09:35	Karen:	No. She must have, she must have studied -
09:37	Ankita:	Yaa but, she doesn't know the term for-
09:39	Karen:	The name. No. She may know.

As we see here, when Karen mentioned rice flowers, Ankita (who has an MSc in biochemistry) said that rice does not have flowers. Maybe this explains why she was not asking Bhimraj about pollination – maybe she did not think that pollination occurred in rice.

But Bhimraj (and later also Smita) told her – and proceeded to show us – that rice does flower.

In her schooling, Ankita had probably once memorised the classification of grasses as angiosperms and the definition of angiosperms as flowering plants, but she had never noticed rice flowers or grass flowers, and it did not strike her that rice is a type of flowering grass. The flowering of rice – probably nowadays the most important flowering that occurs on earth (from the point of view of hungry human beings) is not mentioned in school.

Rice flowers (Fig. 4.1) are tiny, and they do not contain all the same flower parts that we see in a textbook illustration of a flower. There are no petals, which are the main distinguishing features of flowers. The anthers are obvious, but the ovary and stigma are very small, and are hidden inside the bracts (the lemma and palea).

Finally, (at 6:54), Bhimraj started to talk about the role of flowering in rice. But he seemed to be having a hard time explaining it to us, probably because of problems with language (Karen not knowing Marathi and forgetting the terms in Hindi, and Ankita not knowing the terminology in Marathi or Hindi), and also because it was hard to talk about it without having the flowers to show. As Bhimraj was talking, he glanced around to see if he could see any flowers. He knows that the flowering stage is long past (this was during harvesting), but he remembers that there should be some photographs from a month or two ago, when the rice was flowering.

Suddenly he thought of showing the flowers in some grass plants that were right there in front of us. He plucked some flowering grass plants and shows them to us, as he explains the process of pollination and how the seeds develop. He shows how the flowers of grass and rice have filaments with anthers containing pollen sticking out. He tells how the pollen spreads all around, but also lands on the developing seed and makes it 'ready'. Then the long line of seeds in the panicles develop and get ripe. He says that some species of trees have male and female plants, just like animals have male and female, but rice is not like that. Both male and female is in one rice flower.

So, Smita and Bhimraj did seem to have a good understanding of plant reproduction and the process of fertilisation. But they said that they know this not because they learned it in school. They learned it through direct observation in the field.

This shows a contradiction: the people, like the researchers, who study and memorise complicated terminology and biological structures and functions in school probably forget what they learn because it is not necessary for them; while the people for whom the terminology and mechanisms could be useful also may not use much of it, either because they actually do not need it, or because they never got a chance to learn it in school (or learn it in a way that made sense in relation to their work in the field).

4.1.4 Learning About Farming in School

Although most people in India do agricultural labour (at least part-time), there is very little focus on school-level education related to farming. However, textbooks do mention some topics related to farming. For example, in the Maharashtra State EVS Textbook for Class V, there is a chapter on agriculture which states:

> In the traditional methods of agriculture, ploughing, tilling, etc. were done with the help of oxen. A 'mot' (a huge leather bag) driven with the help of oxen was used to draw water from the well. Harvesting, threshing, etc. was done by the members of the farmer's family themselves with the help of oxen. However, farmers now carry out all these tasks with the help of machines. (Maharashtra State Bureau of Textbook Production and Curriculum Research 2015)

The hyperbole escalates as the text goes on to state that despite an increasing population, "the needs of all the people are being met. This has been possible because of modern improved methods of agriculture." Improved seeds, irrigation, fertilisers, insecticides, pesticides, and organic methods are then described. The "success" of the Green Revolution is mentioned.

The contradiction between such absolute statements and the reality that we saw on the farm is jolting. Is this sort of text part of an effort to modernise agriculture by publicising and didactically encouraging the use of modern materials and methods – or by shaming poverty and old-fashioned culture?

This chapter stands in sharp contrast to the Shetkaryaca Asud (Sect. 2.1.2.2), in which Phule emphasises the social aspects of cultivation and takes a strong stand on the side of the oppressed. Outside of the formal education system, Phule's text is still available in Marathi, and is still widely read and appreciated by Dalit activists and people working with farmers and students in non-governmental organisations.

This chapter also contrasts with Chapter 19, 'A Seed Tells a Farmer's Story' from the Class V NCERT Environmental Studies textbook (NCERT 2008). NCERT textbooks are used in Central Government schools throughout India, as well as in some private schools. The textbooks produced by many states are also based on NCERT textbooks, although the present Maharashtra State textbooks are quite different.

Chapter 19 tells the story of how bajra is cultivated, and how cultivation has been changing over recent years. Students are also asked to compare the story of bajra production to their own experiences of grain cultivation and preparation at their homes. Bajra cultivation of the past is romanticised, with statements such as, "When the crop was ready and harvested, everyone celebrated together. Oh! Those wonderful days! With big feasts and lots to eat!" The story continues, telling how new ways came, with canals bringing irrigation and crops like bajra and jowar being replaced by wheat and cotton which "got better prices in the market". The story is told through the eyes of an anthropomorphized bajra seed, who was "forgotten and dismissed". The text states that the new kinds of seeds did not taste as good, but now the son of the farmer who formerly grew bajra made a lot of money from farming and rebuilt the house, bought a tractor, a motorcycle, and an electric pump. But the bajra seed wonders:

Is all this really progress? There is no longer any need for seeds like us, and animals like the bullocks. After the tractor has come, even people who worked on the fields, are no longer needed. How will they earn money? What will they live on? (NCERT 2008, p. 177).

This story provides a basis for suggested classroom discussions on questions such as "What is your understanding of 'progress'?" and "What kind of progress would you like to see in your area?"

But the story does not end here. It continues, telling how negative effects of agricultural modernisation cropped up and led to economic distress: problems due to the use of artificial fertilisers, pesticides and the new seeds, the disappearance of cow dung, loss of soil fertility, increased expenses, and the need for taking out loans. The authors even (briefly) mention how farmers are harassed for not paying back loans on time and are driven to commit suicide. An example of a "new" organic way of farming is presented as an implied solution.

We think this chapter is an admirable example of how the topic of agriculture should be treated in primary school textbooks. Its main pedagogic virtue is that rather than pedantically telling students what is true and what is right and wrong and how they should think and act, the authors ask students to discuss their own situation and experiences. The authors present conflicts, rather than just attempting to present one simple, unconfusing story. And they also do not attempt to present all points of view – which would be specious, since there is no level playing field. The authors are clearly leaning towards being understanding and sympathetic towards oppressed classes – people who are usually underrepresented. This is an example of a pedagogy that attempts to raise confusions in the minds of a student and between students in order to encourage them to engage in authentic questioning and arguing.

Our complaint is that some of the most systemic problems are not addressed, and the authors romanticize the past, while presenting organic farming as if it is a new and unproblematic solution. For example, how and why was farming was modernised if it was really so idyllic in the first place? This question does not appear, and neither is the text presented in a way that might lead students to ask it. And the text does not provide evidence that students may need in order to investigate it. It can be contrasted with the Shetkaryaca Asud (Sect. 2.1.2.2), in which Phule combines discussions of cultivation, food preparation, hunger, and living conditions, presenting a picture that sounds far from idyllic. Phule also analyses the reasons for the oppressive system he describes.

The NCERT textbook, Understanding Economic Development for Class X, includes some discussion of agriculture (NCERT 2006). It states that the agricultural sector produces more food than ever before, and even though most people continue to be employed in agriculture, agriculture produces a relatively smaller proportion of the GDP. Even though the industrial and service sectors produce a greater proportion of the GDP, they do not produce enough jobs. In order to explain this anomaly, the textbook states:

What it means is that there are more people in agriculture than is necessary. So, even if you move a few people out, production will not be affected. In other words, workers in agricultural sector are underemployed. (NCERT 2006, 26)

In explaining why, for example, all the members of one family are working in agriculture, the text continues:

> … [They] work in the plot throughout the year. Why? They have nowhere else to go for work. You will see that everyone is working, none remains idle, but in actual fact their labour effort gets divided. Each one is doing some work but no one is fully employed. This is the situation of underemployment, where people are apparently working but all of them are made to work less than their potential. (NCERT 2006, 26)

The authors are discussing the example of a five-member family that owns 2 hectares of cultivable land (5 acres). They state, "...you do not require five people to look after that small plot."

This is in direct contradiction to what we observed on the farm (which had only 0.4 hectares (1 acre) of land suitable for paddy. We saw that the family was over-worked with the agricultural labour that was absolutely necessary. Having plants and animals to take care of every single day, they could not even take 1 day of vacation or 'sick leave'. The family repeatedly complained about too much hard work. We saw that sometimes they were not able to do rabbing because there was too much other work to do (Sect. 3.2.5.7). They needed to hire extra workers for some of the most labour-intensive work (transplanting, harvesting, threshing), although they could not really afford to do so. Bhimraj had tried to work in the city, but had to return to the farm because they could not manage without his help. Smita said that they could not cultivate more land than what they are already doing because it would be too much work.

The main problem with the textbook is that it does not mention why the farmers are not getting paid more for all their hard work. It does not raise any systemic questions. The textbook on economic development does not mention why there is so much hunger. Nor does it ever use the term 'agricultural crisis' or mention farmer suicides. In fact, this and other textbooks even avoid discussing or using the word 'capitalism' (Haydock 2015).

4.1.4.1   Vocational Training in Agriculture

These days, in the mainstream media as well as in research papers, we hear about the modernisation of agriculture, and how quickly India is developing. We might point to the recent introduction of vocational courses, such as the B. Voc. in Agriculture (Diploma – First Year) at TISS, (for students who have already completed 12 years of school education) as an indication of the importance of modernising agriculture. The stated aim of this course is as follows:

> Agriculture being the prime focus on today's date for Indian economy requires skilled man-power [sic] across India. Further, the sector itself comprises of various sub-industries like Horticulture, Floriculture, Poultry, Organic Farming, and each industry has large number of job Opportunities. Above all, there are huge self employment opportunities in Agriculture Sector. [italics added] (Tata Institute of Social Sciences 2018)

We should note that most people do not complete 12 years of school, so the course is not actually intended to be for the average person. (Although this statistic is of dubious veracity, according to the World Bank, in 2013 the average Indian spent only 5 years in school.)

The first thing we noticed was that although the course is being offered in various parts of India, the medium of instruction and all the textbooks and course material are in English. Since most students are not very proficient in English, the teachers and students actually use various local languages in the classrooms, although students are expected to use only English for all written work. Since the exams are in English, the exams actually function to sort students according to their proficiency in memorising written English.

The other problem we noticed, upon looking over the syllabus and textbook that is used for this course, is that there seems to be too much concentration on 'bookish knowledge'. The topics taught in the course include: the importance of agriculture, the economic importance of the major crops, cropping systems (e.g. crop rotation, mixed cropping, intercropping), the types and objectives of modern methods of soil tillage, types of manuring, methods of nutrient management, types of irrigation, types of pest management, relationships between climate and soil, etc. For each of the topics, the textbook gives lists of bullet points or a couple of paragraphs of didactic 'facts'.

The students are required to memorise terminology and answers to information questions that do not require any higher-level thinking. It is not clear why this information should be memorised (unless it is to discourage higher level thinking and questioning by omitting it). It is very general – not specific to the area in which the students live. The connection to local problems is missing. There is no connection to what students may or may not already know or do. There is no discussion about economics – even the costs of methods are not mentioned or compared! There is no mention of the agricultural crisis or how the structure of the economic system determines the nature of agricultural production. It seems that the practical aspect of the course is very minor. Therefore it is not clear how the course could really help people do farming. It is not even clear how it could help industrialise agriculture. Without practical experience, it would probably not even be possible to understand or retain much of what is to be memorised.

The vocational education courses that we have seen seem to have the objectives of teaching students about modern technology and methods of farming, in order to 'modernise' agriculture. Modernise means converting 'traditional' agriculture to capitalist agriculture. But there is a conflict between this objective and the present situation that we see on the farm. We do not see that the reason that the farming family is not using more modern methods or more modern technology is because they do not have sufficient knowledge of them. They are not using tractors because they cannot afford them. They are not using more artificial fertilisers because they cannot afford them. They are not using more highly advanced and mechanised planters, harvesters, and threshers because they cannot afford them. The farming family had heard about many different kinds of new methods and new technology. They are widely advertised. The family knows that rope and plastic baskets are available in the market, and they do buy them at times. The reason that they also make their own rope and baskets is not that they have a lack of knowledge. They make their own in order to save money (and perhaps also to have more control over their own lives and the sorts of materials they use). Although modern technology seems to be available, it is actually not available to the family.

Many educationists do not seem to realise that this simple contradiction between availability / non-availability exists, which is just a matter of the contradiction between wealth / poverty. They continue to design teaching syllabi, from primary through to vocational levels, in order to teach about modern methods, as if students have a need to be taught such things, There is a contradiction between what is taught in schools regarding agriculture, and what is actually happening in the fields.

According to the school textbooks, present-day agriculture in India does not seem to be very traditional. It seems to be 'scientific', if by 'scientific' we mean that relatively modern technology is being used. But this contradicts much of our experience in the field.

And even if and when modern technology is being used, does this also mean that the people practicing agriculture are doing science or being more scientific than traditional? Do people need to be taught science in order to become more modern agricultural workers? Let's take a closer look at the learning and teaching that is actually happening in the field in order to investigate these questions.

4.2 How Are Methods of Cultivation Learned and Taught in the Field?

As we discussed (Sect. 2.3), the family said that farming is something that hardly needs to be learned. They just pick it up if they live on the farm. They also said that they hardly learn anything at school that helps them do farming, except for general literacy.

This reminds us of what Padma Sarangapani reported in interviews with some children in a village near Delhi (Sarangapani 2003):

Interviewer:	So why do you go to school?
Child:	Because there is no farming left. They used to farm.
Interviewer:	If you want to do farming, why is studying/schooling unnecessary?
Child:	Because it comes by itself. You can sow seeds by yourself.
...	
Interviewer:	What are the things we need to study for? ... Do we need to study for agriculture?
Renu:	We don't need to study (for that).
Amit:	[to Renu] We need to- for reading and writing the name of fertilisers.
Interviewer:	Then why don't they teach you that?

Amit: Oh they will, later.
Renu: Those who fail do agriculture.

Children do not go to school in order to learn about agriculture. They go to school to get out of agriculture (see also Sect. 8.1).

4.2.1 Learning from the Older Generation

We saw several sorts of learning going on in the farm: the farming family was learning from each other; the researchers were learning from the family; the children were learning in school; and (although the researchers did not intentionally emphasize it) the family was learning from the researchers. As mentioned in Sect. 2.3, initially it superficially appeared that older, more experienced family members teach younger members, who do not have as much authority. But as we look closer, this is not really the case.

At one point, Karen was watching Bhimraj ploughing, and Namdev came and took the plough from him. To Karen, who expected that the older generation would be teaching the younger one, it appeared that Namdev may have thought Bhimraj was not doing it very well, and he wanted to show us a better ploughing technique. And at first Karen thought that Namdev did seem to have a better technique because he seemed to be faster and more sure-footed. But she changed her mind after hearing the following discussion:

11:04	Swapnaja:	आता तुमचे बाबा नसतील तर तुम्ही सगळं करू शकता? (Now if your father is not around then you can do it all by yourself?)
11:08	Bhimraj:	हो. (Yes.)
11:08	Smita:	हो. (Yes.)
...		
11:13	Kalpana:	बैल जास्त तुयांचं ऐकतात? (Do the bullocks listen to him [Namdev] more?)
11:14	Bhimraj:	नाही, नाही. आपलंच ना, तसं काही नाही! (No. no. nothing like that, me more!) [Smita, Kalpana, and Ankita laugh]
11:16	Smita:	ते कसे गरम आहेत, हे सावकाश! (He [Namdev] is a little hot headed. He [Bhimraj] is calm.)
11:18	Bhimraj:	मी कसाय, एकदम सावकाश. एकदम व्यवस्थति नांगरकी सगळं व्हायला पाहिजे. (I am really calm [steady, level-headed]. Ploughing and everything ought to be done really properly.)
11:19	Smita:	असू द्या, चालू द्या. (Let it go the way it is.)
11:22	Smita:	ह्यांना व्यवस्थति पाहिजे. तसं तुयांना काय कशी तरी झाली म्हणजे – (He [Bhimraj] needs [work] done properly. He [Namdev] does it just like that [any old way] -)

11:24	Bhimraj:	त्यांचं म्हणजे कसंय – (His is how -)
11:25	Smita:	त्यांचं एकदा इथून नांगर गेला की एकदा इथून जातो. तसं त्यांचा इथून गेला कि एकदा कडेनेच दुसरा जायला पाहिजे. ([pointing at the ground] [For Namdev] the plough goes this way and then it goes that way. [For Bhimraj] when it goes this way, then next time it has to come overlapping the edge.)
11:28	Bhimraj:	आपली-माझी कशीये - सारखी सगळी नांगरकी झाली पाहिजे. (How is my ploughing – it's really even.)
11:30	Smita:	मग आता तुम्ही बोललात ना एका-एका ठिकाणी? एका ठिकाणी नाही फिरत. हे एवढं जमीन असेल ना इथून- एकदा इथून जातो – इथून – म्हणजे पहिल्यांदा गेला ना नांगर, त्याच्यातूनच दुसरा अर्धा पण जातो. असं. ह्यांना चांगली पाहिजे नांगरकी. तसं त्यांचं नाही. एकदा इथून. एकदा इथून. एकदा इथून. एकदा... करायचं नी जायचं. असं त्यांचं. ([pointing to Ankita] You just asked whether they do it over the same place. So that does not happen. If this [pointing to the ground] is the plot of land, his plough goes once in this direction and once in that direction. Like, when the first time it went [this way], the second time it goes halfway [overlapping]. This way. It should be top-notch quality. This is not the case with him [Namdev]. Once this way. Once that way. Finish the job and leave. This is how it is with him [Namdev].)
11:47	Ankita:	म्हणजे त्यांचं कदाचति आधी नसेल. आता थोडं… Like, he might not have been this way earlier. Now maybe a little… [may be softening Smita's criticism of Namdev]
11:50	Smita:	हां… (Yes. [nodding head in agreement])
11:51	Ankita:	…त्यांच्या वयामुळे पण होत असेल… (… because of his age…)
11:51	Smita:	हां. हां. हां. ते पण आहेच. (Yes, yes. [smiling and laughing] Even that's there.)
11:55	Bhimraj:	अं, Madam ला सांग, ज्यांच्याकडं tractor नसेल - त्या - म्हणजे नांगर नसेल- त्यांना नांगरकी करून दिली जाते. (Ah, tell madam, like, those who don't have a tractor, that – I mean, no plough – then ploughing is done for them.)

Here we see that there is a difference between the way Bhimraj and Namdev do the ploughing. Bhimraj has not just learned to plough the way his father ploughs, as one might think would happen if traditions are to be preserved. There is a conflict: Bhimraj has his own way of ploughing, which he thinks is better. Interestingly, even though women do not do ploughing, Smita understands the details and explains these different ploughing techniques to the researchers. She explains and justifies Bhimraj's way with evidence: he is more careful, and does not leave any gaps unturned in between the rows. Smita mentions that the differences are not just technical – they are also related to attitudes and emotions. She did not try to project ploughing methodology as being objective. This contrasts with the way some people try to project scientific methods as being objective – even when the effects of attitudes and emotions are obvious in doing professional science.

What Smita says seems to be criticism of her father-in-law, which one may think is not very respectful. Swapnaja suggests the justification that Namdev is just getting old. Maybe Namdev used to do the ploughing the way Bhimraj is now doing it, but in his older age he has changed. Bhimraj changes the subject at 11:55, when he points out the neighbouring field, that has been ploughed with a tractor by people who have paid someone to plough it with a tractor. In this area, families do not own their own tractors.

4.2.1.1 Children Learning by Playing, Observing, and Experimenting, Not by Being Taught

We mentioned that Namdev was telling the boys to bring things during rabbing. In this way the youngest generation learns from the eldest generation. One might think that Smita would play the main role in informally educating the children and passing on tradition when they are young, just because she is the mother, but with regard to cultivation we did not see specific examples any of the adults explicitly teaching the children. Usually the boys did not seem to be directly involved in the work. They did not participate in transplanting, harvesting, or threshing. At times the work was occurring while the boys were in school.

When rabbing was going on, Pratik and Pranay were running around, and they did occasionally go get something when requested. However Karen noticed Pranay playing on the side-lines, making something with a brick and some sticks and leaves. She thought he was making a toy rab, and she took a photo. She asked him if he was going to burn it, but he ran away without answering. When a few months later she showed the photo to everyone, Bhimraj said, "नही, नही वो ऐसा लकडी खडा कर दिया है उसने, और उसके उपर वो मांडव के जैसा अपुन ये करते है, ना? मांडव खडा करते है, वैसा उसने-" (No, No, he has made these sticks stand, and on it something like a mandav [a kind of shack] we do, no? We erect a mandav, like that he has-). But then we called Pranay and asked him. He remembered right away, and said, 'आग लावण्यासाठी!' (To burn!).

Thus, Pranay was playing, rather than helping to do the work. But he was also observing and imitating the work that the others were doing. But his imitation was very different from what the others were doing. He was experimenting, trying out doing things differently. And there was no one standing over him commenting, explaining, or correcting. By so doing, he was also learning how to do the rab. But he was learning without explicitly being taught, and without his family even being aware of what he was doing.

This incident brings us to ask deeper questions about how learning occurs. Do people learn how to do things very differently than they learn to remember statements?

Learning to remember a statement a person made (e.g. a formal answer to a question), is often done by hearing or reading the statement, copying it orally and/or in writing, repeating it, testing whether the memory is correct, and relearning if it is different. In other words, copying is an important aspect of learning to remember. Understanding could also be important, but even without much understanding, remembering may be possible.

So do people learn how to do something in a similar way – by watching someone do it, copying what they do again and again, testing whether they have done it the same way, and modifying the way they do it if they are wrong?

Maybe there is another way to learn how to do things from other people, which is suggested by what Pranay did. It also involves watching other people do it. But in cases like this, the thing that is being done is quite complex – more complex than just hearing and repeating a few statements. So maybe it is not possible to copy every move, step by step. Instead, the learner analyses what they are observing, tries to understand it, and remembers some basic essence of it. But the actual learning of how to do something cannot occur without actually trying to do it. So does the learner take the stuff in hand (if learning how to do something with stuff) and try to remember each detailed move, copying each move in sequential order? This is not what Pranay did. Our interpretation is that Pranay remembered the general idea, the essence of what was to be done – that a number of layers of various waste material had to be made. Then he started collecting material, and probably at the same time he kept observing what the others were doing. Maybe he came across things that were different from what he saw the others were collecting – like a brick. But a brick might be too good a find to pass up. So the brick also fit into the general idea of making the rab. He was experimenting. Trying to do things this way or that way, observing the results, modifying what he had done, becoming surprised by seeing results that were different than what he expected, etc.

If this is the way Pranay learns, doesn't it imply that old ways may not be passed on unchanged to the new generation?

4.2.2   Learning 'Disabilities'

When the researchers were trying to learn how to do work on the farm, they did things differently than the farmers, sometimes by choice, but more often inadvertently. Some of the differences were due to lack of experience or to some form of 'disability'.

At times, the researchers were confronted by self-doubt, and were afraid that they might not do something correctly, or might make fools of themselves if they failed. Sometimes this made them not even give something a try. For example, Karen felt that she might not be athletic enough to carry a load of material for rabbing, or that the others would laugh at her for trying, so she did not try. But during harvesting, she decided to try carrying a load of the harvested paddy on her head. She was helped to carry a somewhat smaller load than the others, and she did manage to do it. In the process she also learned that she had at times drawn a faulty picture of someone carrying such a load: she had drawn the load turned 90 degrees from how it is actually held on the head. By actually trying to hold the load, in a process of trial and error, she realised that it could not easily be balanced in the wrong orientation. She learned more about the physics of balance and movement through trial and error than she could have learned through instruction or theoretical study prior to practice.

Abhijit has some disabilities with his hands and feet which may not be immediately apparent to others. He worried that the others (both researchers and the family) might have expected him to do something that he would not be able to do. He was not sure whether or how to try to explain his disabilities. He finds that even his friends sometimes make fun of him and do not understand or believe that he really does have certain problems. When he is not sure if he can do something, he would rather try privately than risk failure in front of others, but there would probably be little opportunity to do anything in private on the farm. This got him thinking that what if, with his current disabilities, he was born into a farming family instead of a semi-urban, non-farming family. He thinks that his family never made a major issue of his disability, and they only encouraged him to pursue his studies as much as he can. He notes that his family was over-protective about him yet they saw to it that he lived a normal life.

Gurinder is left-handed, although he was taught to write with his right hand. In the field, he started doing the transplanting with his left hand. The farmers, who may or may not have realised that he was left-handed, asked him to try to use the right hand. Bhimraj, in spite of being left-handed, uses his right hand for transplanting. When Abhijit saw this on the video tapes, he started wondering how he would do it, since he is also left-handed. He said that he would try to find out a way, without disturbing the prevailing order of work, as much as possible.

These examples of people having a 'disability' to do something are defined in relation to other people who are more successful at doing the thing. The shame or embarrassment occurs because of comparisons between different people, and because of the way 'success' is socially defined. Perhaps this is also connected to a socially ingrained competitive spirit that the researchers did not manage to escape. For the researchers who came from a position of power and privilege in terms of class, education, or some other form of identity, it was difficult to realise that they were not as knowledgeable, experienced, or skilled in areas needed to do agricultural labour. This shows how the disabilities were actually socially generated, and they created conflicts that made it difficult to learn.

4.2.3   Teaching and Learning by Showing and Doing

When we saw the rabbing, we were confused. We did not fully understand why it was being done. Initially we thought it was done in order to add fertiliser in the form of ashes to the soil. We had not expected that there would be so many layers of different materials, or that the process was so complex.

Knowing that in theory heat rises, we were sceptical as to whether rabbing would heat up the soil underneath. We doubted that the soil would get very hot after such a brief fire, and that this could be a justification for rabbing. But rather than just talking about how the rab roasted the soil, Bhimraj bent down and dug into the soil underneath where the rab had just been burnt. He felt the heat, and asked others to feel it. He made us feel its temperature, and then we were convinced.

Weeks later, if we had not picked up the rabbed soil in our own hands, we would not have understood what Namdev and Bhimraj meant when they said that the rab was to make the soil 'light'. We might have thought this 'lightness' was some mystical thing. But we found that the soft lightness was really striking.

In the beginning, we imagined that gobar may not burn easily, so we thought it should not be placed on the bottom of the pile, as it was. When we picked up pieces of gobar in our own hands we realised that it was very dry and light, and when we saw how easily it burned, we understood why it is put on the bottom. Then we also understood what the farmers meant when they explained that the gobar is spread out on the sunny field a couple of days before the rabbing and the soil draws the moisture and other substances out of it. Thus, we learned the importance of learning by doing.

Whenever the family was explaining something, they tried to explain by showing and handling the material they were talking about and by doing the process they were explaining.

At the beginning of threshing, the researchers saw the family members take each bunch of paddy plants and hold it against the rotating electric threshing machine. The grain was then swept into a pile on the floor for further separation and winnowing. We also saw a peddle-operated threshing machine in operation (useful when there was no electricity, or to save expenses). After observing, we also tried both methods, both of which required more skill than we at first realised. Thus, we were taught how to do something by showing and doing. Very few words were used to describe what was being done.

Even when we were being taught information, concepts, and reasons, we were usually taught by doing – observing and manipulating actual stuff – rather than just by being told didactically.

4.2.3.1   Selecting Seeds

Sometimes the researchers asked how something was done even though it was not being done at the moment. For example, after harvesting, Karen asked how the seeds are selected for the next season. She had imagined that after threshing they may sort through a pile of seeds, picking out those that are large, or maybe they put seeds in water and select ones that do not float. But Smita explained that the selection is done before threshing. After starting to explain in words, and seeing that Karen and Ankita were completely confused, Bhimraj decided to show them, and he brought them over to a stack of harvested paddy:

01:43 Bhimraj: [to Karen, interrupting Ankita, who was asking them to show
us later on] अभी, madam देखो. मैं देखा रहा है, आपको. (Now,
look, madam. I am showing you.)

[Bhimraj leads everyone over to the stack of bundles at the opposite side of the
veranda, glancing back to be sure they are following. He picks out a large bundle
and turns it over.]

01:58 Bhimraj: ये देखो, (Look at this,) [moving aside three handfuls of panicles within a bundle] ये weight-wala है, ना? (This heavy
(weight-wala) one, nah?) [selecting a fourth hand-full with
his left hand and holding it up]

02:00 Karen: हाँ. (Yes.)

02:01 Bhimraj: इसका ज़्यादा weight होता है. ये देखो. (It has more weight. Look
at this)

02:02 Karen: हम्म्म. हम्म्म. (Hmm. Hmm)

02:03 Bhimraj: क्या अच्छा weight है. (What good weight it is.) [turning
thepanicles with the right hand to show the ones on the
bottom]

02:04 Karen: हाँ. (Yes.)

02:05 Bhimraj: इसके लिए वो उसका, ऐसा, साइड में करके, निकालने का. (That's
why, like this, it is kept aside, and removed off.) [holding it to
the side in the left hand and gesturing downwards with the
right hand indicating the removal of the grains]

02:07 Karen: हाँ. हाँ, हाँ, हाँ. (Yes. Yes, yes, yes.)

02:09 Bhimraj: [holding with left hand] ऐसा, निकालने का. (Like this, that was
removed.)

02:10 Karen: हाँ.

02:10 Bhimraj: उसके बाद मैं, ये निकालने का, (after that, what was kept aside,)
[holding with right hand, then left hand] और, ये निकाल दिया,
[gesturing with right hand] ऐसा थोडा-थोडा लरके. अच्छा,
अच्छा धान हो, ना? ये देखो, (and, that was removed, like that
only some with good grains, good grains, no? Look at this.)
[holding the handful of selected panicles in his left hand and
showing a few of its panicles with the right hand] इसका,
strong है और ये (these are "strong", and these,) [selecting
another flimsy panicle not in the selected handful] हल्का है. ये
हल्का हैं (are light. These are light,) [gesturing to the second
panicle] और ये, (and these) [squeezing the panicles in his left
hand] बड़ा वाला. (are big ones.)

He shows that the selection is made not just on the basis of what the seeds look
like – the characteristics of the entire plant are considered. They even take into
regard the position in which the plants grow in the field, taking only plants from the
centre of the field, because they know that they will be more apt to breed pure. The
best panicles are selected and threshed. The seeds are winnowed in the wind or in

front of a fan so that the light, unfilled husks blow away and are discarded. Only the heaviest grains are used as seeds for planting. Thus, they select the best grains from the best panicles from the best plants from the best part of the field. Bhimraj showed that the entire plant should be strong and have a larger panicle, with a larger number of tillers and grains, fewer empty (unfertilised) husks, and heavier, larger, better grains.

The selection is not simple. Certain qualities conflict with each other. There may be more panicles but fewer grains. There may be more grains but they may be smaller. The grains may be smaller but tastier. Smita explained that the plants should not be too tall and the stems should not be too lightweight or long, to avoid lodging (meaning, the plants fall over before they are ready to be harvested). Too much stalk and leaves are also not wanted. Bhimraj explained that they apply a small amount of urea mixture (chemical fertiliser) at the time of transplantation so that that the plants grow tall, but they do not apply too much because then the plants will grow too tall.

Bhimraj did not describe the selection using very technical jargon for the names and qualities of the parts of the plants when he explained the selection process. This may have been partly because he was speaking in Hindi rather than Marathi. But, more important, he did not need to use complex terminology because he had the stuff that he was explaining in his hands and he was showing it directly to us.

Thus we see that many kinds of conflicts arise from problems of language and communication. There is a conflict between communicating verbally and communicating by showing things. There is a conflict between knowing, or not knowing, each other's languages. There is a conflict between ordinary terminology and the specialised terminology of professional science.

Often when everyone was talking about something, Bhimraj would look around (searchingly or impatiently?) and then go off to show, asking others to come and take a look. As we discussed (Sect. 4.1.3), after harvesting, long after the rice flowers had turned to seed, Bhimraj told Ankita that she is wrong in saying that rice does not have flowers. He took the initiative to go and find some grass flowers nearby, and he showed them to us, explaining that rice flowers are similar. But 10 min later, without saying anything, he went back to the field to find some rice flowers – even though it was the wrong season for them. As he explained afterwards, he knows that, like any grass, a rice plant produces new shoots when the top is cut off, and he remembered that one time the buffaloes had started eating the paddy at the edge of the field, so it grew back very late in the season. He hypothesised that it might be flowering now, and he went to check his hypothesis. He returned triumphantly, rice flowers in hand. With the evidence in hand, the claim was proven. This indicates the importance he places on showing actual physical evidence to prove his point and to teach us. Maybe he thought we would not take his word seriously because he is not in a hierarchal position of authority over us. But actually it shows greater scientific temper to prove a point by observation rather than just relying on the word of authority. It is also a better teaching pedagogy.

Thus, we are led to question our initial assumptions that the family members were relatively uneducated and did not have access to constructivist teaching pedagogies that modern science educationists might advocate, but that are rare in Indian schools.

We should note one other difference between pedagogies in school and in the field. In schools, teachers tell students what to do, when to do it, and how to do it. In the field, the family never asked the researchers to try doing the work. They just did the work themselves, explaining what they were doing as they were doing it, and answering questions the researchers asked. After watching for some time, a researcher might try to help. Sometimes a farmer would notice that the researcher was doing something wrong or having some problem, and they would offer a suggestion. Sometimes the researchers would themselves ask for help.

4.2.3.2 Learning how to Make Rope

When Bhimraj showed Karen how to make rope from paddy stalks he sat down and started to make rope himself. Dnyaneshwar and an Aunty were also sitting on the ground around the pile of paddy stalks making rope. Karen was filming and observing Bhimraj (Fig. 4.3).

Then, after watching him for three minutes, Karen decided to give it a try, so she handed the camera to Kalpana:

Fig. 4.3 Bhimraj making rope from paddy stalks

03:00	Karen:	I will try now. [Karen goes and starts doing the rope-making. Bhimraj doesn't pay attention to her. She seems to be counting exactly how many stalks, lining up the bottoms, fastening the bunch together by winding another stalk around the bunch. Then she puts it under her foot and starts trying to make the rope, but gets lost and looks at how Bhimraj is doing it. Bhimraj is left-handed, and she tries to do it the same way, even though she is right-handed – that is causing a problem, but she does not know why she is having trouble.]
03:42	Bhimraj:	अश्या पद्धतीने झाली की नाही, की मग गुंडी बांधायची. (So once you do it this way, you tie it by wrapping it around.
03:47	Kalpana:	हुम्म.

[At 03:56 Karen looks at how Bhimraj is doing and tries to copy him.]

04:02	Kalpana:	[to Karen] Only two, maam. Maam, only two, I think. [Bhimraj looks at what Karen is doing, sets his rope down, and turns to help her. But she continues having trouble.]
04:06	Karen:	[to Bhimraj] ऐसे? ऐसे? (Like this? Like this?)
04:07	Bhimraj:	हां. (Yes.)
04:09	Bhimraj:	वो हाथ में ले, ये हाथ में लेगा तो भी चलेगा. [Gesturing] ये ऐसा, और उसको अभी ये-घुमालो. इसको. इसको ऐसे-इसको-इसको! (Take in that hand – although it also goes when people do in that hand. Like that, and now twist it. To this, to this, like this, to this.)
04;18	Karen:	अच्छा ये वाला? (Ok, this one?)
04:20	Bhimraj:	इधर ऐसा है ना! इधर से-ये हाथ पे लिया ना, तो अभी ऐसा उप्पर-उप्परसे जाने देनेका.. (Here it is like this, no! [When] from here-taken on this hand, then let it go from this upper – upper side...)
04:30	Karen:	ओह! अच्छा ऐसे! फिर ये वाला? (Ohh! Ok, this way! Then, this one?)
04:33	Bhimraj:	घुमादो! घुमादों सर्फ! (Twist it! Just twist it!)
04:42	Karen:	ये वाला, ऐसे?! (This one, like this?!)
04:44	Bhimraj:	वही, ये ऐसा पकड़ो और ये- (That one, Hold this one like this and this-)
04:45	Dnyaneshwar:	ऐ आपण वेणी घालतो तसं सांग ना, वेणी घालतो! (Hey, tell as if we are making a braid, that we are making a braid!)
04:50	Smita:	Madam, बसल्या खाली. (Madam, sit down.)
04:52	Karen:	Okay. फिर, ये वाला, इस तरफ से? (Okay, then, this one, from this side?)

04:55	Bhimraj:	हां. ऐसा. घुमादों. नेनू उजव्या हाताने दाखव. तुझ्या उजव्या हाताने दाखव. (Yes. Like this. Twist it. Nenu, show using the right hand. By your right hand.)
05:00	Kalpana:	तू एक side ला करत रहा मग Madam पण बघतील. (You keep doing it on one side then Madam will also see.)
05:03	Bhimraj:	वो ऐसा पकड़ा ना, तो उसको ऐसा घुमानेका. (When held like this, then twist it like this.)
05:07	Ankita:	Ma'am, You just observe him. And then try.
05:14	Bhimraj:	हां ऐसा. (Yes. Like this.)
05:15	Ankita:	Ma'am see he is showing!
05:18	Bhimraj:	हां. अभी-अभी घुमादों. (Yes. Now twist.)
05:19	Karen:	ये वाला? फिर ये वाला? फिर ये वाला? ऐसे? (This one? Then this one? Then this one? Like this?)
05:24	Bhimraj:	हां. हां. वही हाथ को- वही-वही हाथ को ये करनेका. (Yes. Yes. That hand itself-That hand itself has to be doing this.)
05:28	Karen:	अच्छा. ऐसे? (Ok, like this?)
05:30	Bhimraj:	हां. ऐसे. इसको. ऐसे. (Yes. Like this. This, like this.)
05:34	Kalpana:	Ma'am, See. See here.
05:35	Bhimraj:	हां. वो देखो. वो देखो. वो देखो. (Yes. See that. See that. See that. See that.)
05:37	Dnyaneshwar:	हां, ऐसा. ये देखो, कैसा आएगा ये. (Yes, Like this. See this how it will come.)
05:42	Kalpana:	ते- Dnyaneshwar, काकी fast करतो भरपूर. (That-Dnyaneshwar and Aunty do it very fast.)
05:45	Bhimraj:	हां. आता ते-ते दाखवतात ना, ते-ते. (Yes. Now she-she is showing na, she-she.)
05:47	Kalpana:	हां. समजलं Madam ना, आता समजलं. येत नाही अजुन. (Yes. Madam understood. Now understood. Still doesn't get it.)
05:49	Bhimraj:	हां. नाही ते, येत नाही. (Yeah, not coming.)
05:51	Kalpana:	हां. ते बघ. काकीने किती पटकन बनवलं. (Yes. See that. How fast aunty did.)
...		
06:29	Kaaki:	[Watching Karen doing it] हा. येती. येती. येती. येती. (Yes. Getting it. Getting it. Getting it. Getting it.)
06:35	Bhimraj:	वो ऐसा एक second. मैं नही आताय. वो धीरे-धीरे से आताय. (It doesn't come like this in one second. It doesn't come now. It comes slowly.)
06:47	Karen:	It's getting difficult.
06:50	Smita:	[In amazement] Madam ना बसायला तरी जमते का बघा! कुठं बसल्यात! (Madam can't even sit [properly]. Look where she is sitting!)

06:51	Karen:	[to Kalpana] You try? You want to try?
06:52	Kalpana:	Yeah.
07:02	Karen:	[Pointing to the rope Bhimraj made, comparing it to the poor quality one she had begun making]: Look how nice this is!
07:06	Kalpana:	It's your first try no. You can compare with mine. [Laughs]
07:18	Bhimraj:	अह Madam, वही रस्सी हो गया, ये ऐसा, उसके होने के बाद, ऐसा, ये देखो. (Uhh Madam, that itself is the rope, this like this, after it is completed, this, see this.)
07:29	Karen:	हां. ऐसे ऐसे. हम्म. (yes. Like this, like this. Hmmm.)
07:34	Bhimraj:	ये हो गया, और इसको, ये गट्ठा बांध के ऐसा रख दिया. (This is done, and this, tied this bunch and kept it like this.)
07:39	Karen:	हां. हां. अंदर? (Yes. Yes. Inside?)
07:42	Bhimraj:	वो बाद में use करनेके लिए लगेगा ना अपुन को, वो ये है. (That is for later use, that is what this is.)
07:48	Bhimraj:	जभी वो खेत मे गया, वो पेंढा बाँधने के मौसम मे इसको काटने का और फिर, पेंढा बाँधने के लिए, चालू. ऐसे. (When it went into farm, cut it during the season of hay stacks and then use it for tying hay stacks like this.)
07:52	Karen:	Oh yeah. इतने ज़्यादा है. (Oh yeah. Is it this much?)
08:02	Bhimraj:	एक पेंढ़ी को डेढ़ घंटा जाताय- (It takes 90 minutes for one hay stack.)
08:05	Karen:	ऐसे? (Is it?)
08:06	Bhimraj:	[Gestures]: ऐसा ये करने के लिए. (To do it this way.)
08:07	Karen:	कब तक सीखते हो? बचपन से? (When do [you] learn? Since childhood?)
08:09	Bhimraj:	अह? (Huh?)
08:10	Karen:	कब सीखे? (When did [you] learn?)
08:11	Bhimraj:	ये ऐसे, ये कर रहे है. ये देखने का, वो कर रहे हैं. वो देखने का. ऐसा करके, सीख जाताय. (This is how they are doing it. To see this, they are doing it. To see that. By doing this, they learn.
08:17	Karen:	वो काफ़ी मुश्किल! [Laughs] (It's pretty hard!)

Our interpretation is that Karen was confident that she would be able to make rope because she thought of herself as being good at working with her hands. When she started trying to make rope, she picked out a handful of stalks, but then carefully counted them to be sure there were 4, remembering that Bhimraj had said to take 3–4 stalks. Bhimraj continued making his own rope, hardly glancing at Karen, even when she was unknowing poking him with the stalks. Kalpana, who was now filming, started to advise Karen, saying that she was taking too many stalks in one hand. Kalpana had never made rope before, but she noticed by observing that approximately the same number of stalks were being held in each hand – something that Karen had missed.

Then Bhimraj put his rope down and tried to help her by placing the right number of stalks in each hand. But then he realised that she was trying to copy him, but this was a problem because he is left-handed and she is right-handed. His rope was also left handed, which cannot be made with right-handed manipulation. So he rearranged the stalks in her hands in the correct positions for right-handed twisting. Especially since she had initially been watching from the other side, Karen had a more difficult time trying to figure out what a mirror image action would be. Bhimraj went back to making his own rope, while keeping an eye on what Karen was doing. Karen was not very agile at getting up and down from sitting on the ground, and Smita remarked (to the embarrassment of Karen) how she kept sitting in an awkward position. It was obvious that Karen was having a hard time learning. It was more difficult than she expected, and she soon gave up, asking Kalpana if she wanted to try. Karen laughingly compared the beautiful long rope Bhimraj produced to her ugly stump of rope. But instead of criticising her 'abilities', the farmers explained that it just requires experience – it takes time. Bhimraj explained how they learn by doing.

We found that prior verbal instruction on how to do something was practically no use – and actually that was not the way any of the farmers tried to teach us. We were given suggestions only after we were already trying to figure out how to do something with our own hands.

But also, it was not possible to just blindly copy hand movements without having some understanding of what one was trying to do.

From the beginning, one observes and understands that rope is made by dividing the stalks in two bunches and (if you are right-handed) twisting each bunch into a right-handed helical strand, and twisting the two strands into a left-handed helix. Thus we realise that the objective of our hand movements is to twist one way to make each strand, pass the newly twisted strand into the left hand and take the other strand in the right hand and wind the strands around each other in the other direction to make the rope. Both of these opposing movements make for a kind of circular overall motion. From trial and error we adjust our hand movement to achieve the desired effect. Observing and understanding that from the beginning, all the stalks need to be bound together and held in place, we keep the end under one foot. As we work, from trial and error we realise that we need to keep adding more stalks to each strand and readjusting the position of the rope under the foot, pulling the new rope back and coiling it up. In case we twist right-handed helical strands into a right-handed helical rope, we would learn right away by trial and error that the rope would just unwind.

This is just one example of how whenever the farmers were teaching us how to do something, they would let us try doing things ourselves before they would offer suggestions or help. This also occurred when we were learning how to transplant the paddy seedlings and how to reap the plants with sickles, as we have discussed above. More than by following orders, we learned by trial and error. When we tried something and saw that it did not work, we were presented with a conflict. We asked ourselves (or others) questions in order to figure out why it did not work. We investigated and tried to understand the process we were trying to do. We manipulated the stuff in our hands and tried alternatives. Sometimes we just chanced upon a solution. Less frequently, farmers offered help without being asked to do so, when they noticed that we were having some difficulty.

Thus we claim that the process of learning how to do something like this, that requires skill and experience, occurs through a process of trial and error by the learner. Through direct observation, the learner gets a general idea of the process. But the details have to be worked out by the learner through trial and error. Without realising what the objectives are, and without trying, the learner cannot just copy hand motions and hope to achieve the same results.

4.3   The Dialectics of Conflict and Problem Solving by the Researchers

Karen had grown up in a family in which several major projects were going on: constructing the family's own house, making furniture, implements and clothing for their own daily use, making art, growing vegetable gardens, and even gathering mushrooms and wild plants for consumption. This is in addition to the sort of work that goes on in most homes: preparing food, cleaning and maintaining the home, childcare, etc.

Thus, Karen's understanding of these kinds of production processes were formed partly as a result of her own experience. When observing the farm in Rudravali she could not avoid sometimes comparing and contrasting with her own previous experience.

In Karen's family the father was the authority and usually directed these processes (but less so for cooking, cleaning, sewing, and childcare), continuously shouting commands, such as "Gimme the wrench! ... Not the pliers, you idiot – don't you even know the difference between a wrench and a pair of pliers!". The shouting and highly emotional level of activity of the entire family was maintained partly by various mistakes and unexpected problems that inevitably arise whenever anyone tries to produce anything – whether in carpentry, building stone walls, painting, or whatever. Karen also witnessed a conflict between the self-assured competence of her father and the lack of self-confidence of her mother. Even the occasional presence of guests could not eliminate the verbal and physical expression of conflict.

So one of the first things that struck Karen about the farming family in Rudravali was that they seemed to be carrying on their cultivation work in a very peaceful manner. It could be that things were more subdued in the presence of the researchers than they would ordinarily be. But as we have seen, the interaction in the family does not seem to be polite or contrived. People are blunt and expressive, interrupting and contradicting each other without hesitation. But at the same time, they are relatively cooperative. And it is not that the conflicts are necessarily more common when people are working together to produce something. Even when people are working individually, they inevitably encounter conflicts between what they are doing, or what they are trying to do, and the stuff with which they are working. Plants never grow exactly the way you think they will, or exactly the way other plants of the same type grow, or exactly the way they grew last season. Tools never work the same way. Mistakes are inevitable. Nature is not very predictable.

Upon observing and interacting with nature we come to understand that nature is actually dialectical (see how we define dialectics in Sect. 7.2.7.2). Nature is a process (not a collection of things) in which inner conflicts keep everything continuously changing. For example, the atmosphere is a process in which various types of gases, water vapour, and dust are vibrating, pushing, pulling, blowing, sucking, swirling, expanding, contracting, rising, and falling, due to opposing forces of motion, inertia, acceleration, gravity, electrostatic charges, magnetism, etc. All these conflicting forces influence the weather and whether it will rain today or tomorrow. Rain may cause the soil to get wetter and wetter, until a point is reached at which it can no longer hold more moisture and it begins to flood. Soil may get drier and drier until a point is reached at which a paddy plant can no longer survive. Human beings are actually an inseparable part of nature, and social relations as well as natural relations are dialectical.

Gurinder was impressed by other aspects when he was learning from the family on the farm. He found that the family would not interfere much when he was trying to do something. They were not even paying a lot of attention to what he was doing. Sometimes from just a quick glance they could see that he was not doing something right. They might suggest that he should try doing it differently. Their suggestions might also be a process of trial and error: suggesting that he try doing it this way or that way.

This relationship between the learners and the teachers was very different from the usual relationship between a teacher and the students in a school. Their relationship with the researchers was also very different from how they interacted with labourers that they had hired on daily wages. The relationships – and conflicts – between any teachers and learners always depend on the power structures in which they are embedded.

4.4 Learning What Hard Work Is

One of the most important things that the researchers learned is that paddy cultivation is very hard work. This is one of the main reasons that it is unsustainable. We learned this only because we were trying to do it ourselves and we experienced the difficulties, even if only to a very limited extent.

If we had any romantic conceptions about the ease and tranquillity of a farming life, it was dispelled by our experience and observation of the drudgery of hard labour. Through experience, we have also learned about the difficulties and risks that are involved. We have friends who have never even grown a kitchen garden (or even potted plants), but imagine that cultivation is just a matter of tossing seeds into soil and waiting for the rain. But some of us have tried growing vegetables – and paddy – for ourselves in small plots and pots, and this has taught us that we do not understand much or have much skill, despite reading books on how to garden. We have grown an entire garden plot that did not yield any vegetables after we had used a huge pile of leaves and grass clippings for manure. We have grown tomatoes that produced lovely leaves, and even flowers, but not a single fruit. We have grown bhindi and baigan that died off due to infestations of pests.

We learned that the process of cultivation is much more complex and difficult than we had first thought, and that learning paddy cultivation is a continuous process that can be learned only through practice over time. Communication (which may include reading and writing) with skilled workers and agriculturalists is also needed, but we concluded that without actually practicing cultivation, it is not of much use. Didactic learning through communication is actually not possible without being in the field, handling the real stuff. It is not even possible to understand the terminology without observing the stuff and experiencing the process. Our experiences in the field led us to become more aware of the problems that occur when work with the mind is separated from work with the hands, as it is in modern agricultural research.

Unfortunately, the researchers did not end up learning as much as we had hoped about how to do paddy cultivation, because we did not spend sufficient time in the field. It was difficult for us to travel to the farm by train due to the insufficient number of trains and over-crowding. Travel by automobile was expensive, alienating, and difficult due to road construction and diversions, and also because it causes some of us to have motion sickness. And also, some of us (i.e. Karen) were not unhappy to avoid hard work!

As we will discuss in later chapters, the family kept complaining about the hard work that cultivation required. But, interestingly, they did not find that cultivation entails too much time pressure, or at least not the same kind of time pressure that other kinds of work – and education – requires. This was discussed in the following conversation:

18:02	Abhijit:	... तुम्हाला माहिति असेल की, काम करतो आपण बाहेर, तर बोलतात ना, 'एवढ्या-एवढ्या ह्याच्यामध्ये हे काम झालंच पाहिजे, नाहीतर तुम्हाला कामावरनं कमी करू.' तसं शेतीमध्ये तुम्हाला अशी काही 'Deadline' असते का? (... you may know that, when we work outside [they] say nah, 'This work has to be in this specific time, otherwise you will be fired!' Like that do you have any sort of 'Deadline' in farming?
18:16	Bhimraj:	नाही. असं काही नाही. (No. Nothing like this.)
18:16	Abhijit:	म्हणजे असं, म्हणजे, एखादी - आता समजा कापणी आहे, – (Means, like this, means, something – for instance reaping-)
18:20	Smita:	हम्म. (Hmm.)

18:20	Abhijit:	-तर कापणी ह्यादरम्यान झालीच पाहिजे- (-then reaping has to be done by this time only-)
18:22	Smita:	नाही. असं काही नाही. (No. Nothing like this.)
18:23	Abhijit:	असं काही नाही? (Nothing like this?)
18:23	Smita:	नाही. (No.)
18:24	Abhijit:	Pressure नाही तुम्हाला? (You have no pressure?)
18:24	Smita:	Pressure असं नाही. (No pressure like this.)
18:24	Bhimraj:	Pressure नाही. Pressure नाही. Pressure नाही. (No. Pressure. No Pressure. No Pressure.)
18:26	Smita:	मग मी काय म्हणते- (Then what do I say-)
18:26	Bhimraj:	शेत-शेत-एक main आहे, शेतकऱ्याला pressure नाही. (Farm-farm-one thing is main, a farmer has no pressure)
18:28	Smita:	- शेतकऱ्याला pressure नसतं! म्हणून या मुलांना easy काम- (-a farmer has no pressure! That is why for these kids, easy work-)
18:32	Bhimraj:	कोणत्या ही कामाचा pressure नाहीये. (There is no pressure of any work.)
18:32	Smita:	या मुलांना easy वाटते शेती. का? पूर्ण ते pressure नसतंय ना! 'आज-आज-राहू दे-उद्या करू आम्ही!' परवा करू' असं म्हणू शकतात ना! (These kids feel farming is easy. Why? There is absolutely no pressure. 'Today-today-let it be – we will do it tomorrow!' Day after tomorrow will be like this they can say!)
18:41	Bhimraj:	आता ही झोडणी आहे ना आमची, झोडणी असं वाटलं की आज नकोय, – (Now this threshing of ours, threshing, if it was felt not to do it today -)
18:44	Smita:	मग आम्ही आज राहून देतो. (Then we let it be.)
18:47	Bhimraj:	-आज नाही करायची! मग उद्या करू!! मग असं- (-not to do today! then let us do tomorrow! Then this-)
18:49	Smita:	नाहीतर परवा! चार दिवसांनंतर करणार! (Otherwise the day after! Will do after four days!)
18:51	Abhijit:	म्हणजे तसं, काढायला उशीर झाला तर नुकसान फार होत नाही? (Means, in the sense, there is not much loss if the work is done late?)
18:54	Smita:	नाही. नाही. (No. No.)
18:55	Bhimraj:	नाय. (No.)
18:57	Smita:	हां, एखादा महिनिभर उशीर लागला तर! पण आपण एवढं तर नाही ना थांबणार! (Yeah only if we are late by a month! But we would not be waiting for so long now!)
19:02	Bhimraj:	एक चार दिविसांचा फरक राहील. एक चार दिविसांचा- (There would be a difference of about 4 days. About 4 days-)
19:03	Abhijit:	नाही, म्हणजे आता कसं असतं, बाहेरच्या offices मध्ये वैगरे नाही का, हे काम ह्या वेळेत नाही झालं- (No, means, now how it is, in the offices outside, if this task isn't finished in this particular time-)

19:08	Smita:	हा (Yeah)
19:09	Abhijit:	-तर तुमचा पगार कमी करू. कविा कामावरून काढू. मग तो माणूस कसा तुया pressure मध्ये असतो ना- (-then your wages would be reduced. Or you will be fired. Then how, that person is under that pressure-)
19:12	Smita:	हां. (Yeah)
19:13	Abhijit:	-घरी गेला तरी तो ऑफिसिचं काम करत असतो. (-Even in home he is doing the office work.)
19:15	Bhimraj:	नाय. (No.)
19:16	Abhijit:	माझ्या वडलिांचं आहे ना तसं- (My father has it that way-)

The cultivation that the family was doing involved a different kind of demand on their time, as compared to a job (or course of study) in which people are required to adhere to fixed daily timings and deadlines. The main difference is that, in cultivation, time requirements are implicitly set by the family, the plants, the animals, and nature itself, and the 'timings', such as they are, are usually not very rigid. In a job, or in school, people are required to adhere to explicit timings that are set by authorities, and may not be set according to any logic or reasons that are known or understood by the workers or students. There are rigid deadlines that may be quite unnatural or against the interests of the people. For Abhijit (as a student), or his father (who holds a government office job), there is great time pressure in their studies and jobs. Even if the work is not very physically demanding, deadlines and timings may make the work very strenuous and undesirable. Although one might think that an urban lifestyle would be easier, and that as capitalism develops, people will have less hard work and more free time, this is often not the case. Thus, because the family has more control (and more rational and 'natural' control) over their own timings and the activities of their daily life, their work was less alienating than the work of Abhijit and his father.

4.5   Summary of Dialectical Conflicts in Learning Paddy Cultivation

As summarised in Table 4.1, the process of learning and teaching paddy cultivation that we observed and practiced is more complex than the superficial appearances in Table 4.1 had suggested. Dialectical contradictions emerge, and this raises additional questions and complicates our understanding, indicating that what appeared to be a simple passing on of traditional ways unchanged is actually not occurring. But neither are modern ways simply being acquired from authorities. What is actually occurring is obscured by appearances.

Table 4.1 Dialectical conflicts in learning paddy cultivation

(1) Although family members state that they learn (or 'know') from tradition and just pick up cultivation from the older generation, they actually learn by engaging in the following, which do not appear to be what people usually think of as 'traditional':
(a) Questioning (including asking 'how?' and 'why?')
(b) Hypothesising
(c) Investigating, experimenting, and doing things differently
(d) Comparing, categorising, and testing
(e) Directly observing
(f) Analysing, arguing, reasoning, and collectively solving problems
(2) Old/new: They use primitive, home-made materials and methods, as well as manufactured materials and modern methods. The newer generation is not always learning or using the same methods as the older generation.
(3) Family members are being pragmatic / abstract, and they are being specific/general.
(4) They are acting locally, but they also are interconnected with the global market.
(5) They are having values such as respect for elders and authority figures, but they are also acting cooperatively, without strictly adhering to one set of power hierarchies. Power relations are conflicting.
(6) They are conflicted between oracy and literacy.
(7) They are engaging in rituals, but saying that they do not believe that the rituals will improve the production of paddy. They say that the yield depends on their hard work.
(8) Availability/non-availability and sustainability: They know about many modern methods and technologies that they would like to use but do not use because they cannot afford them. They are aware of both advantages and environmental problems of modern agricultural practices, but their usage is minimal due to their cost.

In order to further investigate how and why the family is doing, learning, and teaching paddy cultivation, we will have to take a look at the history and political economy of the development of education, agriculture, and science in India.

References

Haydock, Karen. 2015. Stated and unstated aims of NCERT social science textbooks. *Economic and Political Weekly* L (17): 109–119.

Kumar, Krishna. 1989. *Social Character of Learning*. New Delhi: Sage Publications.

Maharashtra State Bureau of Textbook Production and Curriculum Research. 2008. *General Science: Book Five – Standard Seven*, 2014 Reprint. Pune: Maharashtra State Textbook Bureau.

———. 2015. *Environmental Studies (Part One) Standard Five*. Pune: Maharashtra State Textbook Bureau.

———. 2017. *General Science: Standard Seven*. Pune: Maharashtra State Textbook Bureau.

NCERT (National Council of Educational Research and Training). 2006. *Understanding Economic Development for Class X*. New Delhi: NCERT.

———. 2008. *Environmental Studies: Looking around, Textbook for Class V*. New Delhi: NCERT.

Sarangapani, Padma M. 2003. *Constructing School Knowledge: An Ethnography of Learning in an Indian Village*. New Delhi: Sage Publications.

Singh, Gurinder. 2020. *Student Questioning in Student Talk: Understanding the Process and Its Role in Doing Science*. PhD Thesis. Tata Institute of Fundamental Research, Mumbai. http://www.hbcse.tifr.res.in/research-development/ph.d.-theses/thesis-gurinder/.

Singh, Gurinder, and Rajesh Khaparde. 2013. *The State of Experimental Activities in Indian School Science: An Investigation into the Existing Problems and Possible Strategies*, Field work report. Mumbai: Homi Bhabha Centre for Science Education (TIFR). https://doi.org/10.13140/RG.2.2.30910.79683.

Singh, Gurinder, Rafikh Shaikh, and Karen Haydock. 2019. Understanding Student Questioning. *Cultural Studies of Science Education* 14 (3): 643–679.

Tata Institute of Social Sciences. 2018. *B. Voc. in Agriculture, Course Description on Website*. https://www.sve.tiss.edu/index.php?p=courses-hub&courseTypeId=1&hubId=100 Accessed 4 Oct 2018.

Chapter 5
The Historical Development of Education Related to Agriculture in India

5.1 School Education Related to Agriculture Before Independence

Before Independence, public investment in education in India was very limited, especially for the 85 percent of the population who were agriculturalists. During the colonial period, the British had set up an educational system in order to train a small elite class of people for careers as government clerks and for other white-collar professions: "The main object of the introduction of Western education into India was the training of a sufficient number of young Indians to fill the subordinate posts in the public offices with literate English-speaking natives." (Chirol 1910, 34). Its effect was to reinforce fragmentation of Indian society along lines of class, caste, gender, and creed. Education was available mainly just for boys of higher castes and higher economic status. This excluded almost all people who were practicing agricultural labour.

British interest in Indian agriculture increased after Britain needed to import cotton from India in the 1860s, and the need for agricultural education was also acknowledged. However, formal agricultural education was very slow to develop. In fact, Edward Buck, who was appointed as the Secretary of the Imperial Department of Agriculture in 1881, was of the opinion that agriculture could not be taught in schools. Instead of trying to teach farmers, Buck recommended that schools should be established to teach landowners the administrative business of how to manage an estate. Nevertheless, in this period there was some effort in introduce the subject of agriculture in school education in some places, and a few agricultural schools or colleges were also started (e.g. in Madras, Bombay, Nagpur, and Kanpur).

Most efforts for agricultural education were not aimed at the school level. Rather than being concerned with improving the lives of ordinary people who were doing cultivation, they were concerned with increasing productivity and commercialising

K. Haydock et al., *Learning and Sustaining Agricultural Practices*, International Explorations in Outdoor and Environmental Education 7, https://doi.org/10.1007/978-3-030-64065-1_5

agriculture. The motivation for increasing productivity was not to feed peasants as much as to encourage them to produce more cash crops to support their colonial rulers. Its relationship with the nationalist movement and radical anti-colonial politics also made rural development less desirable for British rulers. It was not until the late colonial period that institutes such as Khalsa College in Amritsar were set up that were more concerned with agriculture, rural uplift, and economic development as goals in themselves. Perhaps this institute was accommodated because it was seen to be disseminating science education, and 'modern science' was seen to be non-political. It has been argued that those who established Khalsa College did not think of modern science as being 'western' or as being in need of being reconciled with Indian traditions or methods (Brunner 2018). However, their approach emphasized educating elite 'experts' rather than the masses.

In 1889, the British government had assigned JA Voelcker to recommend improvements for Indian agriculture. Voelcker noted that the few people who were becoming educated (even graduates from agricultural colleges) were getting "a purely literal education" and did not want to return to villages to work in agriculture. Voelcker said, "There is no intelligent farming class, nor even a good class of superintendents" (Chandra 1945, 77). He recommended that agricultural education was required. At the primary level he recommended teaching agriculture through illustrated readers, in addition to the basic subjects. At the middle level, agricultural science and elementary botany and physiology could be introduced, together with elementary physical science and drawing. Middle and high schools could be provided with school gardens so that the students could see demonstrations of cultivation. He also recommended that education should be in vernacular languages, and some agriculture textbooks in local languages were produced. What is new at all levels is the use of reading and writing: the emphasis on work with the mind. However, despite reports, conferences, and policy statements, the implementation of agricultural education was left to the provincial governments, and practically nothing was actually done.

Around the same time, in 1881, Jyotirao Phule strongly objected to an education system that, by deliberate design, excluded most of the population on the basis of caste — both students and teachers. He also questioned the entire structure of the schools, which were based upon bookish 'knowledge', separated from handwork:

> All the schools in the village are staffed by Bhat-Brahman teachers; how will those insolent arrogance-parading teachers, who are sluggish and only know how to talk, who make their living off of farmers but whose value is less than that of the Beldars and Kumbhars who work in mud and earth, who have not the slightest idea of how to hold a farmer's plough but consider themselves to be the most superior of human beings – how will those arrogant teachers whose ancestors made all others inferior give systematic and appropriate education to the children of farmers? (Phule 1881)

He recommended that education in India should not delink productive skills (including agricultural skills) from the intellectual, cultural, and humane pursuits. He recommended that a new set of farmers' children should be trained as teachers, and they should become the teachers of farmers' children. He wrote that only then "will the hold the cruel Brahmans have over their minds be lifted." (Sadgopal 2016).

Along similar lines, Mahatma Gandhi envisaged a form of universal 'Basic Education', Nai Talim, in which there was an integrated teaching/learning consisting of work with the hands, work with the mind, and spiritual values, with a focus on education through productive work: handicrafts, cultivation, and animal husbandry (Sadgopal 2016). Furthermore, the schools would be run autonomously by each village or community, and would be self-financed by selling the products produced by children and teachers, and with the government only helping in the marketing. A few experimental schools along these lines were begun after 1938.

5.2   School Education Related to Agriculture After 1947

There were some attempts to implement Nai Talim after independence in a few parts of the country, but it never became mainstream or widespread. Instead, a Brahmanical educational system modelled on the British system continued, in which agricultural education was side-lined or non-existent. Although schools were set up in many villages in addition to urban areas, this system of education was far from universal.

In 1965 the Kothari Commission (Education Commission 1966) lent support to Mahatma Gandhi's vision, saying, "As education is not rooted in the traditions of the people, the educated persons tend to be alienated from their own culture." But they recommended a discontinuation of efforts at Nai Talim, and instead a new form of general education which focussed on preparing students to contribute to national development and 'modernisation'. Their report states that education and research in agriculture were required, specifically to address the need for self-sufficiency in food production. They suggested the inclusion of "work experience" as part of their "redefinition of [Gandhi's] educational thinking in terms of a society launched on the road to industrialisation".

The report advocated a new form of education that will not "perpetuate traditional patterns of behaviour", as it claims Nai Talim did. It explicitly argued against continuing to offer education in agriculture as a 'craft' in primary schools. Looking at their specific suggestions, it is clear that their main goal of 'modernisation' actually meant capitalist development, which included the development of agribusiness. This was in response to the pressure from foreign industries, which needed to expand their markets for chemical fertilisers, pesticides, seeds, and other products. Thus, the report places high priority on "vocationalisation of education, especially at the secondary school level, to meet the needs of industry, agriculture and trade; and improvement of scientific and technological education and research at the university stage with special emphasis on agriculture and allied sciences."

Since British times, formal education was aimed implicitly, and often explicitly, at enabling the development of capitalism. This is particularly evident in agricultural education. For example, while making note that out of 50 million farms, 6 million of them were of 15 acres or more, the Kothari Commission did not object to this uneven distribution of land. Rather, as Krishna Kumar pointed out, they explicitly suggested that the role of education in agriculture was "to enable the bigger landowners to enhance their material opportunities" (Kumar 1996). "… the post-independence Indian State continued to support the Brahminical-cum-colonial paradigm of education in terms of its epistemic and socio-cultural character" (NCERT 2007).

New universities and research institutions, including agricultural universities, were set up in India in the 1950s and 60s, but they became engaged in research and intellectual pursuits that were not very closely linked to farmers who were working on the land, who remained largely uneducated and illiterate.

The idea of a socially transformative education involving productive work was watered down to become proposals and hollow policy statements for vocational education for oppressed classes and castes, or just 'activity-based' learning, "work-experience", or "Socially Useful Productive Work". None of these were ever taken seriously in practice. The majority of the population continued not to have access to education, and most people, especially women, did not even learn reading, writing, or basic literacy.

The National Education Policy of 1986 (Government of India 1986) stated that efforts would be made to devise a parallel stream of higher-secondary vocational courses based on agriculture (and other vocations), with an emphasis on "development of attitudes, knowledge, and skills for entrepreneurship and self-employment." These efforts were to be Public Private Partnerships. The policy also stated that "It is proposed that vocational courses cover 10 per cent of higher secondary [Class XI and XII] students by 1995 and 25 per cent by 2000. Steps will be taken to see that a substantial majority of the products of vocational courses are employed or become self-employed." However, the implementation of this policy was limited. Vocational education programmes were started in senior secondary schools, which included courses related to agriculture. However, there was a lack of linkage to employment opportunities, and in fact employment opportunities in agriculture dropped significantly after 1990. Even in 2006 it was reported that less than 5% of students were opting for vocational education in Classes XI and XII (NCERT 2007), and probably only a small fraction of those were in agricultural education. Even the aim to "extend science education to the vast numbers who have remained outside the pale of formal education." was not achieved by most of the states. Instead, the gap between the educated and uneducated classes continued to expand, along with the gap between the rich and the poor.

The NCERT produced national curriculum frameworks in 1975, 1988, 2000, and 2005 which included a few attempts to recommend improvements in school level education related to agriculture. They highlighted work-related education, including agricultural work. But in practice, work experience in schools was, at best, limited to maintaining a few extra-curricular hobby classes for "Socially Useful Productive Work". Even these activities very rarely involved students doing even simple gardening, to say nothing of farming. School flower gardens, if they existed, were more apt to be maintained by servants from the mali caste, who received nominal wages.

In 2006, the National Focus Group on Work and Education (NCERT 2007) emphasized the pedagogic role of work in building a national system of education. This was an attempt to break down the artificial dichotomy between work with the hands and the mind, which is supported by the caste system. This Focus Group noted that, in effect, the education system certifies and validifies the caste system. They recommended establishing institutions for Vocational Education and Training that were outside the school system, but might be in collaboration with agricultural universities and other already existing institutions. These institutions were to offer certificate or diploma courses of varying duration for students who had left the regular schools. But here again, the implementation of this programme was deficient.

5.3   The Need to Teach Skills

The conventional view that has been propagated and almost universally accepted in popular as well as academic discourse in the USA since the 1930s is that "the changing conditions of industrial and office work require an increasingly 'better-trained,' 'better-educated," and thus 'upgraded' working population", and 'the average skill' must be raised (Braverman 1998). Harry Braverman pointed out the contradiction in this belief:

> On the one hand, it is emphasized that modern work, as a result of the scientific-technical revolution and "automation," requires ever higher levels of education, training, the greater exercise of intelligence and mental effort in general. At the same time, a mounting dissatisfaction with the conditions of industrial and office labor appears to contradict this view. For it is also said – sometimes even by the same people who at other times support the first view – that work has become increasingly subdivided into petty operations that fail to sustain the interest or engage the capacities of humans with current levels of education; that these petty operations demand ever less skill and training; and that the modern trend of work by its "mindlessness" and "bureaucratization" is "alienating" ever larger sections of the working population. (Braverman 1998, 3)

Despite the passage of years, this belief (or rhetoric) has not disappeared – rather it has spread, encompassing agriculture, and even reached policy-makers in India.

Thus, the Kothari Commission report stated that farmers needed science education so that they would adopt techniques that increase yields:

> The skilled manpower needed for the relevant research and its systematic application to agriculture, industry and other sectors of life can only come from a development of scientific and technological education. (Education Commission 1966)

'Skill' is nowadays in India being advertised in connection with nationalism, and seems to have overtaken, and subsumed, the simple drive for 'literacy'. Over the last 10 years the Government of India has been launching various skill development efforts – the latest being a national campaign called 'Skill India', ostensibly to train millions of people in various skills through government as well as Public Private Partnerships, a Skill Loan Scheme, and by creating vocational institutes, scholarship programmes, training schemes, etc. Skills related to vocational training in agriculture are included in these programs.

But what is 'skill', and how does it differ from 'training' or 'education'? All these terms are vague, and carry various implications despite whatever precise definitions educationists may offer. Thus, people sometimes insist that we should use the term 'teacher education' rather than 'teacher training', because the latter is belittling to teachers. The ordinary definition of 'train' is 'to teach a skill', whereas 'educate' seems to be much more complex, controversial, and 'higher level'. Most definitions of skill refer to the 'ability' to do something. 'Ability' is another questionable attribute, especially in its relation to disability (Oliver and Barnes 2012). Ability may appear to be an individual trait (or even an innate trait), but it may actually depend upon opportunity, and be socially and historically determined. The term 'vocational education' is often used interchangeably with 'training', and here, the emphasis is on learning skills that will help people get jobs, in particular jobs that involve manual labour. Some consider this to be a rather lowly task compared to the mind expansion that occurs when one becomes 'educated'.

Perhaps the difference between training and educating is related to the difference between doing something with the hands, and doing something with the mind. This might explain why 'education' is considered to be 'higher' than training – working with the mind is traditionally considered to be 'higher' than working with the hands. But, can we really separate doing things into these two types? We may sometimes think about something without using our hands, but don't we also use our minds whenever we do anything with our hands – especially when we are learning to do something?

Thus, we can roughly define 'having skill' as using the mind and body to produce something that is physically observable: paddy, baskets, a meal, a dance, a poem, a medical treatment, etc. We will avoid defining it in terms of having an ability to produce something, or making a prediction about whether someone *can* produce something. For the purposes of our study, we do not see a need to try to make quantitative assessments of skill. We will try to avoid overly competitive comparisons. In order to qualitatively judge skill, we will evaluate it through our subjective analysis of the things that were produced.

We saw on the farm in Rudravali that cultivation involves a lot of manual labour, and it was not, and could not, be learned without doing cultivation. People learned cultivation by doing cultivation. Using the term 'skill' as defined above, to refer to doing something that includes manual labour, it is clear that learning-by-doing is necessary in order to learn skills.

However, the term 'skill' is nowadays being applied to a wide range of human activities, as indicated in the draft as well as the final version of the National Education Policy (Ministry of Human Resources Development, Government of India 2019, 2020). Thinking as well as doing are both stated to be skills to be taught in schools. Literacy and numeracy are referred to as the 'Foundational Skills'. There are 'practical skills' as well as 'higher order skills', 'cognitive skills' and 'life skills'. These are collectively termed as '21st Century Skills'. The draft mentions "problem-solving, critical and creative thinking, ethical and moral reasoning, and communication and discussion abilities" as 21st Century Skills. It states that, "Globalisation and the demands of a knowledge economy and a knowledge society call for emphasis on the need for acquisition of new skills … " Of course, IT skills are high on the list, but probably even breathing is being seen as a skill – to be taught in yoga sessions. Ironically, the policy also mentions the need for efficient deskilling:

> Disruptive technologies [such as Artificial Intelligence] will make certain jobs redundant, and hence approaches to skilling and de-skilling that are both efficient and ensure quality will be of increasing importance to create and sustain employment. (Ministry of Human Resources Development, Government of India 2019, 2020).

In competition with China, the Indian government is nowadays attempting to capitalise upon what they call its biggest asset: 'human capital' and the youthful demographic of its population. Skill is being seen as something that needs to be enhanced in order to turn people into valuable 'human capital'. This is an absurd understanding of the meaning of capital – making it seem as if each worker is a capitalist, embodying capital by investing in education and the acquisition of 'skills', which are supposed to be encashable in the form of higher wages (Harvey 2010, 430).

The concept of human capital is very problematic, as has been critiqued by Samuel Bowles and Herbert Gintis (Bowles and Gintis 2002). They point out that people who try to understand problems of employment and education in terms of 'human capital' are ignoring class conflict and the division of human beings into economic classes. Such people try but fail to understand employment problems in terms of 'supply and demand'. This may result in misplaced and unneeded calls for 'skilling'. For example, there may be a need for more labourers in agriculture, and there may be plenty of unemployed people who need jobs, but farmers may not have enough money to pay them because they do not get adequate compensation for their crops. There may be plenty of people who have the skills to do the labour to produce crops or manufacture commodities, but they may not be able to do so if they do not own the means of production. Workers are not in a position to 'encash' their human capital as and when they need employment. And beyond a point, the so-called 'human capital' is of no use to employers either, since they cannot hire all workers. A reserve army of workers needs to be kept unemployed.

5.4 The Dialectics of Cultivation vs Education

Jyotirao Phule stressed the importance of education in agriculture and questioned how agriculture can advance when farmers are deprived of literacy, knowledge of other languages, and other aspects of education:

> "How could he manage to improve his agriculture with the help of books in other languages on agriculture he can't even read the alphabet of his own language? When he is continuously starving, how and on what basis will he send his children out of the village to large cities to study in agricultural colleges?" (Phule 1881).

He was acknowledging the point that advancing in agriculture (or any technology) depends on communication between people, communities, and distant regions, as well as on experience and work with the hands. But he is saying that it is poverty and the lack of basic literacy that is preventing technological advancement.

However, what he wrote in 1881 still has hardly been heeded. Between 1947 and 1980, the Indian government built many agricultural colleges and agricultural research institutions, but very few villagers have been able to become researchers in such institutes. They are also separate from non-agricultural colleges and research institutes. Scientists doing so-called 'basic research' (biologists, botanists, biochemists, molecular biologists, etc) work in institutes that do not have departments devoted to agriculture. Highly educated researchers hardly get their hands dirty, while villagers continue to do the work in the fields.

Although Phule was continuously stressing the need for education, it was not the usual kind of education. Besides learning to communicate in various languages, Phule stressed the need for farmers to learn what their rights as human beings are, to learn to reject blind faith, and to learn not to obediently and fearfully accept exploitation. Even today, it would be hard to find a school with such learning objectives.

Educationists may advocate various sorts of awareness campaigns in order to convince families of the benefits of education. But there does not seem to be a shortage of desire for education at present in India. People go to great extents to try to get their children educated.

References

Bowles, Samuel, and Herbert Gintis. 2002. Schooling in Capitalist America Revisited. *Sociology of Education* 75 (1): 18.

Braverman, Harry. 1998. *Labor and Monopoly Capital: The Degradation of Work in the Twentieth Century*, 25th anniversary ed. New York: Monthly Review Press.

Brunner, Michael Philipp. 2018. Teaching development: Debates on 'Scientific Agriculture' and 'Rural Reconstruction' at Khalsa College, Amritsar, c. 1915–47. *The Indian Economic & Social History Review* 55 (1): 77–132.

Chandra, Satish. 1945. *The Cow in India*. Calcutta: Khadi Pratistan.

Chirol, Valentine. 1910. *Indian Unrest*. London: Macmillan.

Education Commission. 1966. *Education and National Development: Report of the Education Commission 1964–66, Government of India, Ministry of Education, March 1971 reprint*. New Delhi: National Council of Educational Research and Training.

Government of India. 1986. *National Policy on Education*. New Delhi: Government of India.

Harvey, David. 2010. *A Companion to Marx's Capital*. London/New York: Verso.

Kumar, Krishna. 1996. Agricultural modernisation and education: Contours of a point of departure. *Economic and Political Weekly* 31 (35/37): 2367–2369.

Ministry of Human Resources Development, Government of India. 2019. *Draft National Education Policy, 2019*. New Delhi: MHDR. https://www.mhrd.gov.in/sites/upload_files/mhrd/files/Draft_NEP_2019_EN_Revised.pdf.

———. 2020. *National Education Policy 2020*. New Delhi: MHDR. https://www.mhrd.gov.in/sites/upload_files/mhrd/files/NEP_Final_English_0.pdf.

NCERT. 2007. *Position Paper, National Focus Group on Work and Education*. New Delhi: National Council of Educational Research and Training.

Oliver, Michael, and Colin Barnes. 2012. *The New Politics of Disablement*. Houndmills/Basingstoke/New York: Palgrave Macmillan.

Phule, Jyotirao. 1881. *Shetkaryaca Asud*. 2017 Marathi edition. Translated from Marathi to English by Gail Omvedt and Bharat Patankar. Pune: Mehta.

Sadgopal, Anil. 2016. 'Skill India' or deskilling India: An agenda of exclusion. *Economic and Political Weekly* LI (35): 33–37.

Chapter 6
The Historical Development of Paddy Cultivation

Our effort is to understand learning, teaching and doing paddy cultivation on one small family farm. We want to understand whether, or to what extent, the family is following tradition or doing science. In order to investigate this question, we need to relate our observations to the historical development of paddy cultivation. Only then can we really understand what is traditional and what is science, and why cultivation is being done as it is. We need to investigate how and why the processes of paddy cultivation have differed over time.

What is meant by 'traditional' agriculture? Is it unchanging agriculture? We know from archaeological and historiographic evidence that agriculture has not always existed, and that it has changed over the centuries. In fact, agriculture has existed for only a very small fraction of the time that human beings have existed. As we will discuss, there is always some amount of change in agriculture, as all plants and other forms of life are always evolving and the environment is always changing. Actually, there is an interdependence: human society influences nature as nature influences human society. The process of cultivation changes the crop. Rice evolves even without people, and rice also evolves due to the influence of the process of cultivation. So the identification of 'traditional' cultivation may depend on the degree of change or the rate of change. Maybe by 'traditional' cultivation, we just mean that it changes more slowly than what we consider to be non-traditional or modern cultivation.

Our concern is therefore with the nature of this change: how and why change occurs, and how and why particular changes do or do not occur. Many questions arise:

- What are the differences in the process of paddy cultivation over the centuries?
- How does rice differ over the centuries?
- Why has rice changed?
- Are the reasons for change different in ancient times, as compared to modern times?

K. Haydock et al., *Learning and Sustaining Agricultural Practices*, International Explorations in Outdoor and Environmental Education 7, https://doi.org/10.1007/978-3-030-64065-1_6

- In cases where there is not much change or change is relatively slow, then how are agricultural methods and materials passed on from generation to generation?
- If there is change, then how and why are agricultural methods and materials developed?

Thus, our investigation in this chapter is basically focussed on questions of how and why the development of methods of paddy cultivation occur, and how paddy itself changes. This will help us see how changes in methods of cultivation depend on how people learn and teach paddy cultivation.

6.1 The Origins of Rice

6.1.1 Variation and Change in Paddy

Variation exists within all populations of organisms, and variation also exists in rice. For example, if you compare any paddy plants, you will see many differences: the lengths, sizes, shapes, weights, smells, tastes, and colours of any two plants, tillers, leaves, panicles, florets or grains of rice are different.

Some of the differences are partly because random mutations arise in the genome (DNA) of individual plants, and offspring can inherit such mutations. Throughout the lives of the individual plants, their genomes interact with their environments to cause variations in traits. The environment of a genome includes the molecular biology in the cell, other cells, other individual plants, populations of paddy plants, different species of organisms, as well as the abiotic environment (water, light, temperature, weather, soil, etc). Because individual plants with certain variants of the trait tend to survive and reproduce more than individuals with other, variants, the population of paddy plants evolves. The population changes over time. This is called evolution by natural selection, and it occurs without people being around. It occurs without any plan or design – the plants do not evolve because they try to evolve or try to adapt to an environment. In other words, it is non-teleological. It is also important to realise that changes in environment are always occurring, and they affect and are affected by the continuous evolution of organisms such as rice.

However, if people are cultivating paddy, then the population of paddy plants also evolves due to artificial selection. In artificial selection, a population of paddy plants evolve because each season farmers select only the seeds from the types of plants they prefer for planting next season.

In order to cultivate paddy, each season people chose the seeds to use for planting the next season's crop. If they chose seeds from plants that had particular characteristics that they liked, the next crop might have more plants with those characteristics. Of course, some of the characteristics might be just due to the amount of water or sunlight the plants got, or whether they were infested by pests. If the differences are not passed on to the next generation through the seeds, then the selection for those traits will not be successful.

For example, one group of farmers may have noticed that some plants produced rice that was fatter, darker, and with a different aroma than the rice from other plants in their field. If they chose those seeds for growing the crop each year, after many years they might find that their field consists mainly of plants that give fatter, darker, rice with this kind of aroma.

Another group of farmers may have liked rice that was longer and yellower and grew on shorter plants that would not fall over easily, so by selecting the longer, yellower rice for planting, their fields gradually produced rice that was longer, yellower, and grew on shorter plants.

Another group of farmers may have selected seed from plants that grew well with less water. Another group of farmers may have selected seed from plants that had longer panicles and more grains of rice (Fig. 6.1).

No matter how much selection people do, there are still variations between plants, and the rice keeps changing over the years, partly because of the way farmers select seeds. This process is called artificial selection, and occurs during traditional cultivation.

Evolution by artificial selection may be less gradual than evolution by natural selection, or it may be more frequently punctuated by sudden changes. But the biggest difference that people have introduced through cultivation and artificial selection is that, at least to some extent, people can direct and produce the sorts of changes they need or want. Through this process, farmers have developed thousands of different landraces of rice. Some examples of these different kinds of landraces of rice are shown in Fig. 6.2.

Without continual selection for the desirable traits, the cultivated types of paddy tend to revert to less desirable forms. Therefore, cultivation requires continual selection. Paddy seeds do not remain viable for long. After a few years of storage, the seeds will no longer germinate. Especially if seeds are stored in warm, humid conditions, they will not last long. Therefore, a landrace of paddy seeds needs to be continually replanted and reselected.

Fig. 6.1 A tiller of Komal paddy, a landrace produced by the family in Rudravali for their own use

Fig. 6.2 A few varieties of traditional paddy (landraces). (Central Rice Research Institute. 2012:14)

In addition, needs and desires and environments keep changing. Because of this, selection keeps changing and seeds and crops keep changing.

Thus, there is actually a dialectics in the process of seed selection – a dialectics between maintaining a landrace and developing a new landrace. As a result, any landrace keeps evolving. A landrace is thus actually very difficult to identify or define because it is inherently quite dynamic.

6.1.2   The Balance of Nature?[1]

You might think that a stable landrace of paddy could be maintained if it was cultivated in a stable, balanced environment – perhaps if cultivation was done in a traditional way that was in balance with nature. But, does such a kind of cultivation exist? And does a 'Balance of Nature' even exist anywhere or at any time?

The idea of a Balance of Nature has a long history, and can be found in various cultures throughout the world (as reviewed by Kim Cuddington 2001). Its

---

[1] This section on 'Balance of Nature' is adapted from our previously published work (Haydock and Srivastava 2019).

prevalence and persistence is perhaps because people have a longing for a harmonious, unchanging, unending existence. However, there is overwhelming evidence that nature is not, and never has been balanced: (a) there is evidence that nature undergoes continuous change and evolution; (b) there is evidence that nature is not teleological; and (c) there is no evidence for any possible physical mechanism whereby such a balance might be maintained.

Charles Darwin found evidence that non-teleological natural selection is a mechanism of evolutionary change. With an understanding of evolution and its mechanisms, it becomes clear that if nature did adhere to a strict 'balance', life could not exist or evolve. Indeed, interdependent and continuous change and imbalance is an essential characteristic of nature. We will discuss how we see evidence for this in the continuous evolution for rice, both before and after the involvement of cultivators.

The idea of a Balance of Nature in the study of ecology stems in part from a view which dominated the field since the middle of the twentieth century. The belief was that "biological communities are in a state of equilibrium, a more or less stable balance, unless seriously disturbed by human activities" (Campbell et al. 2008, 1211). A community was thought to have a tendency to reach and maintain a particular and stable composition of species. For example, if a forest was converted into farmland and then deserted, it was believed that the vegetation had a tendency to return to its original state of equilibrium (which depended only on the climate). However, researchers have found that this does not actually happen.

A belief in the Balance of Nature is a kind of teleological idealism. It is teleological because it means that natural processes are governed by what is essentially a 'desire' or necessity of nature – a need to adhere to an abstract principle of balance. However, no one has ever found physical evidence for either the existence of a Balance of Nature, or of how a balance is maintained. They have not even been able to hypothesise the existence of any physical mechanism whereby it might be maintained. Although people do make plans and purposely do things (such as paddy cultivation), before people existed, nature, and evolution was basically non-teleological.

The Balance of Nature paradigm can be contrasted to a reductionist paradigm, in which there is no balance regulating different elements of an ecosystem. Instead, each species is regarded as a separate element in its environment, each species reacts to and evolves in response to its environment, and the sum of all the individual species and the abiotic environment determines the ecosystem. However, in this paradigm, observed changes in the environment in response to the species is not given due attention (as discussed by Richard Levins and Richard Lewontin 1985, chap. 6).

Accumulating evidence indicates that a biological community is neither balanced nor reductionist (Kim Cuddington 2001; Frank Egerton 1973). If reasonably long enough time periods are taken into consideration, we do not observe steady state equilibriums in nature. We observe rather a system of interdependent parts (organisms and abiotic factors within and to some extent also outside of the community), in which neither the whole nor the parts completely determine each other.

The whole is more than the sum of the parts, contrary to what an overly reductionist paradigm would claim. The whole is dependent not upon some abstract principle, but upon interdependent processes occurring in physical reality. As Frank Egerton (1973, 347) states, 'Mutations and natural selection gradually change species. Extinctions, species evolution, changes in species composition, and environmental alterations change natural communities'. The vast majority of all species that have ever existed on earth are now extinct. Although the extent and rapidity of environmental disturbance caused by humans is now greater than ever before, even in ancient times entire peoples were wiped out as a result of human actions which caused droughts, extinctions of species, and other environmental problems (Martin Spence 2001).

Of course, humans do disturb nature, and realising that there is no Balance of Nature does not excuse or justify the destruction of nature by humans. It would be a mistake to site the absence of a Balance of Nature in order to rationalise or deny the existence of the environmental crisis. That would be like saying that there is nothing wrong with poisoning a person since their eventual death is inevitable. There is overwhelming evidence that humans have been disturbing natural (unbalanced) processes at an unprecedented extent and rate, and that it is causing environmental destruction that is threatening the survival and well-being of human life.

The problem with believing in a Balance of Nature is that it may lead people to propose faulty solutions to environmental problems, such as just leaving an 'unbalanced' ecosystem to go back to its 'balanced' state. Abandoning a biological community which has been disturbed by human actions will probably not result in it returning to any previous state, or going to a state of equilibrium at any relevant scale of measurement. Since there is no force which will cause a Balance of Nature, the solution to environmental degradation cannot be to just leave things alone and let them go back to their natural balance.

In any case, people do need to eat, and people do need to cultivate food. The process of cultivation does inevitably affect other aspects of nature, just as other aspects affect cultivation. This is what marxists mean when talking about the human need to 'exploit nature'. (For that matter, even hunting and gathering interdependently affect the environment.) People need to question what these interdependencies are and whether cultivation is being done in ways that are desirable, for whom are they desirable, and for whom are they undesirable or harmful.

6.1.3   The Prehistoric Agrarian Transition

Cultivation began with what is called the First Agricultural Revolution, which occurred at different times in different places across the world, with a number of different crops. In many parts of the world, including the Indian subcontinent, archaeologic evidence has revealed that the earliest inhabitants were hunters and gatherers who did not engage in cultivation. Indigenous rice was one amongst many types of grain that were gathered. There are many interesting questions that people are investigating regarding the origins of rice cultivation. Radiocarbon dating indicates that the *indica* subspecies of rice (*Oryza sativa*) was eaten, around 6400 BCE in Lahuradewa, in the upper Gangetic plain, where some of the oldest pottery in South Asia has also been found. Researchers disagree as to whether this rice was gathered from the wild or cultivated (Silva et al. 2015). There is archaeological evidence that rice was cultivated in Harappan sites around 3000–2600 BCE, and possibly earlier (Bates et al. 2017). Some researchers argue that rice was cultivated in the Indian subcontinent as early as 9000–8000 BCE.

Perhaps two or three thousand years after cultivation began, the *indica* subspecies was domesticated – meaning that a genetically distinct variety was developed, which (unlike the earlier varieties) was dependent on humans for reproduction and dispersal (Choi et al. 2017). It is thought that the *indica* subspecies was probably domesticated before 2500 BCE, and was cultivated using rainfed, dry-field methods rather than wet-field methods (Fuller 2011). Different kinds of irrigated paddy cultivation techniques may have been introduced as early as the second millennium BCE.

Domestication of rice is thought to have occurred mainly due to artificial selection – by people, over many generations, selecting seeds from plants with desirable traits (Sect. 6.1.1).

An additional kind of selection is done through weeding: the desired crop plants are allowed to grow in a field and the undesirable weeds are eliminated. Even weeds may be useful for animal fodder or for eating, but unless a field is weeded, the main crops will be impaired by the loss of water, nutrients, and sunlight due to the crowding by weeds. Weed plants pre-existed but evolved in conjunction with the development of agriculture. Weeding may also affect the evolution of the crop by allowing certain varieties to flourish that would otherwise not get adequate conditions for growth. There is evidence that weeding has been practiced in some of the most ancient Neolithic agricultural sites (Bates et al. 2017). Thus, the dialectical conflict between desirable and undesirable plants has given rise to new kinds of plants – both with and without people (artificially and naturally).

Both archaeological and genetic studies indicate that another subspecies of rice, *Oryza sativa japonica*, originated in China, where it was cultivated for a few thousand years before it was domesticated by 6300 BCE. Genomic analysis indicates that there were multiple origins of the different subspecies of domesticated rice from one progenitor, and that the *japonica* subspecies was the first to be domesticated, in China (Choi et al. 2017). Some researchers have reported finding the *japonica* subspecies in late Harappan sites, indicating that this fully domesticated subspecies may have diffused from east Asia to central Asia as early as 2000 BCE. The full domestication of the *indica* subspecies in the Indian subcontinent probably followed contact and hybridisation with *japonica*. This sort of hybridisation must have occurred spontaneously (and possibly very gradually) due to cross pollination, which is infrequent in rice (see Sect. 6.2.2). But how did the *japonica* subspecies get to the Indian subcontinent and how and why did the hybridisation occur? These questions are still being investigated. However, experimentation, communication, migration, and trade were probably important factors in learning, teaching, developing, and engaging in rice cultivation since very early times.

Genetic (and linguistic) analysis and the remains of past cultures provide evidence that a number of crops (e.g. certain grains and grams) were initially domesticated in the Indian subcontinent, and other crops were introduced from other regions, either by trade or accompanying immigration (Fuller 2011). If the evidence indicates that a crop was derived from the local population of a wild progenitor, we can say the crop had its primary origin (or independent origin) in that area. If the evidence indicates that the crop was instead introduced as a cultigen from another region, its origin can be said to be secondary. In case a crop was independently domesticated, there is another, more difficult, question as to whether it was domesticated by people who did or did not have experience of other kinds of food production, or whether it was inspired by contact with people who produced other crops. If a group of isolated hunter-gatherers independently domesticated it, they presumably developed their own methods for domestication. Thus, these different possible scenarios may be characterised by different ways of learning/teaching in which the degree of constructivist methods, or communication with authorities may vary.

As reviewed by Fuller et al. (2016), the important traits that have evolved during the process of domestication include: panicles that do not shatter (dispersing seeds before harvest), fatter grain size, and increase in yield. Labour-intensive cultivation methods also developed, including methods of ploughing and field preparation, and methods of irrigation and manual control of water. In most parts of India, dry rice cultivation was replaced by wet rice cultivation methods, which are more productive.

The extent to which the domestication (and the learning/teaching) was intentional or fortuitous is yet another question. Of course, it is unlikely that cultivation began as a sudden revelation or an 'idea' derived from theoretical analysis that was separate from doing, observing, and questioning. But just as plants evolve through natural selection, it is possible that the seeds and even the methods of cultivation may have arisen to some extent by chance, with little purposive human planning.

Whether there is a causal or dialectical link between the emergence of settlements and the emergence of cultivation is another matter of debate. The evidence from south Asia, in contrast to evidence from other areas, suggests that early cultivation may not have been preceded by the emergence of sedentary, densely populated hunter-gatherer societies. Settlements and cultivation may have developed at the same time. Developments in paddy cultivation are thought to have supported the formation of states, dense populations, and hierarchal social systems, and were also associated with transformation of the environment (e.g. levelling fields, deforestation, construction of water management systems, changes in soil fertility, and increases in the release of methane).

Throughout the evolution and domestication of rice, environmental change has been important. There may be dialectical links between certain climatic changes and the emergence of cultivation and pastoralism (which seems to have generally preceded cultivation). In particular, there was an aridification around 2200–2000 BCE, which probably was one, if not the main factor, leading to the decline of the Harappan civilisation (Fuller 2011). But the same climatic changes may have also influenced hunter-gatherers, through contact with migrating pastoralists, to adopt a more sedentary life-style that included agro-pastoralism. Including the secondary domestication of goats, sheep, and cattle, as well as the adoption or development of cultivation.

How did people first learn to do independent domestication? Most researchers agree that the methods that early cultivators used to develop paddy cultivation must have been largely empirical: based on trial and error. Perhaps foragers collecting a certain species that they found only in scattered patches in particular microenvironments, may have purposely or incidentally concentrated them and/or extended the patches by clearing away other plants. They probably learned through trial and error, communication, observation, and comparison throughout neighbouring communities. Of course, it is very difficult to find evidence for such speculation. However, it is clear that cultivation involves much more than simply collecting seeds from the wild and putting them in soil. Through their agricultural practices, people somehow cause plants to evolve into new varieties. This may happen with or without much conscious planning and specific aiming. But we know that new varieties arise with domestication because domesticated varieties are not found in the wild. When domesticated varieties escape cultivation, they change – usually reverting to something that is closer to a wild type.

Paddy cultivation and the production of rice in an edible form requires more labour than almost any other major crop. However, it is also much more productive than other cereals. The process of domestication and continual artificial selection requires considerable labour. This labour is purposive, and it requires learning and teaching.

There is plenty of archaeological and historical evidence to show that paddy cultivation keeps changing: the seeds, the tools, and the methods of cultivation keep changing, as do the soils, the pests, the weather, the climate, and other aspects of the environment. All of these also vary between different places – even in neighbouring villages and neighbouring fields. Therefore, the learning and teaching of paddy cultivation must keep changing and developing as methods and materials keep changing and developing.

Even this cursory glance at some of the questions regarding the beginning of paddy cultivation cannot fail to leave us with an impression of the complex, dialectical interactions between paddy, the environment, and groups of people, all constantly changing over time. In this context, one wonders how unchanging 'traditions' could possibly arise.

6.2   Diversity and Varieties of Rice

The most dramatic evidence of how farmers developed paddy is the existence of tens of thousands of different types of rice in India (Gurinder Jit Randhawa 2006). Even in one tehsil, farmers have been growing many different varieties, which exhibit a wide range of characteristics and are used to prepare different types of foods (Fig. 6.2).

6.2.1   Traditional Landraces and Commercial Cultivars

There are two different kinds of varieties of rice (and other crops): landraces and cultivars. The term 'landrace' (also called heirloom or folk variety) is used to mean a traditional type of plant: 'a local cultivar' that has been developed by 'traditional' agricultural methods of artificial selection. Landraces have been developed for their use value – for use by subsistence farmers who have developed them and used them primarily for their own use, rather than to sell.

The term 'cultivar' is used for a commercial variety that has been developed by selective breeding done more 'scientifically' by multinational companies or research institutes. The breeding is done through artificial selection and/or genetic engineering. Cultivars are not developed by people for their own use – they are developed in order to be bought and sold. In other words, cultivars are developed for their exchange value. They are developed in a highly competitive manner. There is competition between scientists who try to garner recognition, win prizes, obtain good jobs (in situations in which the number of jobs is limited), and attain positions of social power and dominance. There is also competition between research institutes and companies who compete for the profit they get by selling the cultivars and other products they develop. The companies will not survive unless they keep increasing their profits. Due to all these inevitable kinds of competition, research may not be done through open cooperation and sharing of ideas and findings, even though cooperation may be advantageous to the advance of the research and to the use value of the research. The use value of the science is forced to be subordinated to the exchange value of the science.

High Yielding Varieties (HYV rice) are cultivars. A cultivar is bred to have less genetic diversity – all the seeds usually have more genetic similarity to each other, in order to produce a more uniform crop in which all the plants have more similar characteristics to each other. If everything goes as planned, cultivars should produce a higher yielding or more valuable crop. But in the real world, things seldom go exactly as planned. Since a cultivar is usually bred for widespread use rather than specific local conditions, it may actually be less suitable under some local conditions. Therefore, for the best results, cultivars need to be grown using specified conditions: usually they require more water and more artificial fertilisers and pesticides. If the environment varies too much from the 'average' conditions that are expected, the entire crop may fail. Because cultivars are bred to be more uniform, there may not even be a few individual plants that happen to have differences, such as drought or pest resistance, that the other plants do not have.

Thus, there is a dialectical conflict here: because of the conflict between being uniform and being diverse, a cultivar may either give a good yield or it may fail terribly.

Since a landrace is developed locally by and for peasants or people who are working in the fields, it may be more suitable to local soil, climate, and protection against local pests. But since the selection may not be done according to very precise or rigid specifications, a landrace usually has more genetic diversity than a cultivar. Within one population of one landrace, there may be considerable variation in genetics and in traits of individual plants. Therefore, when there is a drought, flood, pest infestation, or some other unexpected problem, it is likely that not all the individual plants will be affected to the same degree. It may happen that some of the plants have genetics and traits that allow them to survive. Therefore, landraces are considered to be more robust and dependable, even if they may not produce as much yield as commercial cultivars.

Thus, here there is also another kind of dialectical conflict: a landrace may be more suited to local conditions, but also more suited to changes in local conditions. A landrace is more dynamic and less well-defined than a commercial variety. Because they have less genetic diversity, cultivars are less apt to survive various pressures, such as droughts, floods, extreme weather, pests and diseases.

It is important to preserve landraces, just as it is important to preserve diverse wild varieties, because they contain mutant genotypes that produce diverse phenotypes that might be useful, and that might not be reproducible once they are lost. Some landraces may have taken centuries to be developed, or they may have cropped up accidentally or have been found by chance. Scientists rely on landraces, to combine with other varieties in order to produce new breeds that have certain desirable characteristics of a particular landrace.

As we have discussed, our observations of the farming family indicate that the process of artificial selection is done by a surprisingly 'scientific' method in which dialectical conflicts give rise to questioning, hypothesising and investigating. Continuous selection, and the continuous development of varieties of rice, thus suggests that paddy cultivation necessarily involves 'doing science', at least to some extent.

6.2.2 Rice Reproduction and Hybridisation

In some species of plants and animals, there are male and female individuals, but in rice, the male and female are contained within one plant, and also within one flower. As shown in Fig. 4.1, each rice flower, or spikelet, contains the female carpel and the male stamen. The carpel contains an ovary that contains an ovule (egg), which develops in an embryo sac. The embryo sac usually becomes fully developed, and ready for fertilisation, one day before the anthers are fully developed. In the process of 'flowering' (Fig. 6.3), the spikelet opens (the lemma and palea partially separate), the anthers emerge, pollen is scattered, and the spikelet closes, leaving the anthers outside to die and whither up. This entire process takes only 1–3 hours, and usually occurs in the morning. Even before flowering, before the spikelet has opened, mature pollen from the same flower begins landing on the stigma. The flower could be pollinated by pollen from other rice plants, but this is not very likely since it could happen only after the spikelet opens. Therefore, cross-pollination is rare in rice. Within 2–3 minutes of landing on the stigma, a pollen tube starts growing from

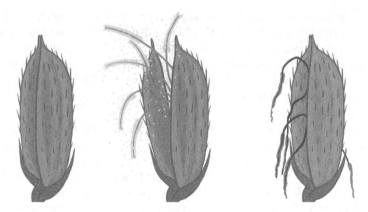

Fig. 6.3 The opening and closing of a rice flower (a spikelet), showing the six anthers producing pollen

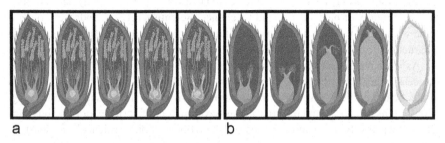

a   b

Fig. 6.4 (**a**) Fertilisation of rice may happen before the spikelet opens. Pollen from the anthers lands on the stigma and pollen tubes grow inside the style, finally reaching the ovule. Sperm from the pollen then goes through the tubes to the ovule. Fertilisation then occurs, in which DNA from the sperm combines with the DNA of the ovule. (**b**) The fertilised ovule then grows into a seed (a grain of rice), which fills the spikelet

each pollen grain towards the ovary (Fig. 6.4(a)). The tubes reach the ovary within 15–60 minutes. Sperm from a pollen grain go down the pollen tube and fertilise the ovule. The genetic material (DNA) from the sperm and ovum become mixed, producing an embryo that contains some DNA from the sperm and some DNA from the ovule. By the next morning, the fertilised ovule begins developing into an embryo, and eventually a grain of rice containing the embryo develops inside the spikelet (Fig. 6.4(b)).

People have developed hybrid varieties of many crops by cross-breeding two different varieties of plants, each having different desirable characteristics. This is done by taking pollen from one variety and using it to pollinate the other variety. For example, hybrid maize can be developed by taking pollen from the anthers of the flowers of a variety of maize that has more starchy kernels and using it to pollinate another variety that has larger cobs. The pollen fertilises the ovaries and produces seeds that, when planted, may give rise to plants with more starchy kernels and larger cobs. This is relatively easy to do in maize, because the male and female parts of the plant are separate: the anthers (tassels) are at the top of the plant and the female parts (silk) are at the bases of leaves. Because of this, cross pollination anyway occurs in maize.

Because each rice flower usually fertilises itself, and each flower produces only one seed, it is difficult for scientists to develop hybrid rice. In order to produce a hybrid, scientists have to stop a rice flower from fertilising itself so that they can artificially fertilise it with pollen from a different variety. This could be done by opening a rice flower before it is fully developed and cutting off all the flower's anthers, then, when the flower is at the right stage, brushing the flower with pollen they have collected from the flowers of another variety of rice. But it is very difficult to do this for every flower. And each flower only produces one seed!

Therefore, until recently, most of the development of new types of rice have not been done by hybridisation. The High Yielding Varieties (HYV) of paddy that began to be introduced into India in the 1960's are not hybrids. They are varieties that have been developed using techniques of artificial selection. In 1956, a semi-dwarf mutant was discovered from which an HYV was developed in China in 1959. In 1966, the semi-dwarf HYV Indian variety IR8 was developed by the International Rice Research institute (IRRI) in the Philippines. Variants of IR8 and other types were developed using methods of artificial selection at the Indian Agricultural Research Institute (PUSA) and other institutions. The HYV grown as a cash crop by the family that we were working with is a variant of IR8.

However, in 1971, a wild rice plant was discovered in China which was a "male sterile plant" – the pollen it produced was not fertile and could not fertilise itself, as rice usually does. The rice flowers of this plant required other varieties with fertile pollen for fertilisation and rice production. This plant provided a way to begin producing hybrid rice. Scientists began trying to produce hybrid plants that had the desirable characteristics of these plants with sterile pollen combined with the desirable characteristics of plants having fertile pollen. By the 1980's they had succeeded, and hybrid rice began being widely cultivated in China. Nowadays various varieties of hybrid rice are also grown in India.

Nevertheless, it is still not clear whether or under what circumstances it is better to grow hybrid rice. Hybrid rice has the same problems as HYV rice, that we discussed above (Sect. 6.2.2). Also, hybrid rice cannot be used to produce seeds that have the same genetics or characteristics as the hybrid seeds. This is because hybrid seeds carry all the genes from both of the types of rice from which they were produced, even though not all of the genes are expressed in the first generation. Some of these genes produce unwanted characteristics if they are expressed. In subsequent generations, they may be expressed, producing crops with larger percentages of unwanted traits in each subsequent generation.

For example, suppose a hybrid was produced by combining a variety that had longer panicles and little pest resistance with another variety that had shorter panicles but was pest resistant. In the first generation the hybrid seeds produced plants with longer panicles and more pest resistance. But if these seeds that are produced are used for the next season's crop, they will produce a new crop that has a mixture of plants with all the characteristics of the parent plants: some plants with longer panicles, some with shorter panicles, some plants that are more pest resistant and some that are less pest resistant. (This example is just to explain the general idea – in real life things are much more complex and ununderstood.)

Because of this, farmers need to purchase new hybrid seeds every season rather than using the rice they produce as seeds. This can be very expensive for the farmers – and very profitable for the seed companies, since farmers do not have the technology or resources to produce hybrid seeds themselves.

But capitalism has an inherent, unavoidable characteristic of commodifying everything it possibly can, and here also, there are attempts to commodify landraces, even though a landrace has been thought to be an "anti-commodity" (Glover and Stone 2018). This is happening in tandem with other efforts to commodify traditional methods of cultivation in order to define 'organic' food: cash crops that can be sold (and perhaps exported) so that capitalists can extract more profit, while seeming to be 'environmentally friendly', 'sustainable', and in support of 'free-trade'. Unfortunately, it seems that people are sometimes more interested in sustaining capitalism than in sustaining life on earth.

6.3　The Historical Relations Between Science and Agricultural Development

In 1870, Lord Mayo advocated setting up a government department of agriculture and opening schools for the study of agriculture in India, in order to spread the application of science to agriculture. He argued, "it is really to the spread of the knowledge of what I call agricultural science that the whole of [the British Agricultural Revolution] is to be attributed" (as quoted by MS Randhawa 1980, 180).

But were technological advances the cause or the result of this agricultural revolution in England? And to what extent was the development and adoption of new technology dependent on some kind of formal education?

Obviously there must be some connection between technological production and science. The development of electric power was based on investigations carried out by scientists. Factories producing chemical fertilisers were designed by engineers who studied chemistry.

But is the absence of science the reason that more advanced technology did not arise earlier in the 'traditional' society of India? And if science is absent, why is it absent? Or is science present but in different forms? And if it is in different forms, why? Or even if the same sort of science is present, are there other, more important underlying reasons why technology does not develop? These questions are important in order to understand the role of science education in the future and in the development of the country.

We need to further analyse the history of science and agricultural development and its connection to the development of capitalism in order to address these questions.

6.3.1   Pre-capitalist vs Capitalist Production

Without investigating the interdependencies between production, money, the labour market, the commodity market, and the credit system, it is not possible to understand how and why the agricultural system works – and why we have the present agricultural crisis.

Long before European capitalism or colonialism, the Indian subcontinent had extensive agricultural production and trade. Besides subsistence farming, surplus products were also being traded and sold as commodities in exchange for money. Labour services were also being exchanged for money. As ways of life keep changing and changes diffuse over large areas, it becomes increasingly impossible to continue to live in the old ways of subsistence farming and/or hunting and gathering, even in small pockets.

There has been considerable controversy as to the nature of the mode of agricultural production and whether or not at various times and places in India it can be called adivasi, feudal, capitalist, or some other type (BB Mohanty 2016). Many people had thought that corporate agriculture would develop faster than it has in India. But instead, in Raigad and in India in general, we see a greater persistence of small family farms and subsistence farming than expected (Athreya et al. 2017). However, some sort of capitalist development is now occurring in which these processes of production and trade are, at least to some extent, being reconfigured according to class relations – and integrated with caste, creed, and gender relations.

Capital appropriates the products of the agricultural labourers, and these products have surplus value embodied in them. What capitalism is all about is the production of surplus value and the circulation and accumulation of capital – and the social relations that enable this accumulation. Technological changes (such as the use of tractors, pumps, high-yielding and hybrid seeds, artificial fertilisers, and pesticides) introduce instability along with the growth of capital into this circulation process. Other problems arise as commodities become integrated into the world market. There is a tendency to transform all possible production which was formerly used by the agricultural labourers for their own subsistence (use value) into commodity production (producing cash crops with exchange value). These commodities assume a life of their own in the hands of merchant capitalists. Even if the commodities are not produced in a capitalist mode of production, capitalist circulation becomes reliant on them, and colonialism, imperialism, and globalisation guarantee continued access to them (see also David Harvey 2013, 153–165).

Agricultural production is inherently very different from manufacturing in several ways. Most importantly, unlike in manufacturing, farmers do not have much control over the amount they produce or the prices they get because of variations in weather, pests, diseases, seeds, water availability, natural disasters, etc. This is what makes agriculture inherently risky, and this is why government support is required, and has always been provided in advanced capitalist countries. To the extent that government support is not provided, capitalist agricultural development is impeded. For example, in the USA the government provides extensive funding for agricultural research, subsidises farming, provides crop insurance, and protects their own agricultural production through trade agreements.

6.3.2 The Development of Science, Agriculture, and Industry in Europe

Do agricultural and scientific revolutions begin with the development of theories and understandings by scientists? JD Bernal argues against this, saying that science does not advance "due solely to the genius of great men, … largely divorced from the effect of social and economic factors" (Bernal 1954, 21). We reject the view that science arises just because a genius happens to be born – i.e. a great man who is capable of doing science. Children are not born as scientists or non-scientists. Neither is science done by individuals working in isolation from society and from their environment.

What are the social factors that are important in explaining the development of science?

Actually science has always been developing dialectically along with the development of technology, modes of production, and social relations. The totality of this human activity occurs in relation to nature and in relation to particular spatial and temporal environments.

Bernal has discussed the importance of practical work in the development of science. He claims that: "the most important and fruitful periods of scientific advance were those in which the class character was at least partially broken down and the practical and learned men mixed on equal terms."

The development of science may depend on some breaking down of this sort of class character, because science requires working with both hands and mind. However, the development of science has also been accompanied by an opposing propensity to vigorously reinforce class character, which is inherent to capitalism. The development of science and technology has led to – and has been led by – the development of capitalism. However, with the development of capitalism, a new type of scientists evolved – scientists who were not just intellectuals, but were more closely associated with technological development.

Note that, according to Bernal, the theory and practice was carried out not by 'intellectual workers', but by different people: practitioners and intellectuals, who "mixed on equal terms" (Bernal 1954, 27). It is debatable to what extent this actually happened. But also note that these new scientists were (almost exclusively) men, not women. This is just one reflection of the social character of science.

According to most dominant academic points of view, science is, by definition, considered to be something that began only with the advent of capitalism in Europe and intensified during their industrial revolution. However, according to Marx and Engels, science was not created but transformed by the advent of capitalism. With the division of society into conflicting classes, the bourgeoisie and the proletariat, the profession of scientist began:

> The bourgeoisie has stripped of its halo every occupation hitherto honoured and looked up to with reverent awe. It has converted the physician, the lawyer, the priest, the poet, the man of science, into its paid wage labourers. (Karl Marx and Friedrich Engels, 1848)

They are probably not referring to farmers working in the fields as people who were previously doing science – they are referring to intellectuals or artists such as Leonardo da Vinci, who were not working for capitalists, but within a feudal system. They are referring to people who were separate from the practitioners, and were looked up to in awe. The profession of scientist – people hired to do science – was created in order to be used by capitalists to create new technologies and commodities and increase their profits.

Capitalist science is inevitably different from any form of science that exists outside of or prior to capitalism. It is necessarily driven by an underlying profit motive. Capitalist science is used to create technology that is commodified and valued for its exchange value when it is bought and sold. But in addition, capitalist science itself (even the most 'basic' research) becomes a commodity to be bought and sold. The process of doing science becomes reified into what appears to be a thing: a piece of technology, a 'body of knowledge', or 'intellectual property': a commodity with monetary value. We will discuss this in more detail, and specifically with regard to agriculture on the following pages.

As we discussed (Sect. 6.2.1), prior to or outside of capitalism, science and technology may be done, for example, to develop landraces of rice. This was and is done by peasants as part of their daily labour. This sort of development is valued for its use value rather than for its exchange value. This kind of science is important in agricultural development, but is relatively slow in developing, due to a number of constraints. Gradually, it becomes eclipsed by capitalist science, which is done for exchange value rather than use value.

In England, technology, education, science and economic development were all interdependent and dialectically related, as both/neither causes and/or effects. Perhaps most significant was the symbiosis of agriculture, industry, and commerce: capital accumulated in industry flowed into agriculture, and the reverse flow from agriculture to industry also occurred. Banks were also involved.

MS Randhawa discusses how the so-called Agricultural Revolution in England in the 17th to 19th centuries was "the result of the application of capitalist methods to the production of crops, the demand for which had been multiplied by conditions created during the Commercial Revolution." (Randhawa 1983, 50) He says, "Improvements in agriculture were made by men with cultivated minds, who adopted it as a recreation, or with the desire of improving what they saw being imperfectly performed." These improvements introduced by the landlord class consisted largely of methods and materials seen while travelling to other parts of Europe, such as new types of crop rotation, and the development of equipment such as seed drills that allowed for row planting and easier removal of weeds by horse-drawn hoes, resulting in improved yields. Perhaps this technology had been developed by practicing farmers, through experimentation and hands-on trial and error. Communication and learning from other places was perhaps more important than invention by the British men with "cultivated minds". Dharampal (2000) has given evidence that agricultural implements and practices were also exported from India to Britain in the late eighteenth century. However, the implementation of new technology in England would not have been possible without the input of capital, which was initially accumulated from mercantile sources (Bernal 1969, 511).

As is well-known, the Agricultural Revolution in England was also made possible by the legally sanctioned imposition of enclosures, the loss of the commons, the evictions of peasants, the "Consolidation of Holdings", and the construction of elite country estates with their gardens and parks. The result was that peasants lost their land, access to the commons, their cattle, and their handicrafts, and were forced to migrate to cities or other countries, some finding industrial employment, and many joining the ranks of the unemployed. In short, the rich got richer and the poor got poorer.

Some have claimed that the agricultural revolution preceded and led to the development of industrial capitalism. However, Utsa Patnaik (2016) summarised the evidence that the development of industrial capitalism in England in the nineteenth century was fostered by colonial exploitation, rather than by an agricultural revolution. The industrial revolution required peasants to be dispossessed from their land. But, contrary to what is often said, this did not occur through an agricultural revolution. The dispossession of peasants from their land [and the establishment of larger capitalist farms] was actually accompanied by a per capita decrease in the production of grain. Also, the displaced peasants were not necessarily absorbed into urban industry – many of them ended up emigrating to other countries. Unemployment was exported. This kept domestic industrial wages from dropping too low, as they would if the reserve army of labour had been larger. If unemployment had gotten too severe, there would not have been enough people able to buy what was being produced. Thus, primitive accumulation in England was accompanied by global flows of capital and labour. The latter was made possible by the genocide of indigenous populations and expropriation of their land in America, South Africa, and Australia. Capitalist accumulation produced mass poverty in the former colonies and affluence in the colonising countries. Industrialisation was possible due to global capitalist accumulation in other countries "through direct seizure of resources by means of force" (Patnaik 2016).

As is also well-known, the development of capitalism, industry, agriculture, and science in England had a direct bearing on India. Imperialism developed and India eventually became a colony of Britain. The effects of colonialism on India included the draining of natural resources and the obstruction of agricultural, industrial, and scientific development, in order to allow British industry (especially textile industries) to flourish. At the same time Britain became increasingly dependent on Indian raw materials. Agricultural products were initially of less importance to Britain than raw materials. Their eventual interest in agricultural development in India was driven by the English textile industry's dire need for cotton following the halt in supply due to the civil war and end of slavery in the USA.

6.3.3   British Colonialism vs Indian Agriculture

Jyotirao Phule gave the following answer to the question of how and why British trade and Imperialism began:

> Since there were mountainous areas here and there in that country, there was not enough land for everyone to survive from agriculture; the cold was extreme; thus all were doing various kinds of artisanship and merchant trades and went to the forefront of all the peoples on the surface of the earth in acquiring education, knowledge and wealth. (Phule 1881)

This is in line with Utsa Patnaik's point that there was a per-capita decrease in the production of grain in nineteenth century England. It is also in contrast to stories the British told about how they possessed some intellectual and technological superiority that enabled and entitled them to go out and colonise the world.

6.3.3.1   Why Was Britain Able to Conquer India?

School children are told that colonialism began after Britain wanted to trade with India because they needed Indian spices. A simplistic definition of trade would be that something is being brought from Britain to India and in return something else is being brought from India to Britain. Children are told that it was spices that were being brought from India to Britain, because Britain could not produce spices. Actually, it is doubtful whether the people of Britain really needed or wanted spices. To this day, British food has the reputation of being very bland. Anyway, the British were taking various things from India.

However, the main British objective was not spices – it was trade itself. To traders, the use value of the commodities they are trading does not matter as long as someone else has a use value for them. And they need not bring the spices all the way to Britain. They could trade the spices to anyone along the way who would buy them. Only their exchange value mattered to the British traders. To traders the spices were, in a sense, immaterial. Through trading, British merchants became rich, enabling capitalism to develop.

But what was being brought from Britain to India? What did Britain have that India did not have? Actually, Britain was hardly bringing anything in return for the things it was taking from India. They had nothing to give. Utsa Patnaik has pointed out that at least after 1857, India did not really get anything in return for all that it was giving to the British traders (Patnaik et al. 2011). The Indians were heavily taxed, and the revenue that Britain 'earned' in taxes from Indians was itself used to pay Indians whatever paltry sums they got for their raw materials and other commodities.

Phule gave an interesting account of how and why the British were able to conquer India. To relish the potency of his writing, it is worthwhile to quote him at length:

> At the same time, the disciples of Hazrat Mohammad Paigambar in Arabasthan annihilated the original political splendor of the Aryans in Iran, and made various forays into Hindustan to take under their control the whole land of those chewed up by Brahmans. Afterwards the Muslim Badshahs became stupefied listening to Tansen's songs by day and carousing at night in their harems, and with this the greatly intelligent English smashed the turbans of the Muslims and easily clasped the country in their arms. There was no great prowess in that, because the ten percent of Brahmans through their fabricated religious writings had denied education, knowledge, courage, sagacity and strength in religious and political affairs to the remaining ninety percent here and kept them inferior. However, after this when it came to the attention of the English that the nine-tenths Shudras and Ati-Shudras were uncivilised in nature and dull in all their work, and carried on their activities according to the dominant Brahman policy, they showed their covetousness to the greatly cunning Brahmans and left all the administration in their hands. With the aim of having sufficient funds to give all the European and Brahman employees whatever high pay and pensions they wanted so that they should at all times have valuable clothes, horses, vehicles, food and drink, the farmers who ate stale bhakri and toiled in the fields night and day were forced to pay revenue rates raised every thirty years according to the whims of the administrators. Not only this, they made a noisy show of giving education to their ignorant children and imposed on their heads a second tax burden known as the "local fund." And they (the farmers) toiling night and day with their children in the fields produce grain, cotton, opium, linseed and other crops with great labour to earn money to pay the agricultural tax and the local fund instalment; and when they go onto the national highway to bring all of these to the market, they find octroi stations every six miles collecting lakhs of rupees. When to overcome their adversities they go to the nearby forests to cut grass or wood or to feed their cattle and sheep, they find that all of these forests have been swallowed up by the government. There is even an octroi tax on the salt they use to make their simple bhakris palatable. (Phule 1881)

This is in line with Utsa Patnaik's claim that Indians were in effect not getting paid for the raw materials that Britain was extracting from India, because Britain was using the money Indians paid them in taxes to pay for the raw materials.

6.3.3.2   Colonialism and Hunger

During the British rule of India, famines were more common than they had been during Mughal rule. Famines are often blamed on natural calamities – in particular on droughts and floods. But droughts and floods do not necessarily cause famines. The famines during British rule were the result of India's exports to Britain, which drained the Indian economy, while helping to finance the 'Industrial Revolution' and raise the standard of living in Britain. Although Indians demanded that exports of grain should be stopped in times of famine, British rulers did not agree, using arguments that acclaimed the glory of 'free trade' (as if peasants had freely traded their grain to the British, purposely starving themselves?). They continued to collect taxes, which peasants had to pay with grain, since that was all they had. Thus, the grain went into the 'free market', meaning, peasants would need to buy back the grain they had produced – with what money?

The British government was forced to do this in order to control prices and food supply for their own residents in Britain – otherwise they would have to confront striking workers at home. Indians had to starve in order to stabilise the British economy.

The Bengal famine of 1943, in which millions of people perished, was one of the most horrendous results of British colonialism (see Fig. 2.10). India's agricultural exports (especially rice and cotton) were crucial to Britain's war efforts during the 2nd World War. Madhusree Mukerjee (2010) has described how Churchill and his government had instituted several policies even though they knew they would cause widespread starvation in Bengal. In 1942, in order to prevent Japan from being able to take rice from Bengal, should they invade, Britain confiscated, destroyed, or forced peasants to sell off what they called 'surplus' rice in east Bengal. The British also ordered boats and other means of local transport to be destroyed in order to deny their possible usage by the Japanese. This also ended up blocking the transport of grain into Bengal. This was part of what the British called their "Denial Policy", which focussed on Bengal because that was where that they thought the Japanese – or the Indian National Army, supported by Japan – would be most likely to invade. Even after a cyclone struck, in October 1942, and then the monsoon failed in Madras, British military contractors and speculators bought (or confiscated) rice and other food, to be used for the army and for feeding people in cities. This caused the price of rice to further escalate, and caused shortages for the peasants who had harvested it and lost their homes in the cyclone. Rich landlords and traders were also hoarding rice. While this was happening, suspects were being rounded up, raped, beaten, and sent to jail in order to suppress calls for independence from Britain. By March of 1943, the famine had reached horrifying levels. Export of grain to Britain continued. This is when Churchill (who was vehemently opposed to agreeing to give a dominion status to India after the war) responded, "They must learn to look after themselves as we have done. The grave situation of the U.K. import programme imperils the whole war effort and we cannot afford to send ships [to bring grain] merely as a gesture of good will." Wheat was available in Australia, but ships were not available to transport grain to India, since they were being used for the war. Britain had drastically reduced the number of ships going to India. Maintaining British food stocks was the priority (and Britain was actually exaggerating its own needs).

However, even nowadays the myth persists that hunger is caused by shortages and deficiencies in the production of paddy and other grains, or by 'natural calamities': vagaries of the weather, or diseases. And starvation may be renamed as 'malnutrition', or simply hidden from 'mainstream' statistics, which focus instead on the 'GDP'. Hunger may be naturalised, or the victims may be blamed. It is easy to assume that hunger is due to a population explosion, forgetting to think about how capital will keep increasing without increasing numbers of workers and consumers.

We even see the persistence of the myth that Britain was helping India to develop, rather than undevelop. DD Kosambi puts it very plainly:

Most of us fail to ask why our countries are underdeveloped, when we go abroad for financial aid and technical experts. The reason for underdevelopment is precisely that our raw materials and our great markets were exploited by the foreigner to his own advantage. Our products were taken away for the price of the cheap labour needed to take them out of the earth, and we paid the highest prices for the finished goods. In a word, the developed countries with very few exceptions are developed precisely because they made profit both ways from us; we were never paid the actual value of the things taken away. It is our resources that have helped in the development of the great industrialised nations of the world; yet we have to go to the same nations as suppliants, not as people demanding return of what is rightfully our own. (Kosambi 1994)

With acerbic insight, Phule discounted the role of the British in modernising India:

Our current ultra-principled truly monotheistic English Sarkar Bahadur extracts the farmer's wealth through various means on behalf of the municipality and from this wealth, apart from finishing the above-mentioned works [e.g. highways and irrigation projects begun by the Mughal rulers], gives them only the kind of education that deprives them of merit, and the strength of the farmers to earn their livelihood honestly decreases day by day. It has become openly known that in these times of peace and plenty, four crores of farmers don't get sufficient food to fill their stomachs even twice a day, and they don't see one day pass without experiencing the affliction of hunger. (Phule 1881)

If 'traditional' refers to the sort of agriculture that was occurring during colonial times, it is hard to advocate 'traditional', unchanging agriculture after reading Phule. How can we romanticise this kind of life?

6.3.3.3 Why Did Agricultural Science and Technology Not Develop Further in India?

Many historians have noted the seemingly unchanging methods of agriculture in different periods of Indian history. In Sect. 2.1.1, we noted that this is reflected in artworks that show cultivation.

Irfan Habib (2008) remarks,

An observer of the rural landscape a hundred years ago might well have been struck by the contrast between the intruding railway lines and steam engines and trains, symbols of modern technological achievement, and the seeming agelessness of the fields, with the half naked peasants working their primitive ploughs, that he saw from the train windows. (p. 51)

We could say that this shows that agriculture has remained in a primitive state over a long time. But according to another way of looking at it, this may also show that relatively advanced forms of agriculture were developed very early in India, and due to their success, they have been maintained over the centuries.

For example, systems of irrigation have been mentioned in very early stories, such as the Jataka tales, which originated as early as the fourth century BCE (Nirmal Sengupta 2018). Archaeological studies have confirmed the presence of dams and irrigation systems (probably for growing rice) in Buddhist sites near

Sanchi dating from the third to first centuries BCE. We have already mentioned that there is archaeological evidence of irrigated cultivation from the Harappan civilisation in the second millennia BCE. The irrigation systems shown in the painting made in 1830 (Fig. 2.7) are fairly advanced compared to European methods of the same period.

Thus, some argue that the usage of seemingly primitive techniques and tools may have persisted because they were effective and efficient. Others oppose this view as being a chauvinistic romantisization. Surely, more efficient and less back-breaking methods should have been developed, and perhaps would have been developed if India had not been colonised. Peasants could have continued to develop further innovations if they had the time and resources to do so.

However, Irfan Habib goes on to discuss how agriculture did change during British rule, in part due to the introduction of railways and steamships. There was an extension of the commercialization of agriculture and an increase in the production of cash crops – including non-food crops such as cotton and jute, as well as oil seeds and paddy, for export. The production of paddy increased compared to other grains, such as jowar (sorghum) and bajra (millet), which were not exported (as paddy was). The increase in cash crops was facilitated by requirements that taxes as well as rent had to be paid in cash, rather than in kind. Rather than producing all of their subsistence needs, peasants began to purchase part of them with the cash obtained by selling their crops. Increasingly, India was sustaining Britain not just by providing raw materials, but even more so by providing a market for British commodities (Habib 2008, 35). This was in addition to land tax, the major source of revenue for the British government. Peasant proprietors in the Bombay Presidency would typically be paying 20–40% of their total yield as tax, and were living in extremely wretched conditions. Peasants had very limited resources to develop their own agricultural science. Thus, colonialism was supporting development in Britain while preventing agricultural development in India.

On top of all the negative effects of colonialism on agriculture, the peasantry also suffered due to the effects of global economic problems, such as the Great Depression of the 1930's. While the British government protected its own farmers from fluctuations in global prices, trade was not regulated to protect production in India. In fact, trade was regulated to impair Indian producers. And we have already mentioned how, during WW II, Britain took grain from India to feed its troops, and restricted the import of grains, resulting in the Bengal famine of 1943 in which millions of people died of starvation.

During British rule, there was also an increase in land used for agriculture, and in deforestation. The decrease in firewood as well as grazing land led to an increase in the use of cattle dung for fuel and a decrease in its use as manure.

As agriculture becomes industrialised, crop development becomes more and more restricted to professional scientists and corporations. Agricultural development increases, but those who work in the fields become less active in its development. This in turn, restricts the kind of development that occurs. As we will discuss, the reasons for development change. Under capitalism, the evolution of paddy is not necessarily just according to people's needs and wants for a tasty, nutritious, plentiful crop – profit for the capitalist must necessarily be the bottom line. Therefore, even if capitalism provides great technological and social improvements over feudal agriculture, the gap between rich and poor keeps increasing.

Another major reason why science and technology related to agriculture did not develop further in India is that the caste system created a huge barrier to its development:

> Over a period, with every labour-oriented activity treated as undignified, the educated sections of society forgot the fact that labour is the source of life.
>
> Because of the denial of education and writing to the labouring castes, science did not develop to a more advanced stage in India. (Kancha Ilaiah and Durgabai Vyam 2007)

The caste system, which is an integral part of Hinduism, has put the people who do labour or work with hands and the people who do intellectual work (researchers like us) into different hierarchical categories. In talking about this division, Kancha Ilaiah [Shepherd] (Ilaiah 2009) has argued that in India, Dalitbahujans, who always worked with their hands, have been regarded as being more lowly than Brahmins, who had more power and did literary work. Dalitbahujans were never involved in the literary kind of work and were thought to be incapable of and ineligible for that sort of work. Kancha argues that historically Dalitbahujans were engaged in various sorts of work like agriculture, metallurgy, ceramics, carpentry, weaving, etc., and developed various techniques and technologies. He says that many of the technological advances in science in India were actually based upon the works of Dalitbahujans and Tribal peoples. However, their work was never recognised by the elite and privileged who controlled science. He sees this non-recognition of the work of Dalitbahujans and Tribal peoples as one of the main reasons for the lack of development of science in India as compared to some other countries.

Agriculture especially requires work with the hands that is integrated with work with the mind. But when, on the basis of caste, people are deprived of reading, writing, and other intellectual pursuits, agricultural development suffers. Agricultural development does not occur through individual effort. Throughout history agricultural development has always been a social effort that requires communication between people who learn and teach each other and work together to develop new methods and technologies. Because the caste system creates barriers between people, it thwarts communication and therefore also thwarts technological development. Because of the imposition of the caste system, people's science is suppressed.

Debiprasad Chattopadhyaya (1989) has written about how even before the time of Buddha, science began developing in ancient India. However, in his view, its development was inhibited due to the political need to maintain a division between "a leisured minority and a toiling majority, that is, in Indian terminology, into the dvijas and sudras" (Chattopadhyaya 2007). Laws were made that "decreed against everything that makes natural science possible" (ibid). Furthermore, he wrote that the general philosophical climate also contributed to the inhibition of science:

> Some of the ablest luminaries in ancient Indian thought used their intellectual resources
> only to inhibit the growth of natural science. From the viewpoint of science at any rate, this
> constitutes the most negative feature of ancient Indian culture. Philosophers with superb
> skill were used as helpless pawns in the grim political game of the vested interests sensing
> danger in science. (Chattopadhyaya 2007)

This was – and is – a struggle between materialist and idealist philosophies. Even nowadays, religious authority figures are venerated to such an extent that it is difficult to express secular views.

According to Chattopadhyaya (2007), the requirements for the development of science are as follows.

1. People need to be aware of the need for a materialist framework, and interested in looking for the material basis for events and processes. If society is dominated by an idealist way of looking at the world, people will instead assume that things are the way they are due to non-material causes, which cannot be empirically verified.
2. People need to be able to free themselves from the grips of mythological-supernatural mystification. Rather, people need to look for "the simple conception of nature just as it is without alien addition".
3. The political situation needs to be such that it does not inhibit the development or spread of a scientific method. Throughout much of history, the scientific method of enquiry has been actively discouraged in order to maintain the division of society into a toiling majority controlled by a leisured minority. Science has been dangerous to the maintenance of the status-quo, and political and legal systems have prevented its spread, by encouraging "myth-making or beneficial falsehood", and by censuring direct observation and the collection of empirical evidence.
4. Philosophers should do scientific explorations of the physical world in order to solve practical problems, rather than concern themselves with "the mystery of the indwelling soul" or pure speculation – or with controlling the toiling people.

He maintains that it is because these requirements have not been met throughout much of Indian history, that the development of science has been inhibited. His third point in particular is important in keeping farmers from doing more science. In Chap. 7 we will discuss more about how this happens.

We have mentioned how Jyotirao Phule described the condition of farmers. He also analysed why the situation was as it was. He repeatedly mentions the ignorance of Shudra farmers and their lack of access to education, as well as religious tyranny and exploitation by Brahmins as what keeps them in such a 'pitiable state'. He says:

> On the basis of all these accounts, one might wonder how farmers could be so ignorant as
> to be looted up until today by the Bhat-Brahmans. My answer to this is that when the origi-
> nal Arya Bhat-Brahman regime was started in this country, they forbade knowledge to the
> Shudras and so have been able to loot them at will for thousands of years. Evidence for this
> will be found in such self-interested literature of theirs as the Manusmriti. After some years,
> four disinterested holy wise men who disliked the prolonged misfortune founded the
> Buddhist religion and campaigned against the artificial religion of the Arya Brahmans to
> free the ignorant Shudra farmers from the noose of the Aryabhats. Then the chief head of

the Aryas, the great cunning Shankaracharya, engaged in a wordy battle with the gentlemen of Buddhist religion and made great efforts to uproot them from Hindustan. However, rather than the goodness of Buddhism being threatened even a mite, that religion kept growing day by day. Then finally Shankaracharya absorbed the Turks among the Marathas and with their help destroyed the Buddhist religion by the sword. Afterwards the Arya Bhatjis, by banning eating beef and drinking alcohol, were able to impose an awe on the minds of the ignorant farmers through the help of Vedamantras and all kinds of magical tricks. (Phule 1881)

Clearly, education needs to play an important role as a route to a better life and emancipation from the trickery and exploitation that upper castes and upper classes impose on others. Education plays an important role in the annihilation of the caste system, which is necessary so that people will work together to communicate, teach, learn, and develop agricultural technologies for their own benefit.

6.4   The Agricultural Crisis and Capitalist Development in India

We saw how the interconnected development of science, agriculture, and capitalism in India was hindered during the colonial period. But why does India not now follow the same path of capitalist development that occurred in Europe?

The situation in India these days is much different from the situation in England in the eighteenth century. There are insufficient jobs to absorb people who are no longer able to even get adequate food doing agriculture labour. But unlike during the industrialisation of England, peasants have nowhere else to go and nothing else to do besides small family farming. With automation and regular economic 'slow-downs', there are not enough jobs in industry. And India has no colonies to send them to, as England did in the nineteenth century. If peasants attempt to leave and fail to find jobs or if some calamity strikes (e.g. if a capitalist/fascist government forces a corona curfew without providing food or shelter), in desperation they can only return to the family and the farm. If some members of the family go off to cities to find jobs, it is very important that they preserve the option of being able to go back to their village as a last resort in case things in the cities do not work out. Peasants know very well the importance of holding on to their rights to the land – that the land is their mode of production and basis for survival – and throughout the country they have banded together time after time to collectively resist dispossession. It is peasants themselves who stand up and ask the rhetorical question, 'Where shall we go if we are forced to leave our land?' Across India peasants keep forming strong movements to resist being dispossessed from the land. They resist because they see what happens when people's land is taken away: they do not get the jobs or compensation they are promised. A common call to resistance is, "Take our lives, but not our land!". Many peasants have lost their lives in these struggles. But some of the struggles have been successful and people have managed to hold onto their land and refuse to let it be taken over by companies and governments for mining, missile test ranges, real estate investment opportunities, highways, Special

Economic Zones, nuclear power plants, and other kinds of development projects (Ranjana Padhi and Nigamananda Sadangi 2020).

Thus, in today's world, developing countries have not succeeded in accomplishing the same sort of capitalist development that happened in other countries in the past. But what is happening is that unemployment is increasing, hunger is increasing, and the gap between the rich and the poor continues to increase. Unemployment and economic inequality have been increasing despite increases in GDP (TK Rajalakshmi 2018). Actually it is due to high unemployment that wages are low, the rate of extraction of surplus value is high, corporate profits increase, and therefore the GDP is high. Nowadays, there is no 'open frontier' to expand into. Therefore, as the GPD rises with rising exports, domestic food grain prices rise and the nutritional status of the population in India declines (Utsa Patnaik 2016). With increases in monoculture and the production of cash crops – especially cash crops for export – farmers can hardly afford to eat what they grow, or grow what they need to eat. Scientific development, whether by farming families or by corporates, is thus impeded.

6.4.1 The Commodification of Agricultural Science

Why does capitalist development in agricultural occur? There is a common myth that the purpose of agricultural research is to make improvements in people's food, clothing, and shelter by making production more efficient. A side effect of this is that farmers supposedly become better off, and are freed from back-breaking labour. Thus, 'scientific' agriculture supposedly benefits both producers and consumers.

Richard Lewontin and Jean-Pierre Berlan (1986) expose this myth. They write that under capitalism, "…technological change drives and is driven by commodity production". While technological development is also dependent on basic research in science, agricultural research has been commodified under capitalism:

> Modern science is a product of capitalism. The economic foundation of modern science is the need for capitalists not only to expand horizontally into new regions, but to transform production, create new products, make production methods more profitable, and to do all this ahead of others who are doing the same. (Levins and Lewontin 1985, 197)

They have also written that it is only as agriculture becomes industrialised that those who work in the fields become less active in crop development, with crop development becoming more and more restricted to professional scientists and corporations.

In the past, farmers (or groups of farmers) who were highly skilled may have been revered because of their skills. Under capitalism, agricultural labourers are classified according to how much they earn as wage-labourers, with the value of labour depending on their wages rather than the type of labour or skill. Human labour becomes generalised: whether one is carrying sacks or driving a tractor, it makes no difference if the wages are the same. Similarly,

Investing in research, which is one of several ways of investing capital, competes with other ways, such as increasing production of existing products, purchasing more advertising, hiring lawyers or lobbyists, buying up businesses in other fields, busting unions, bribing cabinet ministers of potential customer countries, and so on. All possibilities are measured against each other on the single scale of profit maximization. (Levins and Lewontin 1985, 200)

Since under capitalism profit is necessarily the bottom line, the agricultural research and development that is done may not be in the best interests of people or the environment. Since the financial payback for research may take a number of years, expenditure for research may be inadequate. Agricultural research becomes dictated by the requirements of capital, with the underlying profit motive superseding the needs of society and of farmers, and even compromising 'the advance of science'.

Things need not be like this. Alternative systems are possible. Even in the past, in socialist countries some technologies have been purposely developed in order to help people live better lives rather than just to produce commodities (Soviet children's books distributed in India are one example). People need to purposely develop technology for its use value rather than exchange value.

6.4.2   The Green Revolution and the Introduction of HYV and Hybrid Rice

The 'Green Revolution' in India began in the 1960's and was designed under guidance from the USA, through the Rockefeller Foundation. The motivation sprang from efforts to find markets for their goods, in particular to boost the profits of the USA fertiliser and chemical industries and seed companies (Clara Bozzini 1999). The IRRI developed High Yielding Varieties of paddy (HYV), which have become very widespread. The Green Revolution resulted in significantly increased production and increased yields/hectare of paddy. This was probably mainly the result of increased irrigation and artificial fertilisers. It has been estimated that, worldwide, the paddy productivity gains due just to HYV seeds was very minimal – less than 0.8%. (PL Pingali 2012). Some people claim that similar or greater increases in productivity could have been accomplished with the old landraces, by increasing irrigation and soil quality. Worldwide, paddy productivity per hectare doubled between 1960 and 2000, but the increase levelled off after 1990.

In India the Green Revolution has been confined to only certain parts of the country, and has provided priority attention to rich landowners, rather than small family farmers, who have always been in the majority (DN Dhanagare 1987). As a result, the income gap between rich landowners and landless agricultural labourers has increased.

The Green Revolution was accompanied by many kinds of environmental degradation, which are also having long-term negative impacts on paddy cultivation. Modern agricultural practices result in pollution of land, water, and air – both through the practices themselves and the production of the machinery, artificial fertilisers and pesticides, seeds, fuel, and other means of production and transport. This has become increasingly evident over the past few decades:

> Numerous agricultural scientists agree that modern agriculture confronts an environmental crisis. A growing number of people have become concerned about the long-term sustainability of existing food production systems. ... Evidence also shows that the very nature of the agricultural structure and prevailing policies in a capitalist setting have led to this environmental crisis by favouring large farm size, specialized production, crop monocultures, and mechanization. (Miguel Altieri 2000, 78)

Since the Green Revolution, research has been done that provides evidence for environmental effects. However, especially when it is sponsored by vested interests (e.g. chemical corporations that produce fertilisers and pesticides), the research needs to be critically evaluated. Some old and new alternatives to specific problematic materials and methods used for agriculture are available. However, farmers may not adopt these materials and methods if they are not affordable. Corporations are not in a position to promote sustainable practices or materials that are less detrimental to the environment unless it also entails an increased profit for the corporations. Governmental support and centralised planning are necessary. But after the heydays of the Green Revolution (especially after 1990), governmental support for agriculture keeps decreasing (see Sect. 6.4.4). As Miguel Altieri says, "…ecological change in agriculture cannot be promoted without comparable changes in the social, political, cultural, and economic arena." (Altieri 2000, 90).

6.4.3   Loss of Diversity of Paddy

One of the main effects of the Green Revolution was that tens of thousands of varieties of rice (landraces) have disappeared or are not being grown as much, and HYV seeds are being sown instead (see the definitions of these in Sects. 6.1.1 and 6.2). For technical reasons, seeds of HYV and the newer hybrid varieties usually cannot be produced by the farmers themselves for use in subsequent seasons. These seeds necessarily carry genes for undesired traits that become expressed only in subsequent generations. After 1–3 seasons, new seeds have to be purchased from the seed companies. This turns out to be very convenient for the seed companies' profits, but farmers have to pay the price. Also, it puts seed selection out of the hands of the farmers and into the hands of professional scientists. HYV paddy cultivation is more uniform and thus more susceptible to disease, pests, and changing soil and other environmental conditions. Previously, with more diversity within and between varieties of rice, it was more likely that some plants may, by chance, contain variations that would help them survive such problems. This variety forms the basis for selection and evolution of resistant varieties, either through artificial selection or through the new methods of plant breeding and genetic modification. It took many generations to develop so many varieties of rice, and once lost they are irreplaceable. Thus, the loss of this diversity is catastrophic.

How does increasing communication, transport, and trade effect the diversity of paddy and other crops? One might think that it would increase diversity through the spread of more species and varieties of crops to different areas. To some extent this may be true, especially when a completely new type of crop is introduced. But generally, throughout the world, increasing communication, trade, and transport results in less diversity. For example, "In North America alone, an estimated 3-5000 species of wild plants were once used as food, but today 90% of the world's food needs are met by just over 100." (Meyer et al. 2012).

Particularly with regard to paddy cultivation in India, there has been an immense loss of landraces as commercial cultivars have been introduced. A few commercial cultivars have replaced the thousands of landraces that were previously grown in the same fields. The commercial cultivars are promoted by the companies that sell seeds, fertilisers, and pesticides, in order to maximise their profits. Governments have also promoted them, sometimes under external pressure, through subsidies, advertisement, and work by agricultural officers. Sagari Ramdas et al. have studied cultivation in six districts of Maharashtra and Andhra Pradesh, and they report:

> Earlier, each area boasted of 10 to 30 varieties of food crops, not to mention strains and sub-strains; now, these are down to a mere three or four. At present there is an acute scarcity of seeds for food crops as many farmers have stopped storing foodgrains for cultivation in the next year. In other cases, grains stored have rotted after years of disuse. Some species of food crops are on the verge of extinction. There is a serious threat not only to food security and self-sufficient agricultural systems but also to agricultural diversity. The cash crop economy has also gradually eroded the local systems of sharing, collective farming, celebrations of harvest and the cultural practices of the community. (Ramdas et al. 2004, 75)

Wild varieties of rice are also becoming extinct due to the destruction of their habitats due to increasing development for agricultural and non-agricultural purposes, as well as due to change in climate or other environmental changes. As we discussed above (Sect. 6.1.1), all species keep evolving and becoming extinct, even without people, but the current rates of extinction are much higher than ever before, because of human activities.

The process of selection, and the types of rice that it produces, has depended upon the social, political, and economic system under which it has occurred. Under capitalism, profit is necessarily the bottom line because rice is grown for its exchange value, and productivity – the amount of rice produced per hectare (and in the shortest time) – is the most important factor. If rice is grown for its use value, various types of rice may evolve that have a wider variety of characteristics, but some of which may not be very productive. Nowadays in Mumbai, a small shop will typically sell at least five to ten different varieties of rice, but many of these are HYV, and landraces are becoming less common.

6.4.3.1   Attempts to Conserve the Diversity of Rice

In response to the extinction of landraces, there are some efforts to conserve the genetic diversity of rice. Amar Kanwar relates how Natabar Sarangi and Yuvaraj Swain have taken action:

> One day, about nine years ago, Natabar Sarangi, at the age of seventy, decided to search for and plant different varieties of the traditional indigenous paddy seed in his village Narisho, around thirty-five kilometres from Bhubaneshwar in Odisha.
>
> He planted rice varieties that were traditionally planted for different uses and in different soil conditions and land levels. The uses were for everyday staple main means, specifically for breakfast dishes with specific recipes, for desert and snack preparations, for special feasts, to accompany meat dishes, for rice cakes etc. Each was prepared in varied specific ways – steaming, fermenting, drying, boiling, fried, cooked with other ingredients, And for storage or travel, for varied uses and requirement, as one could possibly imagine. There was even a variety of rice that was so tasty, you could simply eat it on its own, without having to flavour or accompany it with any other dish.
>
> He asked a young farmer, Yuvaraj Swain to join him, and together they experimented and planted several varieties of rice in the lowlands, medium lands and uplands, in red soil, black soil, sandy soil, and in hilly tracts. They planted drought tolerant varieties, flood tolerant varieties, cold tolerant varieties, and saline tolerant varieties. They stored the seeds in several little bags and boxes, and offered them to all the farmers in the fields around the village.
>
> ... In the recent past, in Odisha, there were over 30,000 varieties of the traditional paddy seed with an assured stable yield of fifteen to twenty quintals of rice per acre. Now Odisha has only 20 high-yielding varieties of paddy that dominate all rice cultivation. These cannot withstand adverse weather conditions, need twice the amount of water, and require heavy and continuously increasing doses of chemical fertilizer and pesticide inputs. (Amar Kanwar 2014)

Amar Kanwar has organised an exhibit in which 250 landraces of paddy from Odisha are on display. Even to the naked eye, the diversity in the characteristics of the rice on display is astounding. However, just keeping seeds of different landraces in pots on farms or in museums may not preserve the seeds. After a few years, depending on the conditions in which they are kept, seeds die. When they are planted, fewer and fewer percentages of the seeds will germinate as the seeds get older.

Therefore, there are various 'genebanks', such as at the International Rice Research Institute (IRRI) in the Philippines, in which tens of thousands of landraces of *Oryza sativa* are dried and stored at very low temperatures (down to -20 ° C.). For long-term conservation, the IRRI also keeps cultivating the landraces under uniform conditions, for rejuvenation and multiplication. They collect seeds from hundreds of countries throughout the world, and distribute seeds to numerous research institutes. Scientists at the IRRI also test the seeds and study their molecular biology. Scientists try to maintain static landraces – and cultivars – through carefully controlled storage and uniform cultivation, but this is difficult. The nature of nature is that it is always changing.

In addition, there are in situ seed banks operated locally by farmers, such as the above example by Natabar Sarangi and Yuvaraj Swain. In these, expensive technologies are not available, and rice seeds are maintained by cultivation. They are collected and distributed locally. All this is done by people who are usually not called 'scientists' – although the leaders of such genebanks may be educated scientists from mainstream science (Debal Deb 2019).

6.4.4   Neoliberalism Deepens the Crisis

In the period between 1947 and 1990, the import and export of agricultural commodities was restricted in order to protect Indian agriculture from fluctuations in global prices. The government also extended some support to protect peasants from variations in production due to unexpected weather, pests, etc. During this period, prices of paddy were controlled through the Public Distribution System, in which the government bought paddy, stored it, and used it to supply to impoverished families and to supplement the supply of rice in case of crop failure. The government set a Minimum Support Price (MSP) that they would pay farmers for this paddy, which was a guarantee for farmers in case of excess production, in case the market prices fell below the MSP. In this period, the Indian government had an expansionary fiscal policy that was to some extent pro-peasant. Government spending on rural development helped reduce unemployment and increase rural spending power, even though this resulted in a fiscal deficit.

Since 1990, global financial institutions have forced the Indian Government to implement New Economic Policies ('Neoliberalism') in order to reduce the fiscal deficit, reduce government expenditure on rural development and social sectors (including education), and encourage the growth of private (and multinational) industry. This meant greatly reducing the protectionism in crop production. Therefore, when the global price of rice falls, rice may be imported and the price of rice produced in India may fall drastically. When global prices rise, rice may be exported, leading to a rise in the price of rice for consumers in India. This may or may not mean much increase in the prices that farmers get, because traders may reap most of the benefit. Global prices are very volatile, and when farmers increase production in response to a rise in the global price, by the time they are ready to sell their crops, the prices may have already fallen.

Neoliberalism has resulted in an increase in unemployment (particularly in the public sector), and a reduction in rural incomes. Neoliberal policies brought about many changes in agriculture. Agricultural colleges are no longer being adequately supported by the government. Agricultural extension services were cut, to be replaced by research and advice from the private sector – resulting in a conflict of interest, since they were concerned about selling their own commodities. The spread of HYV rice increased, but government support to allow farmers to purchase new technologies decreased. The production of cash crops increased. After large increases before 1990, the rate of growth in crop yields began decreasing. In fact, for the first time since 1947, the rate of growth of food grain production fell below the rate of increase in population. The per capita consumption of grains began to decline. Despite a huge build-up of stocks of food grains with the government, people could not afford to buy the grain they needed. Food grains began to be exported. This meant that the nutritional status of rural people declined, since most of their calories and proteins comes from food grains. The average nutritional intake in India is now the lowest in the world, and the nutritional intake of the rural population is now lower than it was in 1947. Now most of the basmati rice that is produced is exported. The export prices depend upon the huge subsidies given to corporate rice farms in the USA (behind the veil of 'free trade'). Although the retail prices of basmati rice varieties are very high in India (beyond the reach of most consumers), farmers are not getting adequate prices for basmati or for other types of rice. Farmers cannot afford to eat the food they produce for export – and they may be forced to switch to cash crops like sugar or soy beans.

One of the ironies of capitalism is its inefficiency. When agricultural technology, materials, and methods are produced by private corporations competing with each other, the overall process is inefficient compared to what it could be if they cooperated with each other. They may not share supplies they need for production. They may needlessly transport supplies over long distances (even globally) rather than using what another nearby competitor has to offer. They may have to keep shifting factories to different locations in order to take advantage of people who are willing to work for lower wages. Farmers are forced to buy materials that are produced far away if they are cheaper than locally produced materials, even though it is inefficient and harmful to the environment.

Over the past 20–30 years (and especially in the last 10 years) the cost of living keeps increasing and the relative price farmers receive for their crops keeps decreasing. Farmers are not being paid the actual value of the paddy they produce. As the gap between the rich and the poor keeps increasing, the class conflict between farmers and capitalists is becoming more and more acute. Government support for capitalists has increased while support for agriculture has decreased. Government support in the case of droughts, floods, and unexpected weather has decreased. Expenses of all types (farming, daily needs, healthcare, education, etc) have increased. Whatever people have acquired from economic development (consumer items, or improvements in electricity, healthcare, water, or whatever) may make them think that their quality of life has improved. But this development also brings resentment because it has been so uneven, with the well-off getting so much more well-off, relative to the poor. In many ways, instead of reducing conflict, change has brought more conflict, of new kinds.

The amount of money farmers get for the crops they sell is no longer enough to purchase the inputs and essential items that they need. The prices of these items have increased relative to the prices they get for their crops, and the needs have also increased. They are forced to purchase all required items at retail prices and sell at wholesale prices. Selling at very low rates means that relatively more surplus value is being extracted from their labour. This has been happening because the government has reduced its support, and increased privatisation. If farmers formed cooperatives, they may be able to sell closer to retail prices, but in Raigad District this is not happening.

There has been an increase in the standard of living, with associated increases in expenses. It is no longer possible to subsist in the same way as 30 years ago. Neither do people want to go back to the way things were. For example, 30 years ago child birth may have occurred at home with the assistance of a dai. Now child birth is likely to occur in hospitals, and the relative cost is much higher. There may be no option of giving birth at home because a dai is no longer available. The relative costs of health care have increased in part because of privatisation and the withdrawal of governmental support. Health care in rural areas has continued to be neglected, but even if better health care is available, it is available only for a high price. The costs of education have increased due to privatisation, and because there is a perceived need to send children to private schools even if cheaper government schools are available. Nowadays electricity is more of a necessity than it was 30 years ago, but the cost of electricity has also increased.

Actually, there are not enough jobs in agriculture or outside of agriculture. Unemployment is very high (and increasing) for both the uneducated and the educated. Wages for the majority are very inadequate. People doing agricultural labour are not getting the true value of their labour – they are being robbed of the surplus value which is leading to the accumulation of capital and making the wealthy wealthier (see Sect. 7.1.1.3). But the robbers genuinely feel that they are doing all they can to be nice to the workers – and they are! If they were to let the labourers keep the true value of what they produce, the capitalists and traders would soon be out of business.

With the financial crisis that began in 2008, there were enormous increases in food prices, which were not reflected in the prices received by farmers. This resulted in a decrease in the nutritional status of the Indian population, which is still persisting. Within the last 10 years there has been even more decrease in government support for agriculture. Government procurement of rice has decreased, and would have been entirely shut down, except for massive protests by farmers. With the decrease in procurement, the official MSP has become largely irrelevant because farmers are not able to sell to the government, whatever the notional MSP may be. The prices are set by traders at levels that are much below the MSP, and much below the cost of producing the rice, especially if the unpaid labour time of small farming family members is considered. As is typical amongst paddy farmers in Maharashtra, where the government does not procure rice, Smita and Bhimraj had never heard of MSP.

Dhanagare (2016) has summarised the main effects of Neoliberalism on agriculture: (a) increase in cash crops; (b) Lack of MSP; (c) Rising need for loans by farmers; (d) a non-functional Public Distribution System; (e) the promotion of contract farming (bonded labour); (f) marketing which serves corporate interests rather than small family farmers; (g) increasing import & export of crops; (h) over- and underproduction of various crops, and hoarding; (i) increasing farmers' eviction to the non-farm economy. As Dhanagare discusses, this has resulted in a very serious agricultural crisis, which is characterised by difficulties in recovering input costs, increasing costs of seeds, fertilisers, irrigation, and electricity, and increasing debt, hunger, malnutrition, and farmer suicides (Nagaraj et al. 2014). There is a loss of rural livelihoods, increase in numbers of landless labourers, and an increase in migration to cities, resulting in decreasing numbers of farmers who own land. Overall, the contribution to the GDP by agriculture has been decreasing. In addition, the number of varieties of rice keep decreasing as farmers grow HYV rice instead.

In this scenario, how can farmers afford to do science to develop paddy cultivation? And, at the same time, how can we rely on capitalist science to develop paddy cultivation?

Even this cursory glance at some of the questions regarding the history of paddy cultivation cannot fail to leave us with an impression of the complex, dialectical interactions between paddy and groups of people, both constantly changing over time. Thus, one wonders how unchanging 'traditions' could ever have existed.

References

Altieri, Miguel A. 2000. Ecological impacts of industrial agriculture and the possibilities for truly sustainable farming. In *Hungry for Profit: The Agribusiness Threat to Farmers, Food, and the Environment*, ed. Fred Magdof, John Bellamy Foster, and Frederick H. Buttel, 77–92. New York: Originally Published by Monthly Review Press. 2008 Indian edition. Kharagpur, India: Cornerstone Publications.

Athreya, Venkatesh, Kumar Deepak, R. Ramakumar, and Biplab Sarkar. 2017. Small farmers and small farming: A definition. In *How Do Small Farmers Fare? Evidence from Village Studies in India*, ed. Madhura Swaminathan and Sandipan Baksi, 1–24. New Delhi: Tulika Books.

Bates, J., C.A. Petrie, and R.N. Singh. 2017. Approaching rice domestication in South Asia: New evidence from Indus settlements in northern India. *Journal of Archaeological Science* 78: 193–201.

Bernal, J.D. 1954. *Science in History, Volume 1: The Emergence of Science*. New York: Cameron Associates.

———. 1969. *Science in History, Volume 2: The Scientific and Industrial Revolutions*. 3rd ed. Harmondsworth: Penguin Books.

Buzzing, Clara. 1999. Johnson's 'Short-Tether' Policy towards India'. In *Seminar: Amerikanische Aussen- Und Sicherheitspolitik Unter Präsident Johnson Im Dilemma Zwischen Great Society Und Vietnam, 1963–1969*.

Campbell, Neil A., Jane B. Reece, Lisa A. Urry, Michael L. Cain, Steven A. Wasserman, Peter V. Minorsky, and Robert B. Jackson. 2008. *Biology*. 8th ed. MNC: Pearson.

Central Rice Research Institute. 2012. *Annual Report 2011–12*, 14. Cuttack: Indian Council of Agricultural Research.

Chattopadhyaya, Debiprasad. 1989. *In Defence of Materialism in Ancient India: A Study in Cārvāka/Lokāyata*. New Delhi: People's Publ. House.

———. 2007. *Science, Philosophy & Society*. New Delhi: Critical Quest.

Choi, Jae Young, et al. 2017. The rice paradox: Multiple origins but single domestication in Asian rice. *Molecular Biology and Evolution* 34 (4): 969–979.

Cuddington, Kim. 2001. The 'Balance of Nature' metaphor and equilibrium in population ecology. *Biology and Philosophy* 16 (4): 463–479.

Deb, Debal. 2019. The struggle to save heirloom rice in India. *Scientific American* 321 (4): 1019–1054.

Dhanagare, D.N. 1987. Green revolution and social inequalities in rural India. *Economic & Political Weekly* 22 (19, 20, 21): AN-137–AN 144.

———. 2016. Declining credibility of the neoliberal state and agrarian crisis in India: Some observations. In *Critical Perspectives on Agrarian Transition: India in the Global Debate*, ed. B.B. Mohanty, 138–163. London/New York: Routledge.

Dharampal. 2000. *Dharampal: Collected Writings, Vol V. Essays on Tradition, Recovery and Freedom*. Goa: Other India Press.

Egerton, Frank N. 1973. Changing concepts of the balance of nature. *The Quarterly Review of Biology* 48 (2): 322–350.

Fuller, Dorian Q. 2011. Finding plant domestication in the Indian subcontinent. *Current Anthropology* 52 (S4): S347–S362.

Fuller, Dorian Q., Alison R. Weisskopf, and Cristina Cobo Castillo. 2016. Pathways of rice diversification across Asia. *Archaeology International* 19: 84–96.

Glover, Dominic, and Glenn Davis Stone. 2018. Heirloom rice in Ifugao: An 'anti-commodity' in the process of commodification. *The Journal of Peasant Studies* 45 (4): 776–804.

Habib, Irfan. 2008. *Indian Economy, 1858–1914*. New Delhi: Aligarh Historians Society : Tulika Books.

Harvey, David. 2013. *A Companion to Marx's Capital*, Volume Two. London: Verso.

Haydock, Karen, and Himanshu Srivastava. 2019. Environmental philosophies underlying the teaching of environmental education: A case study in India. *Environmental Education Research* 25 (7): 1038–1065.

Ilaiah, Kancha. 2009. *Post-Hindu India: A Discourse on Dalit-Bahujan, Socio-Spiritual and Scientific Revolution*. New Delhi/Thousand Oaks: SAGE Publications.

Ilaiah, K., and Durgabai Vyam. 2007. *Turning the Pot, Tilling the Land: Dignity of Labour in Our Times*. Pondicherry: Navayana Pub.

Kanwar, Amar. 2014. In *The sovereign Forest*, ed. Daniela Zyman. Berlin: Sternberg Press.

Kosambi, D.D. 1994. *Science, Society & Peace*. New Delhi: People's Publishing House.

Levins, Richard, and Richard C. Lewontin. 1985. *The Dialectical Biologist*. Cambridge, MA: Harvard Univ. Press.

Lewontin, Richard, and Jean-Piere Berlan. 1986. Technology, research, and the penetration of capital: The case of U.S. agriculture. *Monthly Review* 38 (3).

Marx, Karl, and Friedrich Engels. 1848. *Manifesto of the communist party*. Trans. Samuel Moore in cooperation with Frederick Engels, 1888; Moscow: Progress Publishers.

Meyer, Rachel S., Ashley E. DuVal, and Helen R. Jensen. 2012. Patterns and processes in crop domestication: An historical review and quantitative analysis of 203 global food crops: Tansley review. *New Phytologist* 196 (1): 29–48.

Mohanty, Bibhuti Bhusan. 2016. Introduction: Agrarian transition: From classic to current debates. In *Critical Perspectives on Agrarian Transition: India in the Global Debate*, ed. B.B. Mohanty, 1–40. London/New York: Routledge.

Mukerjee, Madhusree. 2010. *Churchill's Secret War: The British Empire and the Ravaging of India during World War II*. New York: Basic Books.

Nagaraj, K., P. Sainath, R. Rukmani, and R. Gopinath. 2014. *Farmers' suicides in India: magnitudes, trends, and spatial patterns, 1997–2012*. Tenth Anniversary Conference of the Foundation for Agrarian Studies, Kochi, India.

Padhi, Ranjana, and Nigamananda Sadangi. 2020. *Resisting Dispossession: The Odisha Story*. Delhi: Aakar Books.

Patnaik, Utsa. 2016. Capitalist trajectories of global interdependence and welfare outcomes: The lessons of history for the present, Ch.4. In *Critical Perspectives on Agrarian Transition: India in the Global Debate*, ed. B.B. Mohanty. Basingstoke: Taylor & Francis Ltd.

Patnaik, Utsa, Sam Moyo, and Issa G. Shivji. 2011. *The Agrarian Question in the Neoliberal Era: Primitive Accumulation and the Peasantry*. Oxford: Pambazuka.

Phule, Jyotirao. 1881. *Shetkaryaca Asud*. 2017 Marathi edition translated from Marathi to English by Gail Omvedt and Bharat Patankar. Pune: Mehta.

Pingali, P.L. 2012. Green revolution: Impacts, limits, and the path ahead. *Proceedings of the National Academy of Sciences* 109 (31): 12302–12308.

Rajalakshmi, T.K. 2018. Cooking up numbers. *Frontline* 35 (20): 27–29.

Ramdas, Sagari, et al. 2004. Overcoming gender barriers: Local knowledge systems and animal health healing, chapter 2. In *Livelihood and Gender: Equity in Community Resource Management*, ed. Sumi Krishna, 67–91. New Delhi: Sage Publications.

Randhawa, M. S. 1980. *A History of Agriculture in India*. Volume I Beginning to 12th Century. New Delhi: Indian Council of Agricultural Research.

———— 1983. *A History of Agriculture in India*. Volume III, 1757–1947. New Delhi: Indian Council of Agricultural Research.

Randhawa, Gurinder Jit. 2006. *Document on Biology of Rice (Oryza Sativa L.) in India*. New Delhi: National Bureau of Plant Genetic Resources, Indian Council of Agricultural Research.

Sengupta, Nirmal. 2018. *Traditional Knowledge in Modern India*. New York/Berlin/Heidelberg: Springer.

Silva, Fabio et al. 2015. Modelling the geographical origin of rice cultivation in Asia using the rice archaeological database, ed. Ron Pinhasi. *PLoS One* 10(9): e0137024.

Spence, Martin. 2001. Environmental crisis in prehistory: Hunter-gatherers and mass extinctions. *Capitalism Nature Socialism* 12 (3): 105–118.

Chapter 7
A Deeper Understanding of the Process of Doing Paddy Cultivation

We have seen that the actual process of paddy cultivation is different and much more complex than it at first appeared to be. Our initial impression, as reported in Chap. 2 and summarised in Table 2.1, was that the family was engaged in traditional paddy cultivation that they had learned from the older generation, and that it was remarkably unchanged over centuries.

Upon working with the family over one season of paddy cultivation, we found many dialectical contradictions in how cultivation is learned/developed, as summarised in Table 4.1. We saw that the family keeps modifying the process of paddy cultivation over days and seasons, as new problems and conflicts crop up and new solutions are required.

In this chapter, we will show, with reference to the historical development outlined in Chapters 5 and 6, how the dialectical conflicts between 'traditional' and 'scientific' ways of learning, practicing, and developing paddy cultivation determine its nature. This is how the family cultivates cultivation.

7.1 The Family and the Capitalist Agrarian Transition

As we have seen, Rudravali has soil and climate that are very well suited to paddy cultivation – especially notable is the very dependable rain during the monsoon. Also important, is that it has dry seasons in which paddy can be safely harvested. It is located in an area that is not prone to floods, droughts, cyclones, or other natural disasters. For many decades it has had relatively good connectivity to towns and cities by road, and by rail since 1990. Its people have a long history of being involved primarily in paddy cultivation. Over many centuries, the local people have developed their own varieties of paddy, which they use for their own sustenance.

© The Author(s), under exclusive license to Springer Nature
Switzerland AG 2021
K. Haydock et al., *Learning and Sustaining Agricultural Practices*,
International Explorations in Outdoor and Environmental Education 7,
https://doi.org/10.1007/978-3-030-64065-1_7

The family that we are working with has many advantages compared to other families: they own 0.4 hectares of land that is suitable for paddy cultivation (and more land that provides fodder and material for rabbing), they have not taken any loans (we were surprised to hear this), they have two sons (rather than daughters), and they have three adults who work in the fields, and another who works primarily at home (Prabhavati).

But, as we have discussed, things are changing, and the family is no longer able to survive just on paddy cultivation, as they and their ancestors previously did.

As Smita said, 'अभी हम इतनी मेहनत करके क्या मिलता हैं खाने के लिए? इसीलिए! ऐश-आराम तो कुछ नहीं! सिर्फ दिन-रात ये मेहनत करते. (Now we work hard but what do we get for eating? That is why! There is not even any rest! He only works hard all day and night.) And later she added, 'थोडा पैसा-वैसा तो चाहिए, ना? अभी हम मेहनत मेहनत करके करते, हैं ना साहिब? (But a little money – money is just needed, nah! Now we just do work and do work, isn't it, sahib?).

From this and other conversations, we see that Smita and Bhimraj are struggling to overcome the present agricultural crisis: they say that they are having a very hard time surviving because they are not getting enough money and because the work is too hard. This is despite the fact that they are relatively better-off than most.

Why is this happening? As we will discuss, the explanation lies in how they are circumscribed by the capitalist system.

7.1.1 Paddy Cultivation for Use Value vs Exchange Value

In a pre-capitalist mode of production, the seeds are the life force of nature, bursting forth with 'josh'. as Bhimraj said. The farmers' labour is bound to nature – to the raw material, the environment, and the products of their labour, which they themselves consume. But with the development of capitalism, this changes. Whereas previously 'labour appears as an aspect of the land' (Karl Marx, 1844), with the advent of capitalism, land is an aspect of labour. Land, labour, and nature itself – all become commodified, objectified, and alienated from people.

As we have discussed, we see signs of both capitalist and pre-capitalist modes of production on the family farm. The production of HYV rice as a cash crop in one field, and landraces for the family's own use in another field, exemplify these two modes of production. In a pre-capitalist mode of production, crops were grown for the subsistence of the farmers or for landlords and rulers. Although small amounts of surplus were sometimes traded or sold, it is under the capitalist mode that crops are grown primarily in order to be sold.

What do we conclude about the production of the paddy that the family eats?

(1) The paddy is a physical product: the family sees it growing, feels its weight on their heads as they carry it from the field, and measures it as they store it. Paddy has been produced by their labour. They see and experience this labour, and understand how it is occurring.

(2) The value of the paddy depends on the duration of hard work that went into its production (the socially necessary labour time). As Bhimraj says (Sect. 3.3), it depends on "मेरा- जितना आपुन मेहनत कर रहे है, वह उतना उसको उप्पर से." (My-however much hard work you are doing, it's from how much.)

(3) Labour is social. Since the extended family worked together with each other, it is clear to them that their labour is a social process. They even interact with local people who produce some of the tools and materials that they use.

It's not difficult for the family to understand the value of the paddy that they grow for their own use: it's clear that the sustenance it gives them when it fills their stomachs is what makes it valuable. It is a physical thing that satisfies the physical need for food. But its use value is also because it provides people with emotional satisfaction, the enjoyment of a good taste and something of beauty, the satisfaction of having worked to produced it, and the enchantment in seeing a paddy field grow. The particular taste, smell, and beauty are subjective qualities, but they are based on its physical properties, which they can see, question, analyse, and understand. It's also clear how the paddy was produced by their own labour – they understand this because they did the work themselves.

However, the production of paddy for exchange, as a cash crop is quite different:

(1*) Although the paddy is a physical product that has been produced by their particular kinds of labour, by selling the paddy they act as if their labour – and all the different kinds of labour of all those who produced the tools and other things they used to produce the paddy – is homogenous and equal and is expressed just through the value of the paddy, which is represented solely by the price they get for it. It could just as well have been any other crop or any other commodity that fetched the same price in the market. Its taste, smell, and nutrition are not relevant. It has no use value to the family – only exchange value.

(2*) Although the value of the paddy depends on the duration of hard work that went into its production, when the family sells the paddy they are acting as if the value of the paddy is the magnitude of the value. They act as if its value is an intrinsic, material property of the paddy itself, which is represented by the price that they get for it. They act as if the value of the paddy does not depend on the duration of labour that produced it (the socially necessary labour time).

(3*) Although the social characteristic of the paddy is that it is produced by human labour, when they sell the paddy they are acting as if the material relations between the paddy, the plough, the steel, and the paper money are themselves the social characteristics, and that this is natural and inevitable.

These differences, between the paddy being produced for exchange and for use, illustrate the conflict between the pre-capitalist subsistence farming, on which the family can no longer subsist, and the capitalist cash-crop farming, which alienates the family from the land, from their labour, and from the paddy.

There is a dialectical relation between paddy as a use value and paddy as an exchange value. The two are conflicting with each other. Because of the nature of commodities, the value of a commodity is determined by its exchange value rather than its use value. As capitalism develops, commodification increases, farming becomes increasingly concerned with the production of cash crops, and the exchange value of paddy predominates over its use value. Farmers become alienated from the paddy they are producing.

Note that in feudal times, the people who were doing the work of cultivating paddy may not have been the ones who were eating most of it. But they were not exchanging paddy for money: they were being forced to give various percentages of it to zamindars, tax collectors, priests, etc. (see the Krishisukti of Kashyapa, (Sect. 2.1.2.1). This is very different from the process of producing paddy as a commodity to sell for money.

The family understands how and why the paddy that they eat exists – based on their own experience of producing it. They see that this paddy is a product of their own labour. When they use tools they make themselves (baila, baskets, etc), they know that this labour of making the tools is also 'embedded' in the paddy, since they are necessary for producing it. They have met some of the people who made the plough and forged the ploughshare, and even watched them make it. So they see that the paddy is also a product of their labour.

Although they may not know the details about the history of where this paddy came from long ago, they know that the seeds and the methods of cultivation are actually produced by centuries of social development. They see that in every crop there is variation, and that there are different criteria by which they can select seeds in order to get different kinds of paddy. In the past, other people must have been selecting seeds, just like they select seeds. Actually, the development of paddy and the development of methods of cultivation are constantly evolving processes – they keep changing, and are never really static or balanced or at equilibrium (see Sect. 6.1.1). We have seen how they keep changing due to conflicting forces and inner contradictions.

However, the paddy that the family grows as a commodity to be sold is very different from the paddy that they grow for their own use. We saw that the paddy that they sell is IR – an HYV rice that the family does not eat because they do not like its taste, stickiness, and cooking qualities. They grow it as a cash crop because they can get a better yield with it. The value of the paddy they grow for their own use (Komal, a landrace) is in its use as their food – its taste, smell and nutritional value. But they may not care about the taste, smell, or other qualities (pest resistance, drought-resistance, protein content, nutritional or medicinal value) of the paddy they grow as a commodity. These characteristics will be of no importance unless the paddy fetches a higher price. The value of this paddy is (apparently) determined by the amount of money they get when they sell it. They may produce a bad tasting paddy if it fetches a good price. It may as well be tobacco if tobacco will fetch the same price – the paddy and tobacco would then seem to have the same inherent value.

The farmers do not produce the seeds of the paddy they grow as a cash crop, and they do not even know the people that produce them or how they are produced. They purchase the seeds in the market, without knowing much about where they came from. In this sense, they are alienated from the seeds.

The farmers do not know the origin of the seeds, and neither do they know what happens to the paddy after they sell it. How many times is it bought and sold? Where does it go? Who finally eats it? Do the people who eat it like it, or do they have any complaints? Or maybe it is never eaten – maybe it rots in a godown or is dumped into the ocean. The answers to these questions do not seem to be relevant to the farmers. Isn't that strange?

There are actually a number of things that are very mystical about the paddy that the family produces as a commodity – unlike the paddy which they produce for their own use.

We act as if this paddy has an inherent value, which is its price. Its value does not seem to have anything to do with how long the farmers spent in labour to produce it. And neither does it seem to have anything to do with the labour that was done by the network of socially-connected labourers that contributed to producing the paddy. The person who buys the paddy has no idea who made the plough, or how long it took them to make it. The act of exchange proceeds as if the labour that went into its production makes no difference to the value of the paddy. A certain amount of paddy seems to be equal to a certain amount of money.

There is also a lot of trickery and cheating that goes on at all stages of buying and selling. As capitalism advances, the trickery becomes more and more sophisticated. Some of it becomes legal. People come to expect lying and cheating, while at the same time denying that they are cheating others and being cheated by others. Companies are forced to do such things because of the coercive laws of competition and the need to increase profits.

Companies may try to increase their profits by claiming that a particular commodity has some characteristics that are desirable to customers. They may even resort to false advertising. But as capitalism advances, advertising becomes slicker, the meaning of truth becomes hidden under layers of sparkling plastic packages, and value becomes transformed into intangible brand names and indecipherable labelling that appears to be very scientific.

For example, throughout India it has become impossible for consumers to figure out how to buy atta (ground whole wheat flour) that is not mixed with maida (white flour), unless they buy the whole grain and bring it themselves to a chakki (grinder). The companies that sell the atta usually mix it with maida, because maida is cheaper. Maida is cheaper because it has a longer shelf life, and because it may be imported from USA, where the government gives subsidies and makes 'free trade' agreements (protective tariffs) in order to favour their own market. The term 'free trade' is itself a lie which is not supposed to be openly recognised as a lie. Companies in India label the product as 'wheat flour'. After consumer complaints, the government passed a law in 2019 that they cannot call maida 'wheat flour'. But instead of labelling the product as 'atta and maida' or 'wheat flour and refined wheat flour', companies began using a new label: 'wheat'. Most people do not like to have maida mixed in the atta, because of the taste, nutrition, and texture. But, as time passes, and as it becomes more difficult to get atta that does not contain maida, people get used to it, and forget how pure atta tastes and feels. We have already discussed how HYV rice is becoming more and more widespread even though many people do not like its taste. These are two prominent examples of the general phenomenon of sophisticated trickery resulting in decreasing quality of commodities as capitalism advances (Marx and Engels 1894, 171).[1]

7.1.1.1 What Is Money?

The farmers exchange one thing (paddy) to buyers for another thing (paper money). This appears to be a physical exchange of material things: paddy and money. But how can something that is food be equal to something that is just a few pieces of paper?

Actually, money is not just a physical thing. It is not just coins and paper notes. Money is a representation of value, which is socially determined. We act as if money has some inherent value, but actually it doesn't (we realise this when demonetisation strikes, as it did in India in 2016!). Money is a material thing that represents a social abstraction, the value. Thus, money is a fetish: an inanimate physical object that is treated as if it is abstract, and has magical powers.

Money seems to have the magical qualities that enable us to use it as an intermediary to transform paddy into a television set, for example. Our lust for money drives our lives and determines what we do each day. But this lust is actually not a lust for the physical thing but for the power that it represents.

The lust for money power is one of the driving forces for the accumulation and circulation of money. But it is not just paper (or electronic) money that is moving around. Money is just a representation of value. Value is what is being put in motion and circulated in an ever-expanding spiral, and this is what capitalism is all about. Thus, capital is not money – capital is value in motion and the social relations that enable it (David Harvey 2010). It may appear that capital is money because it is only when capital takes the form of money that can we measure the profit, or surplus value, that the capitalist is extracting.

It is important to realise that capitalists are not accumulating for their own enjoyment – to buy luxuries and eat fancy delicacies. Capital could not accumulate if that were the case, and capital must keep accumulating in order for capitalism to survive. Capitalists could not gain more social power (money) by spending everything on

---

[1] Under Capitalism, history repeats itself, but becomes more sophisticated (and tragic) with repetition. See how Marx described the situation in 1860: "The incredible adulteration of bread, especially in London, was first revealed by the House of Commons Committee 'on the adulteration of articles of food' (1855–6), and Dr Hassall's work 'Adulterations Detected'. The consequence of these revelations was the Act of August 6th, 1860, 'for preventing the adulteration of articles of food and drink', an inoperative law, as it naturally shows the tenderest consideration for every Free-trader who determines by the buying or selling of adulterated commodities "to turn an honest penny." The Committee itself formulated more or less naïvely its conviction that Free-trade meant essentially trade with adulterated, or as the English ingeniously put it, 'sophisticated' goods. In fact this kind of sophistry knows better than Protagoras how to make white black, and black white, and better than the Eleatics how to demonstrate ad oculos [before your own eyes] that everything is only appearance.

themselves or by excessive hoarding. As we have seen (Sect. 6.3.1), capitalists need to keep extending production and incorporating more technical advances. Although greed and malevolence may also be factors in some cases, capitalists are not capitalists because they are mean or selfish. Even if they try to be as nice as possible, or to care about poor people or the environment, they are compelled to keep expanding their capital just in order to survive. If they do not act as if profit is the bottom line, they will be out of business due to competition.

This is why there are attempts to sell new technology and push corporate agriculture upon the family in Rudravali. Rather than developing agriculture in order to supply people with food and make the farmers' lives easier, agriculture is being developed in order to make more profits for rich, powerful capitalists. As capitalism develops, the family may take up new HYV seeds, artificial fertilisers, and new tools and methods. But their adoption of this new technology is impeded by their lack of means to pay for it, and by the lack of government support.

7.1.1.2   What Is the Actual Value of Paddy That Is Sold?

What is actually happening when farmers exchange paddy for money? Where does the value of the paddy come from?

A commodity, such as the paddy that is sold, is not just a physical thing. It has a value which is socially determined. The price, which represents the value, is socially determined. Social relations are hidden behind the material exchange of paddy for money. The way in which the value of the paddy and the value of money are determined is mysterious, because it is not what it appears to be. It appears to be a social relation between the buyer and the seller. It appears to be a matter of supply and demand. But actually, the act of exchange conceals the actual social relations between the labour of the farmers, the producers of the plough the farmer used, those who mined the iron ore to produce the steel, those who transported the steel, those who built the railway on which the steel was transported, and so on. It's impossible to even know who all the people are whose labour somehow contributed to producing the paddy. It's a mystery. It's hidden. It's not that some powerful ruler is purposely hiding this – it's hidden by the nature of the capitalist market system. We just see the appearances: the physical things: the paddy and the paper money. These appearances are not illusions, the physical paddy and paper money really do exist and they really are exchanged for each other. But beneath the appearances, the social relations that produced the value of these physical things are hidden.

Paddy is produced by all this labour, and labour is social. But another strange thing is that social relations exist independently from the will of the people – and yet, social relations are produced by people and depend upon historical conditions.

In capitalism, the social relations of the people who did the labour are replaced by the material relations between the physical paddy and the physical paper money. To the farmers and other labourers, "the relations connecting the labour of one individual with that of the rest appear, not as direct social relations between individuals at work, but as what they really are, material relations between persons and social relations between things" (Karl Marx 1867, 48).

This is a strange perversion. You would think that the value of paddy should be that we can eat it – its use value. But instead, under capitalism, its value is that it produces surplus value, which is a social relation. As David Harvey explains:

...the emancipatory possibilities available to human beings through the sensual physicality of the labour process are perverted and dominated by the social necessity to produce surplus-value for others. The result is universal alienation of human beings from their own potential capacities and creative powers. (David Harvey 2010, 48)

This is what we saw with the HYV paddy that was being produced as a cash crop. For maximum productivity, it requires artificial fertiliser. As Bhimraj said, if they had enough money, then instead of rabbing, more [artificial] fertiliser would be put. With resignation, he said, "Yeah, then what else can be done?! That's how it is." They have to do whatever they can do to increase the productivity. They do not have the choice of growing another kind of paddy to sell if it is likely that it would not be as productive. They do not have the choice of experimenting with seed selection to create paddy with different properties, unless it would result in more profit. And the profit does not end up in their own pocket, but in the pockets of those who are extracting surplus value from their labour.

Karl Marx explained how value is based on what is called, 'abstract labour': all commodities are "reduced to one and the same sort of labour, human labour in the abstract" (Karl Marx 1867, 28). This is the socially necessary labour time. Thus, commodities are congealed quantities of "homogeneous human labour, of human labour-power expended without regard to the mode of its expenditure. ... When looked at as crystals of this social substance, common to them all, they are – Values." (Karl Marx 1867, 28).

In other words, the actual value of the paddy that is sold is the socially necessary labour time that was taken to produce it: the amount of labour time taken by workers of average skill and productivity (of a particular historical time and place) to produce a given amount of paddy.

7.1.1.3 Surplus Value

The family gets the true value of the paddy that they produce to eat. Its value is the value they get when they eat it. But they do not get the true value of the paddy that they produce to sell. The part that they do not get is called the surplus value. The profit is made by traders and corporations who buy the paddy. The profit comes from the surplus value that the farmers are not paid, as well as from the profit traders make just by buying at a low price and selling at a higher price (which we usually take for granted, rather than calling it extortion or robbery).

When we buy and sell, we act as if the physical exchange is all there is – as if these social relations do not even exist, even if we actually realise that they do exist. We act as if commodities and money have some magical power – some inherent value, even though we 'know' that they do not. The 'magic' of commodities really comes from our own labour. But the illusion of this 'magical power' takes control of our lives and we act as if we have no power over commodities, money, prices, and value. Our false consciousness lies in our thinking that we do not have an illusion – our thinking that we are not really acting as if money has magical powers. But we are acting this way.

Paddy is not just to be eaten as rice. Paddy is also seed to be planted. The paddy seeds used for planting are not just raw materials. They have been developed through breeding over years of labour by farmers. As they become more and more com-modified, they are becoming valuable not just because of their use value, but also their exchange value. As such, the seeds are embedded with surplus value which has been supplied by generations of farmers who bred them. Even the cultivars that are developed by corporations are developed from landraces that were originally bred by farmers.

7.1.1.4 Paddy Keeps Changing

There is no one moment in time when paddy is just a thing with one form, size, and shape. In one season, paddy keeps changing as it develops and grows. Its develop-ment depends on its own properties in interaction with the environment and the actions of the cultivators. And over the years, a population of paddy growing in a field is continuously changing and evolving as its environment continuously changes and as society continually changes. Even one grain of rice keeps changing: from the process of fertilisation of an ovum and pollen, to the development and ripening of the grain, to the process of its separation from the roots during reaping, to the pro-cess of separation from the panicle during threshing and winnowing, to the chemi-cal and physical changes it undergoes if it is parboiled, to the process of separation from the husk during milling and polishing, to the physical and chemical processes it undergoes as it responds to changes in humidity, temperature, and other environ-mental changes during storage, to the process of decay or resistance to decay as it responds to fungi, other pests, and/or pesticides, to the process of preparing and cooking it – or the process of germination and growth if it is sown in a field. There is actually no one moment when it exists in just one form. It is always in the process of becoming something else.

We have discussed (Sect. 6.2) how thousands of varieties of paddy have been produced by cultivators practicing artificial selection over several centuries. This is no small feat. It is arguably a much greater achievement, having more impact on social development, than any modern achievement of genetic engineering. Artificial selection and other methods that farmers used to develop paddy and to develop methods of cultivation are largely empirical: based on trial and error. We observed that the family we were working with in Raigad continues to practice artificial selection in order to maintain the landraces of paddy that they cultivate. Indeed, this selection is necessary in order to maintain a landrace, as we discussed. But due to the dialectics of selecting seeds for cultivating subsequent crops, any landrace keeps evolving. The dynamic nature of any variety of rice is inevitable, and this is one of the reasons why paddy cultivation – even so-called 'traditional' paddy cultivation – keeps changing. Thus, paddy – and rice – are actually processes, not things.

Therefore, if we ask, 'What is the value of paddy?' we are asking about the value of a complex process. This process cannot really be understood in isolation from the natural/social environment. Each mouthful of the rice that we are eating embodies the entire process that produced it. In order to question or understand its value, we have to try to question and understand the entire process.

If the paddy is being grown for selling rather than for eating, the process is different – and new complexities arise, making it even more difficult to understand the nature of paddy and the value of paddy.

There is a dialectic relation inherent to the paddy that is sold, as there is with any commodity. It is a dialectic relation between being a reification and being a fetish. On the one hand, paddy (or rice) is reified by treating it as though it is just a physical object, even though it is actually the embodiment of abstract social relations. On the other hand, paddy is a physical object, but it is fetishized when it is sold as a commodity, as if it is just exchange value (an abstraction).

7.1.2   The Family Is no Longer Able to Survive by Farming

One of the main aspects of the agricultural crisis in India is a decrease in soil fertility, water scarcity, salinity, and a lowering water table. However, in the farm in Rudravali, soil fertility does not seem to be decreasing. And there has not been any shortage of rainfall during the monsoon. The family land can support only one crop of paddy per year, because there is no rain at other times of year. (As a rabi crop, the family grows मटकी (matki – a legume that does not require irrigation) in these paddy fields.)

But nowadays, as a rabi crop the family also grows paddy on a plot near the canal on the other side of the village – a plot that is owned by the extended family. This HYV paddy crop is not grown for the subsistence of the family. Most of it is sold as a cash crop. A few people may also be paid to help with transplantation and harvest. Money is required to pay for the electricity that is needed to pump water from the canal to irrigate this field. Coal is used for the thermal power plants that produce the electricity in this region. Therefore, cultivation of this field causes more environmental problems than the non-irrigated fields.

Since 2015, when we started visiting the farm, we did not find much 'modernisation' of cultivation occurring. Mechanisation did not increase. One set of bullocks were replaced by a new set. Transplantation, weeding, and harvesting continued being done by hand. Pesticides still were not used for paddy. Rabbing continued.

However, when we revisited the farm in May 2018, we found a change apart from cultivation: the family had opened a small road-side shop to sell packaged items such as chips and chocolates to local people or tourists passing by. Pratik (now in Class IX) works in the shop. Smita described the situation:

अभी देखो, हम खेती में खा-पिकर अच्छे थे, लेकिन अ-अ-थोड़ा अभी-अ-पढ़ाई बढ़ गयी हैं इसलिए वो दूकान भी डाले, ये भी जाते हैं बाहर काम करने के लिए. पहले नहीं जाते थे. (Now look, we used to eat well from farming, but, ah, ah, now, ah, studying [education] has become more [expensive], that is why we have set up this shop, even he [Bhimraj] goes out for [extra] work. He wasn't going before.)

Why can the family no longer eat for an entire year from the proceeds of their farm? It is not because the family is larger or their land holdings have decreased. It is not because their yield is less or the climate has changed, although these factors are important for other small family farms. But this family is not able to sustain itself on farming, as it did in the past, even though these things have not changed.

One might think that with whatever development there has been in cultivation, the family would be spending less time working or that their work would be easier. However, they are actually doing more work, and more hard work, than ever before: in addition to the work in the fields, there is non-agricultural wage labour in construction, working in the road-side shop, efforts to produce and sell more milk, and even more unpaid labour, such as helping the children with their school-work, and doing the work that might have been done by children who did not go to school in previous generations.

There is a lot of variation between small family farmers all over the country, but one common refrain that we hear is that they used to be able to eat for the entire year off of farming and now they can only eat for six months or less. They may be spending more money than before, but not eating as well. In order to survive, at least one member of the family needs to do some non-agricultural work as well. Bhimraj has to find odd jobs in construction whenever he gets some relief from the farm work. This kind of contract labour is grossly underpaid, with no security and no benefits. Perhaps the roadside shop will bring in some profit – if the value of Pratik's labour time is not considered. But opening a small shop or selling pakoras on a street corner (which the Prime Minister suggested as a 'job') is hardly a solution – those 'solutions' are literally being outlawed (by government initiatives such as demonetisation, GST, banning beef, banning plastic bags, razing unauthorised 'encroachments', eliminating hawkers, etc), so that big corporates can prosper.

How and why is it that the family is now not able to produce enough food for their own subsistence, when 30 years ago they did? As we have discussed (Sect. 6.4.4), the reason is that they are not being paid the actual value of the crops they produce. The reason is also that, under the policies of neoliberalism that were introduced since 1990 (according to World Bank directives), the government has reduced support for agriculture. These policies have caused reductions in the incomes of farmers relative to the cost of living and the costs of agricultural inputs. Since neoliberal policies have also resulted in increased unemployment, farming families also have great difficulty in leaving agriculture or in supplementing their incomes through non-agricultural labour.

Why is this happening? As we have discussed, it is the nature of capitalism. A continuous increase in the cost of living is necessitated by the very nature of capitalism. The nature of capital is that it needs to keep expanding. Capital acquires wealth from the surplus value created by labourers. In order to make more profit, things need to be produced more quickly and efficiently. In order to do this, new technology is introduced and automation increases. As this happens, fewer human labourers are needed, and workers are thrown out of their jobs. But in order for capital to keep expanding, it needs to keep producing more and more. Therefore, capital needs to keep acquiring more markets and more buyers of increasing amounts of products it keeps producing. However, if people are unemployed, they are not able to buy the things that are produced. More people with jobs are needed so that they will buy the products. In order to create more jobs, capital needs to keep producing new kinds of kinds of products and services, including products with imaginary value (e.g. a brand-name, a work of art, or 'intellectual property'). Therefore the wants, needs, and desires of people need to keep increasing. But if unemployment reduces too much, wages may increase. In order to keep wages down and profits up, capital needs to rely on immigrant labour (from Bihar or other states), or the employment of people (e.g. peasants, women) who were formerly not in the labour market. And so, we get stuck in a system that is driven by the need for a perpetual increase in capital – a system which is constantly spiralling out of control as the rich get richer and richer. Thus, a capitalist system cannot exist without continuous expansion. Capitalist systems cannot be stable, and throughout history they have been subject to periodic crises that have required non-capitalist quick-fixes.

If capitalist agriculture continues to develop along the same lines as it has been, the future for the family looks grim. The future for the environment also looks grim. The effects of capitalist development of agriculture have been summed up more than 150 years ago, as follows:

> In modern agriculture, as in the urban industries, the increased productiveness and quantity of the labour set in motion are bought at the cost of laying waste and debilitating labour-power itself. Moreover, all progress in capitalistic agriculture is a progress in the art, not only of robbing the labourer, but of robbing the soil; all progress in increasing the fertility of the soil for a given time, is a progress towards ruining the lasting sources of that fertility. ... Capitalist production, therefore, develops technology, and the combining together of various processes into a social whole, only by sapping the original sources of all wealth – the soil and the labourer. (Karl Marx 1867, 330)

What Marx wrote has unfortunately become all too true in many areas over the last 150 years. Fortunately, the problem of soil fertility has not yet been so severe for this family, although, as we saw, now they do need to purchase some artificial fertiliser (urea) which HYV paddy requires, adding to the cost of their inputs.

In line with what the family said about the effects of rabbing (and with what researchers also say), the fertility of the family's fields is being sustained by rabbing, and by limiting the use of artificial fertilisers. But if capitalist agriculture makes further inroads and the family gives up rabbing and relies solely on artificial fertilisers instead, the fertility of their soil may suffer further. If the family is able to find a way to irrigate their own fields during the dry seasons, it might allow them to produce more rabi crop of paddy. But this might also have a cumulative effect on decreasing the fertility of the soil. If the family takes out loans in order to hire tractors to do the ploughing, they might have to give up their bullocks. But this might also cause a decrease in the fertility of the soil due to the decrease in gobar. The family may then become more and more indebted. With increased work, the family may suffer from health problems and more expenses.

If the family follows the path that many families have been following, they may even be forced to sell off their land, and rent land or work as labourers on land owned by others. More of the family members will have to search for non-agricultural employment. Even without selling off their land, they may be forced to grow more cash crops that they will not even be able to afford to eat themselves. For example, they may have to stop eating the mangos from their own mango trees, because they need the money they could get by selling them (as the prices of mangoes keep increasing with increasing export of mangoes). If they use more of their land to grow cash crops, they will have less land for their own subsistence farming, and their nutrition and health may suffer.

7.1.3 The Need for Agricultural Development

Is agricultural development not desired? People who extoll tradition and 'indigenous knowledge' sometimes romanticise past village life and speak out against modernisation and against 'science' – maybe in preference to some sort of 'appropriate technology' for poor people. They extoll the loss of a natural balance that they think traditional methods maintained. School teachers and textbook authors may also suggest that environmental problems are due to people disturbing the 'balance of nature' (Noel Gough 2019; Karen Haydock and Himanshu Srivastava 2019). People may speak of a 'sustainable agriculture' of the past. But what was being sustained? What was this tradition?

We do see some virtue in how 'traditional' cultivation is supposedly passed along (Table 2.1) and in a way of life in which there is less alienation, as in the past. But especially after observing and participating in paddy cultivation, we also see many problems with the gruelling labour, impoverishment, and physical and mental entrapment it entails. As we discussed, the family kept complaining that their way of farming requires too much hard work, without time for rest or vacation – and they barely make ends meet.

Farmers in India are in the midst of the agricultural crisis, and they are demanding that things change. As we mentioned, the family in Rudravali says that they need to modernise, they need to lower their risks, they need to decrease their physical labour, and they need new technology. But they cannot afford it – because it is available only at high prices from MNC's and private companies.

From what we see of the 'traditional' agricultural life in Rudravali, and from what we read about past rural life in India, it is clear that the 'tradition' is a long-standing tradition of poverty, hunger, and hardship. We see the hypocrisy of what we read in newspapers about India's 'fastest growing economy' and impressive GDP. When we meet the farmers – the people who produce this wealth – we are forced to question the entire system in which this occurs.

We cannot justify efforts to retain or return to perceived indigenous traditions for the sake of tradition. We cannot justify efforts to maintain diverse methods of cultivation for the sake of diversity. Neither can we justify efforts to continue on the road to the capitalist development of agriculture, which as we discussed (Sect. 6.4.4), is causing the agricultural crisis. Rather, we need to make demands based on what is best for labourers – what is most just and egalitarian.

As we mentioned in Chap. 6, we do not look back at the past as a time in which cultivators were living in 'harmony' with nature. There have been plenty of natural disasters and famines throughout history – caused by both social and natural problems. We do not see any evidence that a 'balance of nature' ever existed or that nature would 'balance' itself if we just stop disturbing it (Sect. 6.1.2). As we discussed, paddy keeps changing, even without people. Paddy has never existed in a static, 'balanced' state. Its inherent imbalance is what drives its continuous evolution. The problem is that nowadays, companies are patenting paddy evolution, and controlling the development of new varieties for the sake of profit, not just as a food source.

Just as there is no balance in nature, we do not see a balance in society in the past or at present. We see that society is continuously developing – developing in many interdependent but conflicting ways. By 'developing', we do not mean to imply 'improving', but changing. Improvement may or may not come with change. Just as nature develops due to inherent conflict, society also develops due to internal conflict. Since nature keeps changing, society also keeps changing, because society and nature are inseparable. Therefore we question whether there would ever be a balanced, harmonic, unchanging, form of nature/society.

Could nature/society be 'sustainable'? If by sustainable we mean 'being maintained at a certain rate or level', 'continuing on in the same fashion', or 'conserving an ecological balance', we doubt that it is possible. But is 'sustainable development' possible? Or is there is an inherent conflict between sustaining and developing? What most people actually mean by 'sustaining development' is 'sustaining capitalist development', and there is ample evidence that capitalism is not sustainable, because it involves ever-increasing growth of capital (John Bellamy Foster et al. 2010). It is an inherent contradiction of capitalism (David Harvey 2014). Capitalism necessarily involves making decisions based on making profit rather than on what is best for nature/society or on maintaining a 'balance'.

Rather than just looking backwards to some nostalgic image of paddy cultivation and past village life, or accepting what some say is the inevitability of capitalist development, we need to ask deeper questions, and consider newer solutions.

7.1.3.1 Agricultural Development Is Not Just a Matter of Individual Choice

The family does not have an individual choice about whether to use bullocks or a tractor for ploughing. They would love to have a tractor, or at least rent one to do the ploughing. But they cannot afford it.

The family does not have an individual choice about whether to do rabbing or use artificial fertilisers. They cannot afford to purchase the fertilisers. But if they do not have access to land (their means of production) from which they can gather the material needed for rabbing, they will have to give up rabbing.

The family does not have an individual choice about whether to use gobar or artificial fertilisers. They cannot afford the fertilisers. But if they are forced to give up their cattle, they will not have any gobar. Many farmers are forced to give up their cattle because the government has made the sale of certain types of cattle illegal. Farmers need to sell off their bullocks when they get too old to work, because they cannot afford to feed them. But nowadays, farmers in some areas are afraid to even be seen transporting cattle (especially if they are Muslim or Dalit farmers), because Hindu fundamentalists might accuse them of slaughtering cows, and lynch them.

The family does not have an individual choice about whether to use cane or plastic baskets. If they do not have enough time to make a cane basket, it cannot be made. If cane baskets in the market are more expensive than plastic ones, they will not be bought. If plastic baskets are not available in the market, they will not be bought. If there is social pressure and advertising to make people think that brightly coloured plastic baskets are more attractive and prestigious, this may be hard to ignore. And of course, if Namdev gets excluded from the land in which the bamboo grows (his means of production), his craft will end.

The family does not even have an individual choice about whether to keep hold of their land. A small part of their land was already taken away for the construction of the highway that now runs through the front yard of their house. If the value of their land increases too much, or if their financial situation gets too severe, they will be forced to give up their land. As individuals, they will not be able to resist dispossession or survive as subsistence farmers. The process of capitalist development poses a threat to their means of subsistence.

Society doesn't become capitalist just by people explicitly wanting or learning to be capitalist or explicitly learning about capitalism in schools. People do learn – in and out of school – to fit into a capitalist system, but this learning is implicit, and occurs despite a pervasive, enshrouding ideology in which people usually do not realise that this is what they are learning. Some people may even feel themselves to have the individual freedom to believe and become anything they wish.

Neither do people learn to be traditional by explicitly learning traditions or learning about traditions in or out of school. People cannot be expected to give up glittering new technology, become less individualistic, or have more respect for authority just because their teachers tell them to.

A society does not become more scientific and technically advanced just by giving birth to a few geniuses who go on the become scientists, discoverers, and inventors. Neither does it become more scientific and technically advanced just by making schools in which children are taught to be more scientific.

A society becomes more scientific and technically advanced because it becomes more capitalist, and because it becomes more capitalist, it becomes more scientific and technically advanced. But what sort of scientific advancement is this? Under capitalism, it is one in which the bottom line must be profit, rather than the betterment of people. What we need is a society in which people will use science to change the world in a way that will be best for people – that will prevent people from getting rich and poor, that will prevent human suffering and oppression. Obviously, such systemic changes are not a matter of individual choice.

7.2   Dialectical Conflicts Leading to a Definition of Science

Our goal is to find out how the family learns and passes on cultivation, and how traditional and/or scientific their learning and practicing of cultivation is. We will analyse whether the farming family is actually doing science to some extent, and how and why they are or are not doing science. As we do this, we will define what we mean by science. We will show that, according to our definition of science, the traditional cultivation that the family was doing did involve doing science, although the sort of science that they were doing was somewhat different from the sort of capitalist science that is done by professional scientists. In order to understand this, we will have to analyse why the family was doing science and why professional scientists do science.

7.2.1   How Does the Family Actually Practice and Learn Paddy Cultivation?

In Chapters 2 and 3 we discussed that we found many dialectical conflicts between the superficial appearances and the actual ways the family was doing, learning, and teaching paddy cultivation. These dialectical conflicts are summarised in Table 4.1. We need to stress that these are material contradictions, they are not just contradictory beliefs. Now we will discuss them, in reverse order, in relation to the historical development of paddy cultivation and education, which we discussed in Chapters 4, and 5.

7.2.1.1   Availability/Non-availability and Sustainability

We have discussed how policy-makers and educationists may blame the lack of modernisation in agriculture on farmers' lack of knowledge and skill (Sect. 5.3). However, we also saw that the farming family in Rudravali was aware of many modern materials and methods that they would like to use, but they could not use them due to their cost.

There are usually no simple answers to questions about which technologies to use. Each technology has its advantages and disadvantages. For example, the family explained to us how although they knew there were a number of reasons why rabbing is a good practice, there are also disadvantages. It requires a lot of time and hard work. It causes air pollution. It is risky if the fire goes out of control. They did not need to go to school to learn these things. They could see and feel the advantages and disadvantages: the healthy growth of the paddy seedlings, the weariness in their own bodies after preparing the layers for rabbing, the difficulty breathing smoky air, and the dead trees when a neighbour's fire went out of control. They were also aware through their own experience and their observation of others' experiences that excessive use of artificial fertilisers is not good for the soil in the long run.

But they also realised that because of the cost, they did not have the option of using more artificial fertilisers or trying mechanical methods of preparing compost instead of rabbing. They did not have the option of deciding which methods to use on the basis of the method's use value, which would include the environmental impact. The decision was already made for them, on the basis of exchange value.

As we have discussed, the farm in Rudravali is not directly affected by many of the environmental problems that cause agricultural distress in other areas. They do not suffer from drought, salinity or loss of soil fertility. Because their fields do not have irrigation, they have always been able to grow only one crop of paddy each year – during the rainy season. However, recently they have noticed unusual variations in weather patterns, which may be due to climate change: the arrival of the monsoon (previously around 7 June) seems to be getting later and the rainy season sometimes extends further into September and October. This poses a problem, because rain just before or during harvest can ruin the entire crop. In 2019, much of their paddy crop was ruined due to heavy rains extending into September and October.

The area in which Rudravali is situated has been outside or at the periphery of the Green Revolution. The main change brought on the Green Revolution in Raigad District was the building of additional dams, reservoirs, and canals, which brought irrigation to parts of the area. This allows people to cultivate two crops of paddy each year. But as we saw, irrigation did not reach the family's fields – they do not even have a well or piped drinking water.

The fertility of the family's fields has been maintained by rabbing and fertilisation with gobar and other organic wastes, as well as by growing the legume (matki) during the dry season. This has occurred not by choice, but because they have not been able to afford to use tractors, motorised modes of transport, irrigation, or much artificial fertiliser. If they could, they would prefer to use more modern methods in order to reduce the amount of hard labour and to try to increase their yields and get more money for the portion of the crop that they sell. Since they have to rely on bullocks for ploughing and pulling the bullock cart, and because they also maintain two buffaloes for milk (an important source of additional income), they have a source of gobar. Irrigation – or at least a source of drinking water at home – would be very useful, but it is out of the question since they are too far from the present canals and digging a well through their rocky ground would be too difficult and expensive.

But in case irrigation somehow came to their fields, perhaps they would cultivate an additional rabi crop of paddy as a cash crop, instead of matki, which does not require irrigation. Like any legume, matki replenishes nitrogen in the soil. It is only for their own use, and it provides an important source of protein in their diet. If they planted an irrigated paddy crop instead, the fertility of the soil would decrease. Because there may not be enough material available to rab so many fields twice a year, they would probably have to use more artificial fertilisers, which would add another cost to the expense of irrigation. So, in the long run the family might be worse off: even more overworked, with worse nutrition, and less fertile fields.

We have discussed the relations between 'neoliberalism' and the agricultural crisis (Sect. 6.4.4), and how it affects the family in Rudravali (Sect. 7.1.2). The main effect of the agricultural crisis on the family is that they do not get enough money to survive on agriculture, as they did 20 or 30 years ago. Their methods of cultivation are therefore 'unsustainable'. New methods that might alleviate some of their problems are not available to them because of their expense. Many modern methods might actually exacerbate the situation in the long run. Corporations are not in a position to develop better methods, since they need to compete for profits. And the family does not have the financial resources to do much to develop better methods themselves.

This dialectical conflict between what is in some sense available (technological solutions) and at the same time not available (due to the economic situation) will have to eventually give rise to meaningful change.

7.2.1.2 The Role of Rituals and Gods

Our superficial observation of rituals being practiced on the farm, led to our subsequent apprehension of a conflict between doing rituals and doing paddy cultivation. As discussed, we were surprised to hear what Smita, Bhimraj, and Namdev said about the relation between religion and cultivation. Bhimraj explained what rituals are for "वो अपना अपना अपना प्र-प्र प्रणाम प्रमाण कर लेंगे दे, आपुन. ऐसा." (This is for you to do r- r- respect [obeisance], for your own- your own- your own self. Like that.) But both Smita and Bhimraj insisted that the success of the paddy crop does not depend on the gods, but on "जितना आपुन मेहनत कर रहे है, वह उतना उसको उप्पर से (However much hard work you are doing, it's based on that much), as Bhimraj said.

Initially, we wondered about the possible conflict between rituals and science in cultivation. We were aware that there is a tradition amongst foreign academics to see science as non-existent before the advent of capitalism, to emphasize mysticism and religion in India, and to disregard or deny the existence of science in 'traditional' India. As Hartmut Scharfe (2002) states, "Religion has long been regarded as the dominant feature of Indian culture." Suvira Jaiswal, (2018, chap. 8) critiques how Ronald Inden, "in his well-known book, *Imagining India*, attempts to seek the 'rationalities' of ritual practices of early India in terms of a metaphysical perspective that theorises them as integral to a 'theophanic polity' totally opposite to the 'scientifically grounded politics' of the west."

Our interactions lead us to question these assumptions and to question the nature of both religion and science in a so-called 'traditional' farming family – and in our own lives.

This leads us to ask: what is the meaning of a seemingly irrational ritual?

The researchers come from diverse religious backgrounds, and some of us understand very little about the kinds of religion the family may be practicing. Nevertheless, we might learn about each other's views and practices regarding rituals by looking at our own views and practices. Previously Karen had thought that irrational rituals are just useless superstitions. She never prayed or believed in any god or thought of herself as being religious or spiritual in any way. But now she is asking herself whether she has also been practicing rituals that affect her own social interactions. For example, when she was a child she took part in the 'fun' of making a wish and blowing out birthday candles, even while professing that she did not really believe that the wish would come true if she blew out all the candles, according to the custom. Although she 'knew' it was irrational, she acted as if she did not know that what she was doing was irrational. But she did not realise that this was the way she was acting. In another example, when Karen was 3 years old, her brother told her that there is no Santa Claus, and that parents give the gifts that they say are brought by Santa Claus. He proved it by showing her where her Christmas gift was being hidden by their parents. Karen became very angry – not at her parents for lying to her – but at her brother for breaking the myth. Her response indicates that she did not want to learn. She wanted to continue to perform the ritual and play the ideological game of not knowing what she knew.

Maybe this is similar in some respects to what Bhimraj and Smita were doing: they were saying that they did not believe in rituals, but they were still practicing the rituals.

So this leads to the question: if we say that we do not believe in a ritual but act as if we believe in it, are we actually believing in it? Is this what could be called an unbelieved belief? Are we just pretending to believe in it, or are we being what we pretend to be? It is dialectical. Maybe we actually know that what we are doing is an illusion, but still we keep doing it. But if we know the illusion is an illusion, it is not an illusion! It is not false consciousness in the sense that we are believing in an illusion. It is false consciousness in the sense that we do not realise that our actions indicate that we are believing in an illusion – we actually have beliefs that are contradictory to what we believe our beliefs are. Rather than calling this false consciousness, it will be better to call these beliefs (beliefs that are revealed by our actions), our actual ideology. In other words, making a birthday candle wish, or attaching mango leaves to a plough, are not 'false consciousness' or illusory. The illusion lies in thinking that one does not have illusions.

This shows us the need to investigate our actions in order to understand our consciousness. We need to observe and analyse our material activities such as tying mango leaves to a plough and blowing out birthday candles in order to understand our ideologies. Our ideologies may not be what we superficially think they are. This is the approach that Karl Marx and Friedrich Engels advocated in their analysis in German Ideology:

... we do not set out from what men (sic) say, imagine, conceive, nor from men as narrated, thought of, imagined, conceived, in order to arrive at men in the flesh. We set out from real, active men, and on the basis of their real life-process we demonstrate the development of the ideological reflexes and echoes of this life-process. The phantoms formed in the human brain are also, necessarily, sublimates of their material life-process, which is empirically verifiable and bound to material premises. Morality, religion, metaphysics, all the rest of ideology and their corresponding forms of consciousness, thus no longer retain the semblance of independence. They have no history, no development; but men, developing their material production and their material intercourse, alter, along with this their real existence, their thinking and the products of their thinking. Life is not determined by consciousness, but consciousness by life. (Karl Marx and Friedrich Engels 1846)

Bhimraj and Smita say that they do not believe that gods determine the outcome of the crop. They say that they practice rituals for their own well-being, not the well-being of the plants. In other words, rituals are social abstractions which can affect other human social abstractions, but cannot directly affect plants. The belief that gods and rituals directly affect plants is an example of fetishism and reification: treating an icon, coconut, or other object as if it has some magical power, and attributing one's own actions to objects. Karen was stereotyping villagers and theorising that the family would be confusing the abstract and the material, thinking that plants grow due to some mystical powers. She thought that they would not realise that the power comes from our own creative labour (combined with the material forces of nature). But maybe Bhimraj understood the power of creative labour better than Karen since it was produced from his own sweat. Or maybe not – maybe he also had his own ideological blinkers.

Authoritarianism, faith, rituals, and mystical irrational beliefs, may help people cope with their problems, e.g. by providing people with solace and emotional support. But this is not enough to produce a good crop. As Bhimraj said, the crop requires hard physical work. The crop depends on material relations with soil, water, plants, and the rest of nature, which are interdependent with human interactions. Emotional support from performing rituals may help people work hard, but emotions or beliefs without physical work in the field do not suffice. The work on the farm was characterised by this sort of material interdependence, rather than a vague or mystical 'holism' in which moral values or the judgement of a spirit or god determined crop yields. How can cultivators cultivate if they believe that the success of their crop is more in the hands of some gods than in their own hands? Or that life is just an illusion?

Sometimes a religious approach may be necessary in order for people to face particular social problems. This is what Karl Marx meant when he wrote that religion is "the heart of a heartless world", and the universal basis of consolation.

In the present world, it may be unjust to try to make people give up religion. However, there are certain oppressive aspects of religion that need to be opposed, and if they form the very basis of a religion, that religion itself becomes under attack. This is why Ambedkar opposed Hinduism and encouraged others to also leave Hinduism rather than just try to reform it: "You must have the courage to tell the Hindus, that what is wrong with them is their religion – the religion which has produced in them this notion of Caste" (Bhimrao R Ambedkar 1944).

7.2.1.3 Conflicts Between Oracy and Literacy

We discussed how the family relied much more on oracy than literacy (Sect. 3.2.8). We have also discussed how in formal education the emphasis has been on literacy (Chap. 5).

Kancha Ilaiah [Shepherd] (2010) and Anita Rampal (1992) have written about the 'oral culture' of India, and its comparison with 'literate cultures'. As compared to the written word, the spoken word is even more obviously a process rather than a thing. It is more changeable, responsive, and 'living' than a relatively fixed, static piece of writing. Most people find it easier to communicate orally. Written communication is usually more constrained. Because of this, oral communication is more suitable for certain aspects of learning, working, and doing science. Because it is more inclusive and spontaneous, speaking, rather than writing, may facilitate innovation, creativity, and questioning. It also facilitates communication within small groups and is less individualistic and less alienating. It is more immediate – although written communication through social media on the internet may be changing this.

We might learn something about pedagogy from the dialectical methods used in the Jataka tales, which were orally transmitted, since the third century BCE or earlier. Uma Chakravarti (2012) has discussed how they had both fixity of form and broad, fluid structures that allowed for improvisation, and deliberately made room for performers to contextualise and amplify details at their own discretion. Both the content and the style were expected to be improvised to correspond to particular audiences, and incorporate experiences of the performers and the audience. The audience response might also become part of the story, making for a two-way interaction. At the same time, the Jataka tales also used certain devices in order to lead the audience away from unintended interpretations. This was done by having the Buddha, who was revered as an authority, as a protagonist or narrator. Also, the same characters, with consistent behaviours, appeared in multiple stories, providing some fixity of meaning. Cultivation is mentioned often in Jataka tales, and in their illustrations (e.g. Fig. 2.4). Visual arts also offer methods of communication that overcome some of the constraints of writing.

Compared to writing, the spoken word is more subject to evolution as it passes from person to person and group to group. This may be a disadvantage in doing cultivation, in that it may introduce inaccuracies and spurious errors and contradictions between direct observations and reported observations.

However, Lotika Varadarajan (1979) has noted that since ancient times, a distinction was made between "Hindu sacred knowledge" and "traditional craft and technical knowledge". She claims that "the latter has necessarily to permit greater elasticity as new inventions, techniques and processes have, at each stage, to be incorporated within the existing corpus prior to onward transmission." The two were separate, and had different methods of dissemination. Memorization, pronunciation, mastery, and understanding of sacred texts was highly regulated and restricted to certain members of a particular privileged caste, and for this, the oral tradition was preferred:

> Even after scripts had evolved there was always a preference for the transmission of knowledge through oral rather than written means. This was because it was feared that the written word being accessible to a wider public, might fall into wrong hands with the attendant dangers of distortion and vulgarisation. (Lotika Varadarajan 1979)

People may make important modifications and additions as they orally pass on scientific discourse – resulting in the inclusion of more people in orature, as compared to literature.

On the other hand, the act of writing may help people to analyse and investigate questions. With regard to learning literacy, Lev Vygotsky pointed out that,

> The motives for writing are more abstract, more intellectualized, further removed from immediate needs. ...Writing also requires deliberate analytical action on the part of the child (Lev Vygotskiĭ and Alex Kozulin 1986, 181–82)

There is no doubt that the written word is very important in doing science, and in making the implicit more explicit. However, there is actually a dialectical relationship between oracy and literacy, in which they coexist interdependently. Problems arise when the interdependency is obstructed because certain people are not allowed to be literate or have access to written material. To some extent it may also be obstructed when isolated writers and readers suffer from inadequate direct contact that involves oral communication.

7.2.1.4   Gender Roles and Respect for Authority

As we discussed (Sect. 3.4.3), we initially saw that gender roles were adhered to as expected: ploughing the fields and sowing the paddy were done by men rather than women. All other tasks of paddy cultivation were done by both men and women – except Smita did more of the weeding.

However, we were surprised that Smita was not at all passive. Neither was she aggressive or 'nagging'. She seemed to understand what was going on even when she was not directly doing the physical labour herself. And she played an important role in helping to direct the work.

Other researchers have also found a similar situation. In their study of six districts of Maharashtra and Andhra Pradesh, Sagari Ramdas et al. have reported:

> In the traditional agricultural system women used to have equal decision-making powers. Women and men within a family would take joint decisions about the crops to grow for the next agricultural season. While there were clear-cut gender divisions in roles and responsibilities vis-à-vis various agricultural tasks, women were equal participants in all decisions. It was also the woman's responsibility to select and preserve the grains for the coming year of sowing. (Sagari Ramdas et al. 2004, 77)

On the farm, we observed that the important role of selection and preservation of the paddy seeds was done by men as well as women. But, as we reported, the family sowed both landrace seeds that they selected and HYV seeds that they purchased. The HYV seeds deteriorated after a few years of replanting, and then had to be bought from the market. Women were probably not involved in the purchase of the seeds in the market.

Both women and men took part in collecting, storing, preparing, and distributing gobar for fertilisation, but the purchase and application of artificial fertilisers was only done by men. In case pesticides were used (they were used for chilli but not for paddy), men bought and applied the pesticides, not women.

Most important, the selling of the paddy was done by men, not women. Men received the payments, in cash.

However, women are generally doing more of the work on farms than men. This is partly because the men occasionally go out to do non-agricultural labour, leaving women to take care of the farm. Or men may migrate to urban areas to find jobs, staying away for most of the year, leaving women and children in villages. But it is also a general phenomenon throughout the country that women tend to be busy at work throughout the day, while men may spend some time sitting around drinking tea and talking to each other. When women are sitting around talking to each other, they are likely to also be cutting vegetables, cleaning grain, sewing, or doing some other work. However, on this farm in Rudravali, we never saw either women or men sitting around doing nothing. Both men and women were very over-worked.

Thus, we see that the more modern, capitalist aspects of cultivation involved more patriarchal division of labour than the more traditional aspects of paddy cultivation. This is ironic, because patriarchal hierarchies have often been thought to be characteristic of traditional societies. Many intellectuals assume that there is more gender equality with more 'westernisation' and 'economic development' (i.e. capitalism). However, this is in contradiction to what Sagari Ramdas et al. have claimed:

> With the change in cropping patterns, women have lost all their decision-making powers in the production process and have been marginalised. From decision makers they have become mere providers of labour. (Sagari Ramdas et al. 2004, 77)

This 'modernisation' and loss of decision-making powers has not yet happened to a very great extent to the family in Rudravali. Or maybe the hierarchies were more dynamic than we expected. The power relations between people may have depended on context. Those who are less powerful in some contexts do question authority in other contexts. This is similar to what we found in our study of middle-school students in Mumbai: the same students who were quiet and unquestioning inside a classroom were actively questioning and investigating when they were observed in a more informal context outside of the school (Gurinder Singh et al. 2019).

Actually, the family seemed to be working together very cooperatively. They all gave each other suggestions and directions about what to do. They did not just get or follow directions and advice from the grandfather, who might have been considered to be the family patriarch. For Karen this was strange, because she had expected family members to be working in a more hierarchal structure, based on what she had observed in her own family (Sect. 4.3).

We saw that the dialectical relations between gender, power, and authority, and the conflicts that arise from these conflicts play important roles in how the family learns and figures out solutions to problems that arise in the course of their work in paddy cultivation. They affect how and to what extent they do science.

7.2.1.5   Acting Locally/Being a Part of the Global Market

As we have discussed, it has been claimed that traditional farmers' sphere of work is quite localised in both space and time. In comparison, professional scientists' work may be international and interconnected. This was particularly evident in the contrast between production of seeds locally or by multinational corporations (Sect. 6.2.1). But in Rudravali, the work was local/global.

We saw many examples of ways in which the family was 'acting locally'. Their physical mobility was restricted because they did not own a petrol-powered vehicle, they could not afford to travel very far or very often, train service was minimal, bus service was inconvenient, and travel by autorickshaw was expensive. Besides this, they were tied to the farm in order to take care of the animals and crops every day.

Their communication was mainly local, especially when we first visited, in 2015. At that time they did not have a phone, and their TV was out of order. Even later, their communication was restricted because they communicated in Marathi, and to some extent Hindi, but if any material regarding cultivation was in English, it was not accessible to them.

On the other hand, the seeds they used for paddy cultivation were landraces produced locally, as well as HYV cultivars that had been produced by national and international institutions. They used gobar from their own cattle, but also urea (and pesticides for vegetables) that were produced by national as well as multinational companies, using technology that had been developed mainly in other countries by capitalist scientists.

The paddy that they sold as a commodity was finally consumed at some unknown place, which may have been fairly distant. And the money that they got for selling it was of course also not local. The same money is used throughout the country, and its value, and the price of the paddy they sell, depends on not just national, but also international economics. If and when India exports or imports rice, this affects the local prices.

Thus, we saw how the dialectical conflict between local and global economic and political factors affects how the family in Rudravali learns and develops paddy cultivation.

7.2.1.6   Being Pragmatic/Abstract, and Being Specific/General

Researchers have claimed that if and when farmers carry out their own experiments, their questions for investigation are more specific and pragmatic, whereas professional scientists' questions are more general and abstract. Thus, Christian Vogl et al. (2017) claim that farmers' experimentation is less synchronic and more diachronic – their findings may be valid only locally, in their own specific environment and circumstances, but they may be referenced over a long period of time and experience. This, they claim, is in contrast to the sort of experimentation that formal scientists do, which is more synchronic and less diachronic – the questions may be more general, but with reference to a shorter period of time (since, at least in agriculture, experiments are usually of limited duration). On the other hand, the results of the formal scientists may not actually be very applicable to the specific conditions of a particular farm, or to the changes that may occur on a farm from year to year.

However, the farming family in Rudravali is actually restricted both spatially and temporally because their sphere of work is local, and they do not have much access to written communication concerning the past. They rely on oral communication and on memories which may not span more than a couple of generations. For this reason, their work is less diachronic. We have discussed the advantages and disadvantages of oral vs written communication (Sect. 7.2.1.3).

In the Indian subcontinent, communication has been particularly restricted because people have been separated from each other due to the caste system, which has been enforced since around 100 CE (Tony Joseph 2018). Caste solidifies the gap between physical labourers being pragmatic and landowners being abstract. This can be seen in the general account of paddy cultivation written by Kashyapa for landowners (Sect. 2.1.2.1). Caste brings endogamy as well as barriers and restrictions concerning sharing technology and other cultural and everyday activities. All of these hinder the process of learning and developing cultivation by doing science with one's own hands. For example, it has been shown that caste barriers impede trade in water, resulting in inefficient and inequitable groundwater distribution systems (Siwan Anderson 2011). Thus, even within one village, irrigation becomes specific to certain castes. Obviously, class divisions also have similar effects, and cannot be understood in isolation from caste.

In this regard, we have discussed how the landrace paddy seeds that the family was using have been produced by an experimental process of artificial selection that has spanned many generations. The farmers have not had access to written communication or histories concerning these seeds. However, they have their hands on the seeds themselves, and they know through their own practice how the seeds are maintained and developed over generations by selection. The results of the experiments are the seeds themselves, which are most valid, or appropriate, for their specific location. At the same time, the seeds developed in and for one particular location may be very useful in order to develop seeds that will be useful in other specific locations, because they can supply the genetic diversity that is necessary for plant breeding.

Thus, we see that there are dialectical relations between being specific and general – inherent conflicts that allow people to develop different varieties of paddy.

Another example of this specific/general dialectic is illustrated by the way seeds are selected (Sect. 4.2.3). The farmers look specifically at the seeds, but they also look at the entire panicle, the entire plant, and which part of the field the plant came from. The selection depends simultaneously upon the particular and the general characteristics – which may be conflicting with each other. This means that selection cannot be mechanical or clear-cut. It is always a matter of a dialectic between chance and necessity. It is both abstract and concrete.

Similarly, when examining the standing crop to see if it was ready to harvest, the family looked closely at individual grains on particular plants, checking to see if they were full and hard (Sect. 3.2.7). But they also had to consider the general condition of the over-all crop, since the entire crop had to harvested at the same time.

Finally, when the family was describing the benefits of rabbing, they began by making general statements, but eventually they mentioned all the specific benefits listed in Table 3.1.

7.2.1.7   Using Both Old and New

We have mentioned that there are a mixture of old and new materials and methods on the farm in Rudravali. Until very recently, all tools, such as the baila, panja (rake), broom, plough, and sickles, were made at home or by local artisans. Even now, the only ones that are manufactured farther away are a few small tools with plastic handles. We saw many other examples of both old and new technologies being used: HYV seeds as well as landraces were being sown; gobar as well as urea was used for fertilisation; both foot-operated and electric threshers were being used; cane baskets were being made and used, but plastic baskets were also used; plastic utensils and containers for cooking, carrying, and storing water were being used along with ones made of brass, iron, steel, aluminium, and clay.

Clearly, there is a conflict between old ways and new ways of doing cultivation. And the question is: why do the old ways still persist?

We have already discussed some of the possible reasons why science and technology related to agriculture did not develop further in India in the 18th and 19th centuries (Sect. 6.3.3.3). The two main reasons are the economic deprivation due to colonialism and the effect of the caste system in segregating people, hindering communication, learning and teaching, and hindering those who work with their hands from also working with their minds. Once science and technology had developed in Europe, and India became a British colony, the development of Indian science and technology was (at times forcibly) prevented from developing, so that British technology and industry could prosper, and so that agriculture could be under British control. Even nowadays, original research and development of new technology is hardly being supported in India – technology is largely imported and modified from that developed by dominant powers. Farmers do not have the leisure or money to do sufficient research to develop technology that can compete with what the west produces. Banks and the financial system also play increasingly important roles, and end up trapping farmers in debt more often than allowing them a means to develop their farms.

Years ago, Bhimraj had to return to help his father with farm work after attempting to take up a job in a city. This allowed their small family farm to continue to exist according what appears to be feudal relations. It is not uncommon for people to leave their farm to get employment in a city, but then face some sort of problem (e.g. a health problem, unemployment, communal violence). If they have a village to go back to, they do go back if there is no other option. Going back to the village is the last resort. But people realise that it is very important to preserve this last resort. Bhimraj says that his children will also have this option (since to some extent they picked up farming just from growing up on the farm): 'आमची मुलं कशी आहेत,

याच line मध्ये आहेत. मुळात शेतकरी आहे. मग त्याच्यामुळे किती जरी, त्याने बाहेरचं काम किती जरी केलं, म्हणजे कंपनी करू द्या, duty करू द्या, लास्ट ला गावाला आले का नाही, का ते परत वळणार आपल्या शेतीकडेच. (Our children ... are basically farmers. Then because of that even if they work a lot, means let them do company work, duty work, at last when they come back to the village, they will turn back to farming again only.) Maybe Bhimraj and Smita said this because they had become discouraged after hearing Karen asking what is the use of education when even educated people may not get jobs. Bhimraj says that he is sure their children will actually turn to farming, and Smita says, 'आता शेतक-याचा मुलगा शेतकरीच होतो, डॉक्टरचा मुलगा डॉक्टर होतो, हे परंपरा आहे, आमची मुलं शेतकरीच होणार. (Now a farmer's son only becomes a farmer, doctor's son becomes a doctor, this is the tradition, Our kids will also be farmers only.) They seemed to be saying that actually these are still feudal times and children are bound to the occupation of the family into which they are born.

We discussed how as capitalism penetrates agriculture in India, it begins exerting control over all sorts of inputs to small family farms. Tools, farm machinery, seeds, pesticides, fertilisers, and irrigation are increasingly becoming commodified and produced in a capitalist mode of production. Even drinking water is becoming a commodity that has to be bought and is sold by someone for a profit. We have also discussed how and why capitalism has not penetrated to a greater extent into agriculture, and why the small family farm still exists – rather than being taken over by large-scale corporate agriculture, as has happened in some countries. Some had thought that an agricultural revolution would occur in India, leading to an industrial revolution as well. But we discussed how even in Britain the industrial revolution occurred not through an agricultural revolution, but through primitive accumulation and colonial exploitation (Sect. 6.3.2). Acquisition of peasant land holdings and usage rights in England forced peasants into cities where they became industrial workers. But there were not enough jobs for all of them. Unemployment was kept within control because the 'excess' peasants emigrated to British colonies. Clearly, a similar thing cannot happen in India today. There is no place for dispossessed peasants to go (Sect. 6.4).

The family in Rudravali cannot survive just by continuing to do things in the old ways. Bhimraj has to go out to get jobs outside of agriculture in order for the family to survive. But unemployment keeps increasing, and it is difficult for him to find jobs. Who knows how his sons will find jobs.

The most striking 'oddity' that characterises Indian economics today is that although the industrial and service sectors produce most of the GDP, they do not produce enough jobs, and most people continue working in the agricultural sector. As we discussed (Sect. 4.1.4), the Class X NCERT economics textbook authors do mention this problem. However, the authors do not explain why it is happening. They do not even attempt to explain why there are not enough jobs.

Instead, they say that too many people are working in agriculture, and they are not working hard enough – they "work less than their potential". Clearly, they are blaming the victims.

This directly contradicts what we saw on the small family farm: everyone was complaining about how overworked they are.

The textbook authors go on to suggest some remedies, such as getting loans, installing irrigation systems, cultivating more fields, getting a few family members jobs in industry elsewhere. But they never suggest that the farmers should get the true value of their labour in the first place.

When asked about the possibility of expanding their paddy cultivation to another field, Smita categorically ruled it out, saying that they can barely manage all the work as it is. Anyway, we wonder how could they come up with the money to rent another field, seeing that they want to avoid the loan trap. And debt would be a real trap, because in the current system they will not get the true value of their labour, and probably not even enough money to pay back their loan.

But what we are often taught, both through the mainstream media and through formal education, is that poverty in India is decreasing. What is not taught is that this 'reduction in poverty' is an illusion based on an official definition in which the 'poverty line' keeps getting lowered, as it is adjusted by a price index that underestimates the amount of money required to maintain the same nutritional status. If we use a correct method to estimate poverty, we see that poverty remained almost the same between 1973 and 1993, but rose sharply in both urban and rural areas during the 'economic liberalisation' between 1993–94 and 2009–10. What is not taught is the evidence that at a global level, capitalism is predicated on an increase in wealth for a few and a relative increase in poverty, unemployment, and hunger for the majority.

Therefore, there has not been a large increase in corporate farming, as many had predicted. Old ways have not all been replaced by new materials and methods. There are not as many tractors as our textbooks proclaim. No matter how many times we hear educated people say that they have been eradicated, we still see bullocks pulling ploughs and carts and other 'old-fashioned' ways of village life. In Raigad, and in India in general, we see a greater persistence of small family farms and subsistence farming than expected (Venkatesh Athreya et al. 2017). Small family farms have persisted even though many peasants are being forced out of agriculture and being forced to find at least part-time jobs elsewhere.

7.2.1.8   Doing Science or Following Tradition?

When we analyse in detail the paddy cultivation that we observed, we find that the family was engaged in a process in which they were continually confronting conflicts, questioning, and investigating (Table 4.1). This actual process of learning paddy cultivation was very different from the 'traditional' ways of learning that were suggested by the superficial appearances (Table 2.1). The so-called 'traditional' paddy cultivation that we saw did not adhere to the characteristics that the other researchers listed in Table 2.1 had suggested.

For example, when the family was setting up the kanaga in order to store the paddy that they had produced for their own consumption, they carried on a continual process of questioning in order to figure out how to set it up. This was an integrated, collaborative effort in which all the people who were present took part in questioning, suggesting solutions, being creative, presenting and justifying arguments, and testing. This was not done just by thinking and discussing. It was part of doing the work itself.

Why was the family doing this process of questioning, investigating, and experimenting? They were doing it because they needed to do it. They needed a storage container. Their process of figuring out how to make it, making it, and figuring out how to use it, was guided by their own necessity of having to store their grain. They were doing it for their own use – for its use value.

Interestingly, in this case it was not the older generation that was teaching the younger generation. The elder was asking questions and the younger generation was playing the largest role in figuring out the answers and directing their elder, although it was still basically a cooperative process in which even the answers were questioned by the answerers.

Namdev's first big question – how to fill the kanaga, was actually a matter of physics, because the kanaga was just a cylinder with no bottom. It was not clear whether it had to be sealed at the bottom, or how a bottom could be fit to it. The problem and its resolution included emotional involvement. Namdev may have been afraid of going inside the kanaga, or maybe he just did not want to do what the others were telling him to do. He was also worried about how he would get out, or whether he might get buried in the paddy as it was filled.

Bhimraj and Smita explain that "Just as the rice would keep going down, down – at the same time, the person will keep coming up-".

The metaphorical significance of the situation is striking: Do peasants rise with the production of grain? Or get buried, despite – or because of – the amount of grain that is falling upon them? Of course, peasants should rise – they are producing grain to sustain themselves. But it is the convoluted necessity of the spiral of capitalist accumulation of profit that may cause peasants who produce grain to drown in their own labour!

But then, was it the elder who was buried, by the scientific temper of the next generation? No, as shown in Fig. 3.12, Namdev survived his burial, rising up as the grain poured down upon him! (We might metaphorically wonder whether he would have risen so easily if he had been buried in a container of grain that was to be sold for its exchange value rather than used as food by the family...)

As we saw, the family does do some experimentation on the farm. However, experimentation is limited by the economic risks that it entails. The family had tried cultivating both with and without rabbing (Sect. 3.2.5.7), but this was not done on purpose as an experiment to find out which method is better. Normally they cannot afford to not do rabbing, because it is risky. In order to compensate for the risk, they would have to apply artificial fertilisers – but they cannot afford to do that.

Due to lack of money, doing things in two different ways is always risky. Every change in methodology is a risk, and some options are more costly than others. Because of the cost and the risk, they really cannot afford to be very experimental. Whenever two different methods of cultivation are tried, one of them would almost definitely produce less yield than the other. It would usually be less risky to plant all fields the same way, as close as possible to a way that has already been done, in order to get better results, or to maximise profit if the crop is being sold. Therefore, two different methods are usually not done purposely in order to figure out which method is better. Besides, as we discussed, it is difficult and expensive to control variables in cultivation. It has been claimed that one difference between 'modern western science' and traditional ways, is that in the latter not many variables are controlled (Imogen Bellwood-Howard 2012).

If the crop is being produced for exchange value, profit must be the bottom line. The implications suggested by the results of experiments may not be heeded if they will not result in an increased profit. Therefore the value of doing experiments is limited. Extra time is required to do experimentation, and especially if a crop is being produced for exchange value, there may not be enough time for experimentation. Also, due to exhaustion, over-working, and lack of time, experimentation may be limited.

We saw that there are also serious dangers of experimenting. Fires could spread if the rabbing is not carefully controlled. Animals could die if they are not properly cared for. People can fall ill if they are overworked.

Thus, at times the family did experimentation by necessity, and/or by mistake, rather than by design or desire. Experimentation is expensive and especially risky for people who are very dependent on their crop for daily survival.

7.2.1.9 Science Is Done with Variable Sets of Interdependent Aspects

We have seen that the family in Rudravali did seem to be engaged in a process of questioning and investigating. However, this process was not a well-defined, orderly, step-by-step process of identifying questions, hypothesizing solutions, devising methods, testing them out, making observations, analysing the results, and making conclusions. They were not doing science, if that is how science is defined. But we doubt that science is ever so orderly – even though it is certainly useful to write summaries in scientific reports using such step-wise sequences. What we observed in the field is that the family engaged in these and other aspects in a much more complex manner, as they interacted with material realities. Through a complex network of continuous questioning, observing, trying out different methods, and figuring out how to solve new problems, the evolving process of paddy cultivation was simultaneously learned and developed by new generations.

Disregarding for a moment that we have found them to be part of dialectic relations, these aspects (some of which are listed in (1) of Table 4.1), are aspects of the 'Scientists' Toolbox' that scientists use to do science, as claimed by Dan Wivagg and Douglas Allchin (2002). The aspects they mention as being part of this 'toolbox' include: questioning, hypothesizing, predicting, testing, directly and indirectly observing, measuring, experimenting, manipulating, collecting data, analysing and evaluating data, reporting, communicating, reading literature, modelling, tinkering, playing, imagining, being creative, comparing, calculating, doing statistical analysis, etc.

As we have discussed, we have seen that many of these aspects are, at one time or another, part of the process of doing paddy cultivation. If we define the process of doing science as being variable processes consisting of interconnected networks of various aspects from this 'science toolbox' – with the number and order of aspects being variable rather than linear and fixed – then we see that the family is doing science as they do paddy cultivation.

Rather than trying to define a binary of processes which either are or are not science processes, we think it is more appropriate to talk qualitatively about processes as being more or less scientific, depending on how many of these aspects and which aspects are used, and the extent to which they are utilised or emphasized. If we do this for particular cases, we do not necessarily find that what professional scientists do is more scientific than what a farming family does.

The possibility that the family may be doing science arises because of our admittedly broad definition of science, which however does not include everything that people do or every way that people learn.

For example, do people decide when to sow seeds by consulting a horoscope, by asking an authority, and/or by analysing weather patterns and referring to the experience of previous years – or by some combination of all of the above? To what extent do people learn how to make rope through a process of trial and error or to what extent do they learn by obsessively copying some ritualistic or habitual gesticulation? Do people learn to plough by copying exactly what an authority from the older generation does, or do they learn by questioning and trying to figure out how and why it should be done?

We mentioned the conflict between science and religion (Sect. 7.2.1.2). In as much as religion involves having an unquestioned faith in some kind of authority, it conflicts with the questioning that is basic to doing science. However, we concede that not all religious activity involves having unquestioned faith. And professional scientists are not being scientific when they are blindly accepting beliefs on the basis of faith in 'scientific' authority. Of course, we are not saying that what someone is doing is necessarily better just because it is more scientific. We have to analyse why the 'science' or 'non-science' is being done, for whom it is being done, and what its effects will be.

Using our definition of science as a process – a network of aspects of asking questions and investigating – it is possible that even if people are not producing or using modern technology, and even if they are not writing a new 'body of knowledge', or e.g. understanding the complicated biochemistry that explains why plants are the way they are, they may be doing science as they do paddy cultivation. And/or, people may be doing science to a limited extent, or producing technology and knowledge to a limited extent. And/or, people may be doing science (and producing technology and knowledge), but it is not generally known or recognised as such by the dominant foreign research community.

Using this definition of science, it could be that the paddy cultivation we observe is quite 'traditional', but traditional is actually not traditional – it may be more 'scientific' and more dynamic than one might suspect.

But if so, again we come to the question of why science is limited and why science and agriculture have not progressed more than they have in India. We need to reconsider the definition of 'traditional' in relation to the development of paddy cultivation throughout history. Only then we see how farmers are very confined by economic and social circumstances, (e.g. caste, class, and gender). Because of spatial and temporal limitations, the spread and pace of technological change is reduced. However, the basic method which is used to develop materials and methods seems to be similar to what we call the method of doing science.

Still, you might argue that this is not science – it is just an everyday process that is different from the way professional scientists work, since it does not result in scientific development on a scale, extent, or rapidity that is characteristic of 'real' science. Our answer to this is that it may be true that it is not producing such visible technological results, but we still think it is worthwhile to call it science because the process is so similar. Using the term 'science' for what the farmers do, may give them some recognition and credit for what they are doing. And developing landraces and developing cultivation is no small accomplishment, when we look at the bigger picture of human development over all time.

Thus, we claim that the family was doing a sort of science. And they were doing it, and doing it the way they were doing it, due to the dialectical conflicts (Table 4.1) that they were confronting in the course of their work. In the following section we will discuss the family's opinions and recognition that what they were doing was science.

7.2.2 The Family Defines Science

Whether you believe that traditional paddy cultivation is or is not science depends on your definition of science.

When we asked Smita what she thinks science is, she said, "विज्ञान म्हणजे – आता-वेगवेगळ्या प्रकारचे शोध घेणे मग त्याला आपण विज्ञान बोलतो नाही का?" (Now, science means, to do different kinds of discoveries, then don't we call it science?).

We then asked her whether there is science in her own life. She replied, "म्म – माझ्या जीवनात विज्ञान म्हणजे – आता बघा हां, माझ्या जीवनामध्ये विज्ञान म्हणजे मग कसं होईल सांगते बघा. मी तर ह्यालाच विज्ञान म्हणेन, म्हणजे माझी प्रगती झाली, म्हणजे कशी झाली बघा, आता माझी दोन मुलं आहेत, ती शाळेत जायला लागली, आणि त्यांच्या नंतर – त्याच्या अगोदर मी काय करायची? चूल-मूल सांभाळूनच होती! आता मुलं शाळेत गेली आणि, थोडं-मोठी झाली ना मुलं, शिक्षण वाढलं, खर्च वाढलं, आता आमचं शेतीमध्ये – ही शेतीच अशी! नुस्कानी होते वैगरे. तेव्हा माझे – मिस्टर माझे आता कामाला जातात. त्या पैश्यावरती, किती, चारशे नी पाचशे रुपये रोज येऊन – आणि ते कधीतरी काम मिळतं. पंधरा दिवस भेटलं तर आठ दिवस घरातच बसतात, महिनाभर भेटलं तर आठवडाभर ते – महिनाभर ते घरातच बसतात. आता पाउस यावर्षी जास्त पडला, काम नाही, मग मी – ते काय म्हणाले, "चला ठीक आहे,

तुम्हाला थोडंसं, थोडंफार मदत व्हावं म्हणून मी थोडासा छोटासा व्यवसाय टाकला." माझ्यासाठी तर विज्ञान हाच होईल. असं. (Mmm – science in my life – now look, ok, I'll tell you how it is, how there is science in my life. I have made progress – this is what I call science, Now, how did it happen, just see. Now I have two children, they have started going to school, and after that – Before that – what did I used to do? I used to look after the hearth and home! Now the kids went to school and they got a little older, the kids got higher education, the expenses went up. Now we have the farm – this is the farm! There were losses. my Mr. [husband] goes to[construction] work. In that money, how much, four hundred or five hundred rupees daily – and that work is temporary. If there is work for fifteen days, then he sits home for eight days, if there is work for a month then for a week, he – he sits home for a month. The rains were too much this year, no work, then I – what did he say, "It is okay, You should have a little, a little bit of help. That is why we set up this small business." This itself is science for me.)

It is interesting that Smita did not first say that science is some kind of 'body of knowledge'. She said that it is discovery – which is a process or act of doing, not a physical or abstract thing or body. This is very much in line with what we noticed as the definition of science that was revealed by what the family was doing. They were doing science in the course of their everyday work and this science was a matter of questioning, investigating, and discovering.

It was not easy for us to understand Smita's definition of science. But she seems to be saying that in her life, doing science means changing, or making progress through discovery. And she seems to be considering science to be both social science and natural science. She associates science with discovery and technological and developmental progress. She gives the example of recognising a social/economic problem and figuring out how to solve it: changing from working around the house and taking care of the children to taking an active role in figuring out how to bring more money into the household. What she discovered was how to make some money from non-agricultural work by setting up a roadside shop to sell chips and chocolates. This was in addition to the money that Bhimraj started bringing in from construction jobs. The family did not just become mute observers, accepting the economic deprivation that they were confronted with, or resigning themselves to their destiny. Their conflicts forced them to question their position and this drove them to try to find various ways to earn more money. In this process, they were working against all sorts of forces (e.g. the heavier than usual rain) that made this doing of science difficult.

We were surprised to hear that Smita seemed to think that science may point to a solution to the problems that are due to the agricultural crisis, and that the solution is not just technocratic. Maybe science is not needed in order to invent new technology that will help the family out of the crisis. The technology already exists. The problem is that they cannot afford the technology. It is a social/economic problem. So the solution is social/economic – to figure out how to raise their income through non-agricultural labour and through engaging in petty trading. The solution Smita found was to try doing something other than agriculture. And she has questioned and investigated (done science) in order to find this solution. This science was done for its exchange value – in order to figure out how to survive the agrarian crisis.

Bhimraj was not present when Smita gave her definition. When he arrived, a little later, he gave a somewhat different answer to the same questions:

26:54	Karen:	विज्ञान क्या हैं? (What is science?)
26:56	Bhimraj:	[pointing to the paddy field] विज्ञान याने – ये, इसके, इसमें – (Science means – this, in this-)
27:00	Abhijit:	नाही, असं. तुम्हाला विज्ञान म्हणजे काय वाटतं? आता शाळेमध्ये बोलतात पोरं ना, आम्हाला विज्ञानाचा तास होता. मग तुम्हाला असं बोलल्यावर – तुम्हाला काय सुचतं त्याच्याबद्दल? (Not like this. What do you think is science? Now say in school, no, that we had hours of science. Then when said like this – what do you think of that?)
27:08	Bhimraj:	विज्ञानमध्ये कसंय, प्रगती जास्त होते. म्हणजे, की बाबा, ह्याची माहिती, किवा हे किती अंदाजाने होऊ शकतो – (In science, there is more progress. I mean, baba, knowledge. Or how can predictions be made -)
27:16	Smita:	[softly, from the back] असं नाही, असं नाही. (Not like this. Not like this.)
27:17	Bhimraj:	- की हे सगळं जमतो. विज्ञानाने. आणि त्याच्याने आपल्याला – आपल्याला पण करायला इच्छा होते. असं. (– that all this is possible because of science. And because of that – for us – even we want to do the same. It's like this.)
27:25	Abhijit:	शेतीच्या बाहेर पण विज्ञान आहे का? (Is there science outside of farming?)
27:27	Bhimraj:	आहे. (There is.)
27:28	Abhijit:	तुम्हाला वाटतं? (You think?)
27:29	Bhimraj:	हो. (Yes.)
27:30	Abhijit:	कुठे कुठे? (Where all?)
27:31	Bhimraj:	जिथे विज्ञान आहे तिथेच ते सगळं साध्य होतं ना? विज्ञान नसेल तर साध्य नाही. ... असंय ना? (Wherever there is science, this all is possible, right? If there was no science, this is not possible. Isn't it?)
27:40	Smita:	बरोबर मिळालं madam ना उत्तर? (Did madam get the correct answer?)
27:42	Abhijit:	काय? (What?)
27:43	Smita:	[laughs]
27:44	Abhijit:	बरोबर, चूक असं काही नसतं. (There is no right or wrong.)
27:45	Bhimraj:	नाही. (No.)

27:46	Smita:	नाही नाही नाही! माझं त्यांचं same झालं काय? मला मगाशी विचारलात ना? त्यांना आत्ता विचारलं. तेव्हा मी विचारलं. (No no no! Are mine and his [answers] same? [you] asked me earlier, [you] asked him now. That is why I asked.)
27:52	Abhijit:	ते आता madam ठरवतील – (That Madam will decide.)
27:53	Smita:	(laughs)
27:54	Abhijit:	-मी कोण ठरवणारा?! (Who am I to decide?!)
27:55	Karen:	नहीं, पर वैज्ञानिक अलग हैं? वैज्ञानिक अलग हैं या, या – (No, but is a scientist different? Is a scientist different or, or -)
28:00	Bhimraj:	हां! इसमें वैज्ञानिक अधिक हैं, उसके बाद में वो एक करना पड़ता हैं. (Yes! In this there is more science. After this there has to be one.)
28:07	Dnyaneshwar:	नाही. नाही. (no. no.)
28:07	Smita:	अलग आहे का विचारतात ते. (Is it different, is what they are asking.)
28:08	Bhimraj:	हां मग बरोबर आहे, अलगच आहे. (Yes then it IS correct. It IS different!)
28:09	Smita:	ऐका ना, ऐका ना. नीट ऐकत नाही! (Listen nah, listen nah. [you] do not listen carefully!)
28:12	Bhimraj:	आता वैज्ञानिक जो शंशोधन केलेला आहे – (Now the research a scientist has done -)
28:14	Smita:	[whispers] रेकॉर्डिंग चालू आहे. (Recording is going on.)
28:14	Bhimraj:	-त्याच्यानंतरच आपल्याला सगळं समजलं जातं ना? – (–after that only we understand everything, no?-)
28:18	Smita:	[whispers] ते चालू आहे. (It is going on.)
28:19	Bhimraj:	-असं. नाहीतर आपल्याला ह्याची माहिती नसेल तर आपण कसं करणार? (–this way. Otherwise if we do not know about this, what will we do?!)
28:23	Abhijit:	म्हणजे हे तुम्हाला वैज्ञानिकांकडून समजलं? (Which means you understood this from the scientists?)
28:25	Bhimraj:	नाही. नाही. (No. no.)
28:26	Smita:	नाही. नाही. हे आधी जाणून आहे ना! (No. no. this is known beforehand nah!)
28:27	Bhimraj:	हे आमचं-आमचं आजोबा – हे आहेत ना – (This – our – our grandfather – he is there nah -)
28:29	Dnyaneshwar:	Practically हे- (Practically this-)
28:30	Bhimraj:	-पहिलं आजोबांनी हे सगळं प्रदर्शित केलेलं आहे. त्यानंतर आता आम्ही करतोय. आता ते आजोबांनी केलंच नसतं, आम्हाला माहितीच त्यातली नसती, तर आम्ही केलं असतं? (– first grandfather has done this. After that we are doing. Now if this was not done by grandfather, we had no knowledge of it, then would we have done it?)

28:40	Abhijit:	नाही, मग madam विचारत आहेत, की मग, तुम्ही – तुमच्यामध्ये आणि वैज्ञानिकांमध्ये फरक आहे का? (No. then madam is asking, that if, you – is there a difference between you and the scientists?)
28:47	Bhimraj:	हां – ते – नाही – ते- त्यांचं पहिलं हे आहे ना, त्यानंतरच आपण पाठीमागे आहोत. (Yeah-that-no-they-they have done it first, after that only we have followed them.)
29:20	Abhijit:	(to Dnyaneshwar) तुम्हाला काय वाटतं? (What do you think?)
29:22	Bhimraj:	नाही त्याला तेवढा जास्त अनुभव नाही. (No, he does not have much of experience.)
29:25	Smita:	नाही नाही सांगतो तो. ए madam हे करतील बरोबर. – (No no, he will tell. Madam will correct it -)
29:27	Bhimraj:	(to Dnyaneshwar) वैज्ञानिक म्हणजे काय? (What is a scientist?)
29:28	Smita:	- आणि मग ते तिघांचं एकत्र करतील. (– and then they will combine all the three.)
29:30	Abhijit:	म्हणजे तुला विज्ञान म्हटलं की तुम्हाला काय सुचतं? (Means what do you think when someone says science?)
29:34	Dnyaneshwar:	विज्ञान म्हणजे काय? (What is science?) ...
29:50	Dnyaneshwar:	विज्ञान म्हणजे सामान्य ज्ञान. म्हणजे जे इकडे practically असेल ना – (Science means general knowledge. Meaning whatever practically is here, nah -)
30:00	Bhimraj:	म्हणजे पहिलंच जे- (Means first whatever -)
30:01	Dnyaneshwar:	- म्हणजे जे काही आपण असं – आता हे बीज टाकलेलं आहे, त्याच्यापास्न उगवणार, म्हणजे धान उगवणार, आता त्याच्यापास्न जे रोपटं तयार होणार, रोपटापास्न परत ते – त्याला फळ वैगरे येणार. (Means whatever we – now this is seed is sown, from it there will be growth, meaning the crop will grow, now from that whatever plant will grow, from it another – it will bear fruits.)
30:15	Abhijit:	ते विज्ञान? (That is science?)
30:16	Dnyaneshwar:	हम्म. (Hmm.)
30:17	Abhijit:	त्याच्या बाहेर विज्ञान आहे का? (Is there science outside of it?)
30:18	Dnyaneshwar:	आहे. आहे. वेगवेगळ्या ह्याच्यात विज्ञान आहेच ना? (Yes. There is. In different things, there indeed is science, no?)

It's very interesting that in the beginning of this conversation, the first thing that Bhimraj thought of when defining science (before we asked him about science in his own life) was with reference to the paddy fields we were standing in. However, Abhijit thought that this was not an appropriate answer, so he tried to guide Bhimraj towards defining a school kind of science, which is what first came to his mind. But Bhimraj did not entirely give up on the way he had begun to answer, and he still mentioned the application to their own lives on the farm. He said that science is knowledge, through which predictions are made and through which progress is made. When he said that even they want to do so, he might have meant that even they want to do science, or that even they want to do their farming the way science makes possible, so that that they also progress. When he said (at 27:31) that if there was no science, this [farming] would not be possible, he might be thinking that even their farming is based on science. Smita had not mentioned their own farming as being scientific, and on this point she seems to disagree with Bhimraj.

Dnyaneshwar defined science as being the process of how plants grow and reproduce. This is similar to another popular conception we have often come across: that science is nature itself,[2] or the natural laws that (supposedly) govern nature. This conception almost seems to place traditional paddy cultivation as being closer to science than some other definitions allow.

After Smita complained that Bhimraj was not giving the correct answer, Bhimraj said that it is only after scientists do research that we understand things. He is implying that science means not only the existence and description of nature, as in Dnyaneshwar's definition, but also understanding how and why things happen. This is the definition that we have heard from many farmers: that scientists understand the 'why' of things. In this case, science may again assume a more privileged position amongst the highly educated. (Ironically, some scientists claim that science can only deal with 'how', not 'why', and a domain such as religion deals with 'why', as we will discuss in Sect. 7.2.8). Although at first Bhimraj seems to be implying that they do understand, and they do science, afterwards (at 28:30) he says that what they know comes from their grandfather, and they only follow what their grandfather says (which we have shown is not actually true (Sects. 3.2.4 and 4.2.1). Previously also, Bhimraj had told us that they are not scientists because they do not understand the 'why.' However, with regard to many aspects of paddy cultivation, he did explain the 'why', as we have discussed (Sects. 3.2.1 and 3.2.2).

In associating science with modern development, Smita and Bhimraj may also be thinking that science is advanced technology, and what makes more modern life possible. Many farmers that we have talked to say that science is artificial fertilisers, pesticides, tractors, combine harvesters, and HYV seeds. Clearly, this technology is rather different from the technology used by the family, or by subsistence farmers. Using this definition of science, the paddy cultivation we observed is hardly science. Bhimraj seems to be agreeing when Smita (at 28:26) said that they have not learned from scientists (although, her statement was in response to a rhetorical, leading question from Abhijit).

As Bhimraj was explaining his definition, Smita (who was standing to the side) murmured her disagreement. Maybe she was thinking that he should have mentioned discovery rather than knowledge, although actually his definition was not

---

[2] Interestingly, some people define God as being nature. In Hindi and Urdu, the word, कुदरत (kudrat) means God/nature, or the Almighty. So then, does science mean God?

really very different from the one she had given earlier, before Bhimraj had come. Maybe Smita was taking our questioning as the sort of questioning that teachers do at school. She seemed to be trying to get Bhimraj to give the correct answer – meaning, what the teacher (Karen) thought was the correct answer – or maybe her own answer. She seemed to think that if he listened properly to Karen, he would tell the correct answer. She wanted Bhimraj to sound good in the recording (see 28:07–28:18). On the other hand, Smita was laughing, maybe thinking the questioning and answering is amusing, and maybe also doubting whether there is just one correct answer. She seemed to be pleased with her own answer. However, later she said that Karen will combine all the three people's answers – and that is actually what we are doing here!

Most of the science we saw the family do was being done for its use value – to produce food for their own use. They were not doing science in order to develop some technology or commodity that is to be bought and sold, and therefore valued for its exchange value. For example, the science that they were doing as they designed and set up the kanaga was not in order to produce a kanaga for sale. It was to be used for storing the paddy that they eat. Their experimentation in rabbing, development of seed selection, and Bhimraj's deviation from his father's methods of ploughing were not done just in order to increase profits – they were done mainly to produce a type of paddy that had use value as their own food. Their methods of learning by doing may have been less efficient than if they had just followed the methods given by some authority. Their rigorous observing was more scientific than casual glancing, and may have required extra time and energy, but it was done despite any loss in productivity that it may have entailed. All the doing of science that was entailed in making a basket was done despite the fact that a machine could have produced a basket more quickly. Thus, in as much as it was done for their own use rather than for profit, their science was very different from capitalist science.

But the science that Smita and Bhimraj refer to when they define science as being about progress, modern technology, and setting up the roadside shop, may imply a more capitalist kind of science that is associated with 'economic development' and exchange value. This kind of science may not involve them in doing so many of the aspects of science (experimenting, observing, etc) that we observed them doing in the process of paddy cultivation. Many of these aspects require time, resources, and expense, which have to be weighed against the profit they will get in the end. Doing more science would increase the already enormous risks they face. The work needed to do petty trading requires less physical involvement and less freedom and creativity. It is in this sense a more alienated kind of work. The products they sell are mass produced and pre-packaged by multinational as well as Indian companies, who also produce the advertisements, including a signboard for the shop that displays the MNC brand name. The Maximum Retail Prices (MRP) are already printed on the packages. The packaging and the process of production and transport of the products will pollute the environment. The products for sale may not be healthy, and they may not be what people really need. Their use value is not as important as their exchange value. And most of the profit will end up with the MNCs, not with the family. The family will not have a role in deciding all these things. They may try to sell their own products, but they would probably not be able to compete with the large companies.

When the family uses modern technology (seeds, fertilisers, machines), the kind of science that they do is reduced and is more concerned with exchange value. They become alienated from science. They do not do science to create, design, or produce these technologies. Their science is restricted to testing them in the field. But this testing is limited because of the risks and the need to worry about the exchange value of what they will sell in the end. And selling will be necessary in order to try to recoup the increased expenses involved in incorporating modern technology.

The adoption of modern technology involves a decrease in the family's learning, developing, and doing science. It also requires less skill and less hands-on work. For example, the neighbours paid for someone to come and plough their field with a tractor rather than ploughing with a pair of bullocks. The tractor required less skill, less time, and less involvement for both the driver and the family who paid for it. It takes the family months to train a new pair of bullocks to pull a cart, and much more training is needed before the bullocks can pull a plough. Two bullocks need to learn to work together with each other and with the person who is guiding them, using reins, a stick, and their voice. The person also needs to learn how to teach them, and needs to understand a lot about cattle. And this is in addition to the skill of holding the plough with the correct pressure and angle, moving at the correct speed, leaving the correct space between rows, and figuring out how to turn sharp corners. Learning to use a tractor is orders of magnitude easier and faster, since many aspects are mechanised, taking many aspects of control out of the driver's hands. For example, the distance between rows is set by the distance between the row of discs on the plough. The depth is automatically the same for each one. By comparison to a person ploughing with bullocks, a person driving a tractor is very relaxed (if not bored), sitting on a seat, hardly moving, with very few decisions to make. Ploughing with a tractor can be learned in a few days.

Using artificial fertilizer instead of rabbing also requires less skill, time, and involvement. We saw how the days of work – collecting material, spreading it out in layers and burning the rab – can be replaced by one person walking around the field for a few minutes, tossing out handfuls of urea.

In the more 'modern' methods, there are fewer minute-to-minute decisions to be made, and fewer degrees of freedom. Less planning is required. Less designing is involved. Less creativity is possible. The planning, designing, and creating of the methodology has already been worked out by the company that produced the machines and materials. Because there are fewer degrees of freedom, there is less experimenting to do. The more technically advanced procedures are more uniform. All this also brings about more alienation between the farmers and the land. The farmers' actions become more mechanical. A farmer becomes more of an appendage to a machine.

According to a definition of science as the process of doing science, their science may be decreasing with their 'economic development'. The science is instead done by scientists who invent and design the tractor, the plastic basket, or the chemical fertilisers. Ironically, even the relative economic status of the farming family may be decreasing with their economic development, since they are becoming deprived of increasing percentages of the surplus value they are producing.

We have discussed how many intellectuals consider science to have begun with the advent of capitalism in Europe and intensified during their industrial revolution (Sect. 6.3.2). According to this definition, to the extent that the farming that we observed on the farm is pre-capitalist traditional farming, it is not scientific.

This definition is related to the popular conception that science is the end result of the processes of production of 'knowledge': the knowledge of the world, categories, nomenclature and definitions. Students that we meet, or people who have gone through a number of years of formal education, often seem to think of science as being this sort of 'body of knowledge'. It is the lists of facts they are asked to memorise in order to pass exams. Bhimraj, and especially Dnyaneshwar sometimes gave this sort of definition in addition to their other definitions.

But what do we find if we look closely at what is taken to be the 'body of knowledge'?

Those of us who are researchers speak from our own experience. As students, initially we did think there was a fixed body of knowledge, which seemed to be what was written in books and known by authorities. We wished our teachers could just transmit all this knowledge to us. We wished we could somehow know it all – and it seemed that we could if we just worked hard enough (and if we had acquired enough brains from our ancestors). But as we went on to higher education, instead of getting easier, the task seemed to become more and more impossible. What we thought were simple 'facts' seemed to be becoming much more complex and questionable. And as went on to do research in science, looking into one specialised field in even more depth, we realised that so many things in that field raise new questions that never occurred to us before. Of course, many things are not worth questioning, or spending time investigating, but it is not very clear how to tell which questions are worth asking. Ironically, one aspect of doing science that does not usually receive enough explicit attention in school or even in professional capitalist science is the nature of questioning and how and why particular questions are asked and other questions are not asked. How do we choose? The interdependence between social factors and choice of questions is not sufficiently discussed or explicitly investigated. We were expected to just do research on questions we were assigned, or on questions closely related to what others were doing. It was only in writing grant applications or introductions to research papers that such meta-questions might arise, and even as PhD students, we were often not very involved in these things.

On the farm, we found that the reasons for asking particular questions and the interdependencies between questions and social factors were more obvious and more explicit. Questions were asked as part of the process of doing cultivation. It was often obvious that the questions were important in order to do the work that was needed in order to produce rice – and obviously the family needed rice to eat. For example, they asked, 'Where is the baila?', 'What is the difference between this crop and last year's crop?', 'Does this rice taste different from that rice?', 'Why should we plough this field less deep or more deep?' Clearly, these questions were necessary, and directly related to the social nature of their work: who should do which work, how good their rice would taste, how hard and tiring the work would be, etc. They also asked questions such as, 'How much is this paddy worth?' and 'Why does the fertiliser cost so much?' and 'Should our children find jobs outside of agriculture?'. These questions, were also clearly related to social aspects, and it was even more complex to find their answers.

A number of farmers we have met have said that science explains 'the why' of how things are. And they say that they are not scientists because they do not understand 'the why'. But maybe they do not realise that no kind of scientist fully understands 'the why'. It is only in the way science is presented in school that makes it seems to be a fixed body of knowledge full of all the correct answers. And as we discussed, farmers do actually understand, or try to understand, quite a bit about 'the why' of things. But in so doing, just as in any doing of science, they also come to see that there are many 'whys' that they do not understand – i.e. that they have many questions.

Actually, there is no fixed, unchanging 'body of knowledge' that can supply answers. Continuous asking and investigating is needed. The fixed 'body of knowledge' is a chimera. It is actually so dynamic and fuzzy that it is hard to even see a 'body' at any one point in time. The location of the baila keeps varying. Each year the paddy crop is different from the last year. Taste preferences depend on personal/social factors. The characteristics of the soil are different at different times of day and in different months and years. New kinds of pests keep arising. The economic situation keeps changing.

What did we find when we looked for the 'body of knowledge' on the farm in Rudravali? We expected to find a rigid, well-defined conception of what a rab is and how it is prepared. But we found that it depends on availability of materials, how much time the family had, and who was doing each thing. We expected to find one simple, straight-forward method of ploughing with bullocks. But we found that even the father and son were ploughing differently. We expected to find that the family was using the same, well-defined seeds that their ancestors had used for ages. But we found that they were using various different seeds, and they themselves were defining and redefining the landrace in the process of selecting the seeds each season. We expected to find the family using the same methods for storing paddy that they and their ancestors had always used. But we found that they were constructing a new kind of storage container, working out the design and method of filling it as they went along.

Whether we consider what is happening in our own experience of mainstream science or on the farm in Rudravali, we do not see a 'body of knowledge'. What we actually see is not a thing, but a process. Both the 'body of knowledge' produced by 'modern western scientists' and the 'body of knowledge' of 'traditional' cultivators are actually very dynamic processes. And we do not see how they can be considered to be separate from the process of 'constructing knowledge' or doing science.

7.2.3   Truth and the Objectivity of Science

Not having a fixed, unchanging body of knowledge does not mean that there is no absolute truth, as some postmodern relativists claim. The baila was, after all, in a particular place. A member of the family did decide that they really do prefer the taste of one kind of rice over another for a particular dish. It may really be more likely that the clods of soil would be larger and more difficult to break if the plough-ing was deeper. The amount of money the family got for their paddy actually was Rs 35,000, not Rs 70,000. These were truths at some point in time.

However, the truth may not be easy to find – and the truthfulness of our beliefs and understandings is hard to judge. It may be hard to tell how deeply the ploughing was done and whether the clods are larger than last time. It is very difficult to find out why fertiliser costs what it costs. Sometimes even the baila may be hard to find. And in many cases, we will only find probabilities, not yes or no answers to our questions.

According to our framework, the truthfulness of the answers lies in physical real-ity, which is nature. And nature is actually not a collection of things but the process of interdependent aspects of nature (including society) interacting with each other. We interact with physical reality in order to ask and answer questions. For example, physical reality is the changing position of the baila, which we move and ascertain through our processes of observing. Physical reality is our process of eating and tasting rice in our mouths, which we evaluate by comparing the experience to past experiences and past associations. Physical reality is the changing expression on our mother's face, as we wonder whether it is a smile or a frown or indecipherable. Physical reality is the size of the clods of soil that we estimate by observing, holding in our hands, feeling their weight, manipulating, and breaking.

According to our framework, science questions are based on physical reality. There is a physical reality that we are observing, whether directly or indirectly, and with whatever distortions. Doing science also includes making – and questioning – abstractions: concepts, theories, hypotheses, etc. But the abstractions relate, directly or indirectly, to physical reality. Therefore, when we do science, we act as if physi-cal reality does exist. The act of doing science implies that we actually believe physical reality does exist – otherwise we need not act (e.g. collectively cultivate, cook, and eat).

According to our framework, we do not think that physical realty is a figment of our imagination. This is what we mean when we say that there is an objective real-ity. However, we do not definitely know or understand much about what that reality is. There are limitless questions about physical reality, partly because nature itself keeps changing. Therefore, all questions will never be asked and all answers will never be found, although we do often see that there is some progression in our understanding.

We do not believe that the process of doing science can be – or should be – objec-tive. Different people (and even one person) will have different subjective observa-tions of objective reality at different times and places, due to complex interdependencies that are inherent to nature/society. However hard scientists may try to be objective, there will always be some elements of subjectivity in their research. Rather than just trying to overcome (or ignore) all biases, we should try to recognise our biases and points of view, and analyse how they affect and are affected by the doing of science. We should try to avoid biases that we find oppressive, and try to do science for the oppressed.

It is not possible to do an objective agricultural science that will be good for 'all stakeholders', including farmers, landless labourers, traders, consumers, and corporations. Their different objectives and goals are contradictory. The needs of the corporation are inevitably contradictory to the needs of the cultivator because the corporation needs to extract profit from the labour of the cultivator. In a capitalist system, class conflict is inevitable.

We need to take sides. For example, the manufacturer of a pesticide will not want to find that the pesticide they have developed is unsafe, because that will reduce their profit margin. Therefore, the development of pesticides should not be done by private enterprise. Pesticides should be developed for their use value, not for their exchange value. The research should be done by and for the farming family – the people who need to use pesticides and eat the crops they produce. We need to do research for people, not for profit.

To act for the needs of the farming family, we may need to do double-blind studies of the pesticide's efficacy and safety, if this will be more apt to protect their safety. But we do not need to do a diversity of studies, just for the sake of diversity, including studies that are biased in favour of the corporation. For example, we do not need to do studies to find out how to manipulate farmers to buy seeds that they do not really need or want. Neither do we need to include diverse studies that, for example, use the position of the stars in order to declare whether a pesticide is effective. We need to try to figure out which are the most worthwhile questions and kinds of investigations – from the point of view of the needs of the farming family. We need to question answers that are just based on accepting what some kind of authority says – whether the authority is an astrologer, a CEO, a religious pundit, a Nobel prize winning scientist, or an experienced farmer.

We do not need to 'objectively' consider the point of view of the oppressor, the tyrant, the slaveholder, or the richest 1% in order to decide that we need to take sides with the oppressed. We need to try to understand the history and basis for the oppressive system, and the reasons why oppressors oppress, but we can do this without denying that we have a point of view against oppression.

We have chosen to side with the oppressed – with the farmers who are facing the brunt of the agricultural crisis, rather than with the corporations, who must keep profit as their bottom line. We advocate explicitly doing science for oppressed farmers rather than just trying to be objective and unbiased. The stated effort to do objective and unbiased science is anyway a façade for doing mainstream science that is actually for the advance of capital, for national interests, or for military power. Doing science for these things means doing science against the poor, against international interests, and against world peace.

Our stand can be compared to MK Gandhi's famous talisman, in which he suggested what to do when you are in doubt or when "the self becomes too much with you". He said:

> Recall the face of the poorest and the weakest man (sic) whom you may have seen and ask yourself, if the step you contemplate is going to be of any use to him. Will he gain anything by it? Will it restore him to a control over his own life and destiny? In other words, will it lead to Swaraj for the hungry and spiritually starving millions? (Savita Sinha and Mohammad Husain 1994)

This talisman used to be printed at the beginning of every NCERT textbook. It is interesting that Gandhi suggested finding truth by referring to the physical reality of people's poverty, given that he was usually very much an idealist, believing that "morality is the basis of things, and that truth is the substance of all morality" (Mohandas K Gandhi 1927, 29). He seemed to believe in the existence of an absolute truth – what is morally correct for one and all. But in this talisman, rather than saying that one can find this truth by listening to one's 'inner voice' (as he did sometimes say), he is acknowledging that one's point of view may obscure the truth. For that matter, he also wrote that his definition of truth has been "ever widening" (ibid). He was concerned with the process of searching for and experimenting with truth, which, he said, was never ending. So perhaps his search for truth was in some sense similar to the process of doing science, even though it was predominantly a deeply religious search.

Our stand on objectivity in science is similar to what Howard Zinn wrote about objectivity with regard to social sciences, and history in particular:

> [The historian cannot] ... avoid emphasis of some facts and not of others. This is as natural to him (sic) as to the mapmaker, who, in order to produce a usable drawing for practical purposes, must first flatten and distort the shape of the earth, then choose out of the bewildering mass of geographic information those things needed for the purpose of this or that particular map.
>
> My argument cannot be against selection, simplification, emphasis, which are inevitable for both cartographers and historians. ... The historian's distortion is more than technical, it is ideological; it is released into a world of contending interests, where any chosen emphasis supports (whether the historian means to or not) some kind of interest, whether economic or political or racial or national or sexual.
>
> Furthermore, this ideological interest is not always expressed in the way a mapmaker's technical interest is obvious. ("This is a Mercator projection for long-range navigation – for short range you'd better use a different projection"). No, it is presented as if all readers of history had a common interest which historians serve to the best of their ability. This is not intentional deception; the historian has been trained in a society in which education and knowledge are put forward as technical problems of excellence and not as tools for contending social classes, races, nations.
>
> ... My viewpoint, in telling the history of the United States, is different: that we must not accept the memory of states as our own. Nations are not communities and never have been. The history of any country, presented as the history of a family, conceals fierce conflicts of interest (sometimes exploding, most often repressed) between conquerors and conquered, masters and slaves, capitalists and workers, dominators and dominated in race and sex. And in such a world of conflict, a world of victims and executioners, it is the job of thinking people, as Albert Camus suggested, not to be on the side of the executioners." (Howard Zinn 2002)

What Howard Zinn says about historians, cartographers, and social sciences, we can also say about the natural sciences and scientists in general. Science research necessarily focuses on some questions and ignores other questions. We need to ask why and for whom science is being done in order to understand the nature of science in different times and places. Capitalist science that is presented as if it is done for the common interest of all people, as if it is the universal body of knowledge, or the science of an entire nation or society, actually conceals fierce conflicts of interest.

7.2.4 Empiricism and Spirituality in Science

It is not uncommon to find intellectuals in India who take science to be a way of reasoning, deemphasizing the connection to work with the hands. This may be connected to their caste prejudices that take reasoning to be of a higher status than manual labour. Reasoning is also considered to be more modern than traditional work with the hands.

However, multiple ways of defining science are often entertained. For example, the Kothari Commission (Education Commission 1966) seemed to have an epistemology with an interesting mix of science and spirituality. They recommended that the country should move towards a new "age of science and spirituality" that involved moving farmers out of their "age-long conservatism through a science-based education" in which they become interested in experimentation. They say: "In a traditional society, production is based largely on empirical processes and experience, on trial and error, rather than on science; in a modern society. it is basically rooted in science." As Krishna Kumar (1996) pointed out, it was interesting that they did not see the empiricism of traditional trial and error to be science.

Perhaps the authors of the report were considering science to be essentially a 'body of knowledge', which is based on a system of logical reasoning and formulation of theories and laws that explain physical reality. In this view, the 'body of knowledge' is not known to most (uneducated) people who contribute to production: science is done by professional scientists. Even so, it is strange that the authors apparently do not see the role of empiricism in the formation of the 'body of knowledge'. Perhaps they also think that 'basic' science is separate from, and must precede the application of science to the development of technology.

Their conception contrasts with the Weberian claim that in India,

> 'knowledge' is not a rational implement of empirical science such as made possible the rational domination of nature and man as in the Occident. Rather it is the means of mystical and magical domination over the self and the world gnosis. It is attained by an intensive training of body and spirit, either through asceticism or, and as a rule, through strict, methodologically-ruled meditation. (Max Weber 1958, 331)

Whereas Weber sees the limited development of science as due to a lack of empiricism and rational experimentation, the Kothari Commission authors explain the lack of science by the presence of empiricism. We think that Weber fails to observe the actual extent of empirical science. On the other hand, the Kothari authors fail to recognise the importance of empiricism in science – perhaps because they actually see science as being 'facts', concepts, or theory based on logical reasoning rather than on empiricism. This is the sort of knowledge that Weber implies is not science.

Of course, it is short-sighted to think that science is based only on sense perception, or empiricism, or logical empiricism, or deductive reasoning. It is short-sighted to think that one or the other of the aspects of doing science comes first, and the others follow one at a time. As we have discussed, we should not deny the interdependencies between empirical observations, inductive reasoning, and social/natural realities/subjectivities.

7.2.5   What Is Scientific Observation?

Some scholars differentiate between 'everyday observation' and 'scientific observation'. But we question this terminology, if not the entire distinction. We wonder if it's possible to do 'scientific observation' every day, in the course of one's everyday activities. Actually, we would like to encourage the spread of scientific temper, which means doing science every day, throughout our daily activities (see Sect. 7.2.8). In any case, there is no binary between observations that are either everyday or scientific – observations may be more or less scientific.

Some people have claimed that in order to observe as scientists observe, we need to have a considerable amount of 'content knowledge', and we need to be told what to observe (Catherine Eberbach and Kevin Crowley 2009).

How do scientists observe? And what is the role of observation in their research? In discussing how she does her research on maize, the scientist Barbara McClintock explained:

> No two plants are exactly alike. They're all different, and as a consequence, you have to know that difference, ... I start with the seedling, and I don't want to leave it. I don't feel I really know the story if I don't watch the plant all the way along. So I know every plant in the field. I know them intimately, and I find it a great pleasure to know them. (Evelyn Fox Keller 1983)

She said that for each plant, she has to understand

> how it grows, understand its parts, understand when something is going wrong with it. [A plant] isn't just a piece of plastic, it's something that is constantly being affected by the environment, constantly showing attributes or disabilities in its growth. You have to be aware of all of that.... You need to know those plants well enough so that if anything changes, ... you [can] look at the plant and right away you know what this damage you see is from- something that scraped across it or something that bit it or something that the wind did. (ibid.)

In other words, in order to understand a plant, she has to have some historical background knowledge of the plant. But this knowledge does not come from reading about the plant – it comes from observing it intimately and rigorously over its lifetime. Evelyn Fox Keller wrote that in order to observe, Barbara McClintock said that she had to have time and patience to "hear what the material has to say to you," and the openness to "let it come to you." (Evelyn Fox Keller 1983).

Understanding is interconnected with observation, but how can we say how much we need to understand in order to observe? Or how much we need to observe in order to understand? Observing and understanding are interdependent, continuous, ongoing processes. Knowledge or understanding does not come just from reading books and listening to authorities speak. "In science, the stuff is the authority" (Eleanor Duckworth 2012), and the stuff needs to be observed in order to understand.

In summary, we claim that the process of scientifically observing is a dialectical process since it is inherently characterised by opposing or contradictory aspects. The conflict between the following dialectically opposing aspects is what gives rise to scientific observation:

(a) Questioning (not knowing) /Answering (knowing)
(b) Expecting the hypothesized / Allowing for the unexpected
(c) Necessity / Chance, opportunity, having something to observe
(d) Needing / Wanting, caring about the plants, animals, and environment
(e) Specific focus on certain aspects / Consideration of the whole, and interdependencies with the environment
(f) Freedom from fear, indifference, shrewdness, objectivity / Caring, respect, subjectivity
(g) Identifying, naming, labelling, comparing / Questioning of identities, names, and labels, recategorizing, creating names
(h) Qualitative analysis / Quantitative measurement
(i) Accuracy and precision / Approximation and generalisation (using appropriate resolution)
(j) Experience (knowing) / Naiveté (not knowing)

We claim that these characteristics are what make observing more scientific.[3] More scientific observing will have more of these characteristics and/or they will be more prominent. Observation is essential to doing science. Scientific observing is not done independently from doing other aspects of science such as hypothesising, comparing, testing, identifying, categorising, etc. There is no particular order in

---

[3] Observing that is less scientific is observing that does not arise from so many of these dialectic relations, or in which they are less significant. For example, observing is less scientific if it occurs with less questioning/answering, or if it occurs with quantitative but not qualitative measurement, with wanting, but without needing, or with needing but without wanting.

A relevant example of relatively less scientific observation these days is watching TV or videos, scrolling through photographs on Facebook, Instagram, or Whatsapp, or taking photographs, when it is done relatively mindlessly, without much care, without much questioning or answering, without asking why click in this direction at this moment, and without much need or desire (except for the need to escape reality). These observations may include clicking (or not clicking) on 'like' without much thought (which is hardly even quantitative, since the range of response is so limited). Of course, more scientific observation of the same videos or photographs is also possible, and does also occur at times.

which these aspects are done – they form interdependent networks rather than one series of sequential steps.

Based on the evidence we discussed (Sect. 3.2.7), we claim that the farmers were observing quite scientifically.

7.2.6 Capitalist Science vs the Science We Saw on the Farm

Science is actually never done by individuals who are always free to follow whatever whim comes to their mind. And neither should it be. Professional scientists also have social responsibility, which as JD Bernal pointed out, became especially apparent after scientists in the USA made atom bombs that killed thousands of civilians when they were dropped on Hiroshima and Nagasaki in 1945 (JD Bernal 1954, 1). The bombs were not made by individuals, but each of the individuals was responsible as being part of the process of making the bombs. Any kind of science is always a social process that depends on its past history, involving many people. However, modern 'western' science appears to be individualised. It is organised in a hierarchal structure in which individual leaders get awards, credit, fame, money, and power.

Under capitalism, professional scientists have characteristic compulsions. Formal science is usually restricted to a small elite working in separate, highly specialised disciplines. Formal science is done by professional scientists who are getting paid salaries. The science is done for its exchange value. They are selling their labour and doing science for its exchange value. Even if they are doing 'basic research' in a government institution, they need to justify their research based on its importance and possible long-term applications to the development of technology. If they work in agricultural institutions or companies, their work may be much more closely bound and shackled by the underlying profit motive.

Professional scientists often claim or act as if they should be objective and purposely try to avoid including social/political aspects in their studies. We have already discussed how we think objectivity is actually not possible (Sect. 7.2.3). But the small farming family does not even attempt to claim this or work like this. For example, when analysing the maturity of a paddy crop, they need to consider all sorts of aspects: whether the grains are ripe, whether some plants are less mature than others, whether it will rain in the next few days, whether their health and fitness will hold up for the coming days of extreme hark work, whether they can afford to hire labourers, etc. The question of whether the harvesting of the crop should begin does not have one objectively correct answer. The answer depends on the point of view of the harvesters and on many subjective factors. This is the case with many of the crucial questions that arose during cultivation. And as Bhimraj said in deciding how to do the threshing, they need to consider that more energy is needed to operate the threshing machine that needs manual pedalling, but another kind of energy – electricity – is required for the other machine, and electricity is expensive and is not always available. Their economic and social constraints are explicit. Therefore, they cannot work without considering social/political aspects: the relative costs of alternative methods, their own physical and mental strength and exhaustion, their interactions with co-workers who are also family members, their emotions when interacting with the buyers, relations of caste, class, gender, etc. According to Imogen Bellwood-Howard, who has studied farmers in Ghana, "Farmers exist in a world without the disciplinary divisions scientists and students are used to. ... Financial, demographic and biological factors interact, ..." (Imogen Bellwood-Howard 2012).

The family was actually doing some very important experiments by necessity. Probably the most important experiment was done as they selected seeds for the next season from each year's harvest (Sect. 4.2.3.1). Based on their observations and past experience they had to make a hypothesis as to which seeds will be best for planting. The hypothesis was tested by cultivating the seeds and observing the resulting harvest and comparing it to past harvests. The next round of selection might be modified, depending on the results. The only alternative to this experiment is to purchase seeds instead, which was to be avoided due to the cost.

Thus, we saw that seed selection requires cultivators to engage in a process which can be described as a network of interconnected and interdependent questioning, observing, hypothesizing, analysing, comparing, inventing of technique, experimenting, etc. Probably in the ancient past, just as we saw in the present, cultivators needed to keep asking and investigating questions such as, 'What is the difference between this crop and last year's crop?', "Does this rice taste different from that rice?', 'What will happen if we plough less deep?', 'Are these plants producing more tillers because of the type of seeds or because they got more rain after flowering?' In other words, we claim that paddy cultivation requires people to do science.

We were struck by the importance of direct observing in doing all aspects of paddy cultivation, as we have discussed (Sect. 3.2.7). We showed how scientific observation was quite frequent, in the sense that it arose from a number of different kinds of dialectical conflicts (Sect. 7.2.5). It is intimately connected with questioning, which also arises from dialectical conflicts (Sect. 3.2.5.8). Conflicting observations, or conflicts between observations and beliefs are what often gives rise to questioning. There is a dialectics between questioning and answering in which knowing, or thinking that one knows, is in conflict with not knowing. Even if it occurs at a subconscious level, this dialectic gives rise to more rigorous observation. If people think they know the answer, they will have no need to observe in order to check the answer. Similarly, if people are sure that their hypothesis is correct, they may not see the unexpected when it occurs. A hypothesis is not a hypothesis unless it there is some doubt about it, and unless people allow for the unexpected when they are observing. For example, if the researchers had been very sure that their hypothesis that the rab does not heat the soil underneath is correct, they may have refused to test it by feeling the temperature of the soil – or even if they felt it, they may have convinced themselves that it did not feel warm (even if they had measured it with a thermometer, they might have convinced themselves that the thermometer was faulty for some reason).

We also saw how on the farm there is a dialectic between needing and caring (Sect. 3.2.7). Needing and caring are inherently in conflict with each other, and from the dialectic between the two, observation is enhanced and new ways of cultivation may emerge. But what happens under a capitalist system, is that unmet needs overwhelm caring. Profit is necessarily the bottom line, so caring is of value only if it increases profit. When needs are hardly met, it may be difficult to care. The synergy between needs and caring breaks down, resulting in situations in which the farmers may suffer.

The family is directly involved in the process of continuous change and development of the paddy and the materials and methods of its cultivation. This is particularly true for the paddy that they are producing for use value. Unlike the paddy that they grow for exchange value, they are concerned about the taste, smell, nutrition, cooking qualities, and other properties of the paddy they grow for their own use. They are intimately involved in the process of the production of this paddy. Because they literally have their hands on this entire process, they are in a position to manipulate the paddy with their own hands, to experiment, to wonder about it, to ask questions, and to investigate it. They do science as they engage in its cultivation and use: by selecting seeds, trying out different storage methods, experimenting with different ways of cooking it, etc. The paddy that is grown for exchange value is somewhat removed (alienated) from their immediate experience, needs, and desires. Its seeds are HYV seeds, purchased in the market, developed by professional scientists, and produced by agricultural institutions and corporations. Its taste does not matter to them personally, since they are not eating it. They are not experimenting with ways to cook it. Their main concern is with the price the crop will fetch in the market, and the amount they have to pay for the seeds and fertilisers. They have very little to do with determining or controlling these prices. The prices are out of their hands. They do not appear to have the power to manipulate the prices, or experiment with ways of getting a more fair price. How can they do science in order to get the true value of the crop they sell? It could be done, but it would be difficult, and would require concerted social action of some sort. It is also difficult because it seems so impossible – it seems that the price is somehow imbedded in the paddy itself, rather than being socially constructed in a complex manner.

7.2.6.1 Science and Technology for Whom?

In order to understand the nature of science we have to consider why science is done and for whom it is done: for the rich and powerful or for the poor and oppressed (people lower on the kyriarchal scale). There is a conflict between doing science for people and doing science for profit. As capitalism develops it becomes less and less possible to do science for people and develop technologies for people if and when it conflicts with science and technology for profit. And it often does conflict. Therefore, advancing agricultural technologies may have very harmful effects on the environment and on farmers like the family we have been interacting with. They do not have the choice or the freedom to do things in ways that will avoid harming the environment or in ways that will produce food for its use value if those ways are not adequately profitable. The bottom line is the exchange value. Producing and developing varieties of paddy that are useful (e.g.: nutritious, tasty) for the family's own use value (and for the interests of the community, and society/nature) would be the most rational – and scientific – approach. But this kind of science and technology may not be possible if it conflicts with the exchange value of the paddy. The result of this conflict is that a more rational agriculture, may not be possible under capitalism, even though capitalism also results in technological improvements, as Marx and Engels pointed out:

> The moral of history, also to be deduced from other observations concerning agriculture, is that the capitalist system works against a rational agriculture, or that a rational agriculture is incompatible with the capitalist system (although the latter promotes technical improvements in agriculture), and needs either the hand of the small farmer living by his own labour or the control of associated producers (Marx and Engels 1894, chap. 6)

7.2.7 Science Is Historical Dialectical Materialism

7.2.7.1 Dialectical Questioning, Observing, and Investigating Are Basic to Science

One very crucial part of paddy cultivation is that in order to decide when to harvest the paddy, the farmers have to predict that it will not rain for several days. If it rains during harvest, the crop will be spoiled. We discussed how the farmers had said that actually, they cannot predict the rain (Sect. 3.2.5.4). But still they have to make a prediction in order to do the harvest. They did so, knowing that their prediction may not be very accurate, because rain is actually unpredictable. We claim that this is a very good example of what doing science is. We take it as evidence of the farmers' scientific temper (Sect. 7.2.8). If they had less scientific temper, they may have claimed that they know for sure, without referring to observations or evidence to support or refute their belief, or maybe even purposely ignoring evidence that contradicts their conclusions. Or, if they had less scientific temper, they may just mechanically and unquestioningly follow the advice of some authority. If they had less scientific temper, and it did rain despite their prediction that it would not, maybe they would have blamed their karma or their failure to have conducted some ritual. (Cassie Quigley 2009; Gurinder Jit Randhawa 2006).

To recognise that we do not know something is really the essence of science. Science is actually not a list of 'facts' that we 'know'. Science is 'not knowing', and knowing that we do not know, which gives rise to questioning and investigating. In other words, there is a dialectical relation between knowing and not knowing, which gives rise to questioning and investigating.

We have discussed how we have not found any very fixed 'body of knowledge' in paddy cultivation (Sect. 7.2.2). Even if there was an unquestionable, fixed 'body of knowledge', it would not be part of the process of doing science, because it would not be questionable. Questioning implies a lack of knowledge, which is the opposite of knowledge.

Questioning forms the basis of doing science. Without questioning, there is no investigating. The only way that one can be sure that a question has a definite answer is if one knows the answer. But when one thinks that one knows the answer, the question ceases to be an authentic question, and it ceases to be a matter of investigation.

Our point of view stems from our observing and experiencing the importance of questioning in the process of paddy cultivation as well as in the process of doing professional science in which some of us have been involved. In our varied experience we are continuously struck by the absence of a static understanding of physical reality. As professional scientists, we keep questioning the understandings of others and of ourselves. Therefore, we cannot accept any definition of science in which questioning is not given central importance.

However, questioning is not a process that occurs separately from answering – or searching for answers. We have discussed how the family engages in a process that consists of an interconnected network of continuous questioning, observing, analysing, hypothesizing, experimenting, reasoning, comparing, etc. (not in a particular order).

Besides questioning, observing is another aspect that is basic to the doing of science. This is because science is about questioning physical reality, and investigating physical reality requires a connection with direct or indirect observation of it.

We have already discussed how and why we think scientific observing is dialectical. It is inherently characterised by opposing or contradictory aspects, as we have discussed in Sect. 7.2.5.

Also, rather than being a process in which individual scientists objectively seek truth, we see science as a social, collaborative process that inevitably includes subjectivity. Using this definition, perhaps the 'traditional' practice of paddy cultivation by this family is actually quite 'scientific'.

7.2.7.2 What Is 'Dialectical'?

What is the general definition of 'dialectical' that we are using?
As Friedrich Engels wrote:

> ... the world is not to be comprehended as a complex of readymade things, but as a complex of processes, in which the things apparently stable no less than their mind images in our heads, the concepts, go through an uninterrupted change of coming into being and passing away, in which, in spite of all seeming accidentally and of all temporary retrogression, a progressive development asserts itself in the end. (Friedrich Engels 1886)

As explained by Maurice Cornforth (1975) as well as John Bellamy Foster, et al. (2010), this means that in physical reality, everything changes, and therefore things do not exist as things, but as things in the process of becoming something else. Thus we see the difficulties in thinking that a thing is identical with itself: because everything keeps changing and what is 'self' at one time no longer exists at another time. This is referred to as 'self non-identity' (Wolff-Michael Roth 2007a). This forces us to modify our everyday system of logic in which A = A to use an HDM logic in which A ≠ A. In other words, all 'things' are actually processes, and things may not actually be what they appear to be.

In doing paddy cultivation, it is not difficult to realise that everything changes, and that all things are actually in the process of becoming something that they are not. Living things are in the process of dying. A stalk of paddy is in the process of becoming straw for rabbing, which is in the process of becoming ashes and soil.

The rabbing is itself dialectical. It is made by arranging dry material that will burn easily. But it also includes a sprinkling of soil and water on top of all its layers. The rab needs to burn, but not burn too much. It contains opposing forces of dryness and wetness, giving rise to a controlled burning.

We saw that even tools were dialectical. For example, Namdev used a small bundle of paddy straw dipped in a bucket of water as a tool to spread water on the rab. The same sort of bundle of straw was lit and used by Smita to spread the fire to the other side of the rab.

The farmers said that scientists understand the 'why' of things. This implies that things are really not just things, but things that came to be things for some reasons. In other words, they are things coming to be – or processes.

These processes cannot be understood in a mechanical way as being the sum of individual parts. The parts are what they are as a result of being part of the whole. Soil is not just a sum of its individual constituents. Leaves keep changing as they dry up, decompose, burn, and become ploughed into the field. How they change depends on what is next to them and what they are interacting with. At the same time, their decomposition affects the processes that are occurring in their environment. A rab is not the sum of things in layers. Leaves burn differently depending on their position in the rab. Soil is different depending on its position in the rab. The fire burns differently if there is soil on top of it. The soil acts and reacts differently if it is below or on top of the fire. As Richard Levins and Richard Lewontin say,

> What constitutes the parts is defined by the whole that is being considered. Moreover, parts acquire properties by virtue of being parts of a particular whole, properties they do not have in isolation or as parts of another wholeIt is not that the whole is more than the sum of its parts, but that the parts acquire new properties. But as the parts acquire properties by being together, they impart to the whole new properties, which are reflected in changes in the parts, and so on. Parts and wholes evolve in consequence of their relationship, and the relationship itself evolves. These are the properties of things that we call dialectical: that one thing cannot exist without the other, that one acquires its properties from its relation to the other, that the properties of both evolve as a consequence of their interpenetration. (Richard Levins and Richard Lewontin 1985, 3)

7.2.7.3 Nature Is Historical Dialectical Materialist (HDM)

We need to define what we mean by nature, and why we say that it is dialectical. In order to define nature, we have to distinguish between what is nature and what is not nature. In order to do this, we will refer to the terms, 'idealist' and 'materialist'.

By 'idealist', we mean the belief that ideas are basic and matter is derivative from ideas. In other words, ideas can exist independently of physical reality. By ideas, we may mean thoughts, concepts, reasoning, logic, consciousness, a soul, a spirit, or a god. These are abstractions that cannot be observed with our senses of sight, hearing, tasting, smelling, or touching.

By 'materialist', we mean the belief that matter is basic and ideas are derivative from matter. By matter, we mean physical reality, or physical processes – things that we can perceive directly or indirectly through our senses. By 'materialist' we do not mean that only matter exists or that ideas do not exist. We mean that: events occur that humans do not perceive; physical reality existed prior to the existence of human beings and human consciousness; and human consciousness does not exist separately from human bodies, as explained by JBS Haldane (1940).

This physical reality is what we mean by 'nature'. However, for us at present, 'nature' means nature/society: the inseparable unity of the human (social) and the non-human living and non-living environment, in their complex interactions, interdependencies, and movement.

This is in accordance with Karl Marx's (1844, p. 143) description of the dialectical relationship between society and nature. Thus, according to an HDM framework, reality is the inseparable object/subject: we human beings are acting upon, studying, and doing science on a material world which is also acting upon us. This social mediation is discussed by John Bellamy Foster et al. (2010, 215–47), and in a review of their work (Karen Haydock 2017).

We use the word 'historical' to mean that physical reality (nature) changes in time. Furthermore, the character of change is such that gradual quantitative change may eventually lead to radical qualitative change. Thus, one process is replaced by another process which is qualitatively different. Actually, we observe that nothing lasts forever in its same form.

Thus, when we say that nature is historical dialectical materialist (HDM), we mean that there are inherent conflicts and inner contradictions in all things/processes which keep the whole in motion. Also, all things/processes are interconnected and interdependent. Following the interpretation by the marxist biologists Richard Levins and Richard Lewontin (1985), dialectical relations are opposing aspects that do not occur separately in time as causes and effects. They exist as a unity of inherent opposing forces (see the example of lodging, below). However, these dialectical relations give rise to change in physical reality in time, which is why we say nature is 'historical dialectical materialist' rather than just 'dialectical materialist'.

This is similar to the way Friedrich Engels (1876) defined the laws of dialectics as the transformation of quality into quantity, the interpenetration of opposites, and the law of the negation of the negation. However, following Richard Levins and

Richard Lewontin, we will avoid calling them 'laws', in order to avoid the misconception that there is some externally imposed 'law' which causes things to behave dialectically. Rather, we want to emphasize that nature is inherently dialectical – and non-teleological in the case of non-human nature.

Although processes in nature are dialectical, not all descriptions of these processes are dialectical.

For example, the description of a paddy plant falling over (lodging) because the wind blows it, is simply a matter of cause and effect. It is not a dialectical description.

But there are inherent dialectical relations in the same process, for example:

- The wind pushing the plant and the roots holding the plant in place;
- Fibres in the paddy stalks stretching and bending due to their elastic properties and a resistance due to the rigid structure of the layers of cell walls opposing this force;
- A social structure which supports the use of urea fertilisation that makes the plants grow too tall, and thus become prone to lodging, and an opposing social structure which make it too expensive to use too much fertiliser;
- The weight of the top of the plant increasing as the grain fills the spikelets, pulling the top down and the strength of the stem opposing the bending.

These dialectical relations are not causes and effects. Neither do they occur sequentially. The fibres are not flexible because the fibres are rigid. Neither is the rigidity of the fibres causing their flexibility. Neither is there first a rigidity and then a flexibility or vice versa. The wind is not causing the roots to hold the branch in place. The roots are holding the stalk at the same time that the wind is blowing it. The social structure that supports the use of fertilisers does not precede or cause the social structure that makes the price of fertilisers high.

The entire complex, interrelated set of inherent dialectical relations and opposing forces are all part of the process in which the plant sways back and forth, until it suddenly completely falls over and cannot rise again (which is an example of gradual quantitative changes leading to a sudden qualitative change).

7.2.7.4 What Do We Mean by Saying That Science Is HDM?[4]

We have already discussed how we find many aspects of doing paddy cultivation to be aspects of doing science (Sect. 7.2.1). And we have also discussed how we find many aspects of doing science investigation to be dialectical. When we say this, we are saying that the *method* of doing science is dialectical. By 'method' we do not mean a standardised procedure or a set of discrete aspects. We are using the term 'dialectical method' to refer to the type of analysis that Marx used throughout Capital: "to use material evidence and find dialectical relations in it – relationships which reveal contradictions that explain how motion occurs and why processes

---

[4] Parts of this section have been adapted from what has previously appeared in Singh et al. (2019).

proceed the way they do" (Karen Haydock 2017). This definition of 'dialectical' is what we think both Marx and Engels meant by it. In addition to using a dialectical method, they also wrote about how nature/society is dialectical.

Realising that physical reality is HDM, we think it makes sense to use an HDM method of science to investigate (and change) physical reality. The HDM world view as well as the method is what we call marxism.

When we say that the doing of science is HDM, we mean that the various interconnected, inseparable, and interdependent aspects of doing science are dialectical and arise as dialectical conflict. We have discussed how questioning is dialectical, and how scientific observation is characterised by opposing or contradictory aspects. Hypothesizing, experimenting and testing are dialectical in that they involve contrasting and comparing alternative (opposed) methods. Analysis may be dialectical if it includes the analysis of how dialectical conflict reveals how and why processes occur. We should also stress that this entire process of doing science is done collectively by people interacting with each other.

We have already discussed how we saw evidence of this happening occasionally, and to some extent, on the farm in Rudravali. The family did not always explicitly refer to what they were doing as being dialectical, or as being science. But their actions revealed that they were doing HDM science to some extent.

We think that science should be a historical dialectical materialist (HDM) process. And when genuine science is done, it often is HDM because it is based on observing, questioning, and investigating physical reality, and nature/society is HDM, as we explained above. However, when the capitalist structure imposes a profit motive, people are prevented from doing HDM science.

We realise that science is not always done in a very dialectical way, or in a very HDM way. For example, if we use an HDM method to investigate the wind blowing and the paddy plant falling over during lodging, we would investigate the sorts of the dialectical relations which we listed above. If we used a less HDM method, we might investigate just one aspect separately, say the amount of force needed to bend the stem by a certain amount, without adequately considering opposing forces, nonlinear dynamics, or interconnections to other factors. We might ignore the social, political, and economic relations that contributed to the use of fertilisers and to environmental problems. A more reductionist approach is less HDM.

Since nature itself is dialectical, even if we do not consciously try to find dialectical relations – we may find them when we do science. Therefore, and also since we recognise that our preconceptions and our ideological framework affect us as we do science, and that doing science is a social process, it is useful to use an HDM framework.

The depiction of 'the real process of science' (The real process of science 2017), which is given for students and teachers in the Berkeley University website is in line with our definition of science. An HDM method of doing science is similar, but includes a few additional aspects and helps us to see the dialectics of nature. As discussed by Levins and Lewontin (1985, 267), this method is not positivist – it is necessarily subjective, influenced by emotions, and interdependent with social, political, and economic systems. These interdependencies are complex, conflicting, and require attention from science educationists. However, this does not mean that we deny the existence of an objective reality or the existence of misconceptions, truth, lies, or right and wrong. The point, rather, is that, as Eleanor Duckworth (2012) says, 'In science, the stuff [physical reality] is the authority'. Furthermore, we cannot understand physical reality as isolated 'things' independently of past things/processes: we need to study the historical relations between them.

When we use an HDM method to do (natural/social) science, understandings can be challenged, questioned, investigated, requestioned, and reinvestigated. They keep changing as contradictory or new evidence is identified. New evidence keeps being found, partly because physical reality itself keeps changing. In doing science dialectically, we need to investigate how conflicts – inherent physical opposing forces – drive processes, rather than thinking that processes follow basic, abstract 'laws'.

We have shown how conflict is inherent to nature/society. We have also shown how conflicts give rise to questioning. It is not possible to abolish all conflict. Therefore, living a life without any conflict is not, and should not be a worthwhile goal. Conflicts need to be recognised and resolved, if possible. But new conflicts and new questions will then appear.

Due to interdependencies in physical reality, it is problematic to do science in an overly reductionist manner, looking individually at separate aspects of things or processes without simultaneously or subsequently reconsidering interactions with the larger context. We need to study interdependencies and historical relations (changes throughout time). HDM science is a socially mediated process of investigating physical reality, which is historically contingent, emergent, and probabilistic (Foster et al. 2010).

Dialectical analysis involves persistent questioning, without taking anything for granted, and it reveals distinctions between appearance and essence. But we do not define 'essence' in an idealist, logical empiricist, or mechanical materialist sense. We do not think an essence is necessarily teleological, or is immediately apparent to an individual and can be directly perceived by the senses. We use the word 'essence' to mean the HDM essence: "a struggle of opposites, which objectively exists but can be revealed only gradually, through observation, inductive reasoning, and social and historical critical analysis" (Karen Haydock 2015). We collectively (not individually) observe and analyse processes in order to understand their basis in inner conflicts in the material world. Thus, an essence is not an inner *thing*, but a dialectical *process*. And we learn about the essence by ourselves engaging in a dialectical process of questioning and investigating the process. By investigating, we reveal deeper meanings.

Thus, we see science not as a set of 'things' or a bounded or independent 'body of knowledge', but as a process of doing HDM science. We see meaningful learning as being concerned with using this process and developing 'scientific temper'.

7.2.8 Developing Scientific Temper

"To develop scientific temper, humanism and the spirit of inquiry and reform" is one of the Fundamental Duties in the Constitution of India. Scientific temper, as defined by Jawaharlal Nehru (1946, 509–15), is similar to but more broad than most definitions of 'scientific literacy', in that rather than just relying on authority, it includes the practice of using a scientific method throughout our daily lives to ask and search for answers to all sorts of questions. It must go beyond the domain in which science is professionally done, including the consideration of ultimate purposes, beauty, goodness, truth, and socio-political aspects.

However, The Nehruvian conception of scientific temper is often taken to emphasize science as a method of reasoning, with less emphasis on the interdependence between reasoning with the mind and doing with the hands. The materialist basis and ways of finding empirical evidence may be lost, or de-emphasized.

The word 'temper', as Nehru used it, does seem to refer more to a way of thinking or an attitude than a way of doing. Throughout his book, The Discovery of India, he keeps referring to different types of temper: the temper of the people at a particular time, the temper of the government, the emotional temper, the natural temper, revolutionary temper, vain temper, and the temper of the country.

Nehru contrasts the scientific method with the method of religion, which he says, relies on emotion and intuition. And because it is concerned with its own vested interests, religion produces: "narrowness and intolerance, credulity and superstition, emotionalism and irrationalism." He says, "It tends to close and limit the mind of man, and to produce a temper of a dependent, unfree person."

On the other hand, he says that science has been taken to be an "objective method", concerned with 'fact'. He writes that the scientific approach is: "the search for truth and new knowledge, the refusal to accept anything without testing and trial, the capacity to change previous conclusions in the face of new evidence, the reliance on observed fact and not on pre-conceived theory".

He argues that this scientific approach should be combined with temper, connected with the ivory tower of philosophical thought, logic, and reason. He writes:

> ... [The scientific method] is necessary, not merely for the application of science but for life itself and the solution of its many problems. Too many scientists to-day, who swear by science, forget all about it outside their particular spheres. The scientific approach and temper are, or should be, a way of life, a process of thinking, a method of acting and associating with our fellowmen. (Nehru 1946)

But even worse than thinking that scientific temper is concerned mainly with reasoning, many people interpret the propagation of scientific temper to be a matter of disseminating 'scientific information' and orchestrating knowledgeable scientists to answer the questions of ordinary people. This approach almost makes science appear to be a dreary body of knowledge to be transmitted, unchanged, from one generation to the next. In other words, science becomes just another form of traditional knowledge.

These conceptions of how people should inquire and search for truth can be compared to the approach advocated by Ambedkar in his interpretation of Buddhism. According to Ambedkar, the Buddha said that having correct views requires "the abandonment of all doctrines which are mere speculations without any basis in fact or experience." (Ambedkar 1957) He also wrote: "[The Buddha repudiated the Brahmanical] thesis that the Vedas are infallible and their authority could never be questioned. In his opinion, nothing was infallible and nothing could be final. Everything must be open to re-examination and reconsideration, whenever grounds for re-examination and reconsideration arise" (ibid.). The Buddha even said that his own message "was open to anyone to question it, test it, and find what truth it contained" (ibid.). And Ambedkar himself did do this, developing his own interpretations and applications of Buddhism. He stresses that the search for truth is done through experience, testing, and questioning everything. Ironically, this may make Ambedkar's Buddhism appear to be more of science than religion. It may even be a more genuine science than the kind of faith-based science through the 'dissemination of scientific knowledge' that is advocated by some professional scientists.

There is another sense in which the definition of scientific temper is in contradiction to what some mainstream scientists believe science is. They sometimes say that the methods of science can only be applied to certain areas. Some say that science does not deal with questions of morality, ethics, or sometimes even social problems in general.

Some people have tried to define and delimit the sorts of questions that science can address. For example, Jerry Coyne (Professor Emeritus of Ecology and Evolution at the University of Chicago) claims that the difference between science and non-science is that science can only deal with 'how', not 'why', and a domain such as religion deals with 'why'.

Some science educationists believe that teachers should only tell students 'about' environmental problems, focussing on the 'science', without telling students what to do or what is right and wrong – and without focussing on social aspects or engaging in activism in the science classroom (Karen Haydock and Himanshu Srivastava 2019; Annette Gough 2006). Other educationists now agree that acquiring 'skills, attitudes, motivations and commitment to work individually and collectively toward solutions' should be a goal of environmental education, as suggested by the Belgrade Report (UNESCO-UNEP International Environmental Workshop 1975). However, even though such policies may be advocated, they are rarely implemented in schools in India.

Our view of the method of HDM science is in agreement with a sort of scientific temper that is concerned with the process of asking and investigating any sorts of questions, using both mind and body. It rejects a separation between social and non-human natural sciences.

The kind of scientific temper which we think is important, according to an HDM framework, is not concerned with individuals remembering concepts or performing science process skills such as remembering how to use scientific instruments. These are things that people learn as and when the need arises – if they have sufficient motivation and scientific temper. Learning and developing scientific temper is done by doing a sort of science that includes both experimentation and communication (e.g. discussion and reference to literature). Scientific temper is a collective, interdependent characteristic of people in particular environments with particular needs. Similar to Wolff-Michael Roth's (2007b) definition of scientific literacy, we agree that it is thus an emergent process.

Following Paulo Freire (Freire 1970), we see the value of (both natural and social) science education in its possible role to encourage students to work together in order to question the status quo and become more active participants in trying to create a better, more just and equitable world. Countering this, there is a conservative effort to relegate the process of science to professional natural scientists, and to aim science education more at sifting and selecting students and teaching them to remember 'the (sic) science content', using cognitivist theories that prescribe some universal laws of development based on the cognitive structure of the mind (e.g. see Paul Kirschener et al. 2006). However, according to an HDM framework, and in agreement with Anna Stetsenko and Igor Arievitch (2002), we do not find any convincing evidence for such universal laws of development, and anyway we are more concerned with those who are being sifted out. We see less need for people becoming walking encyclopaedias and more need for people to question, critically analyse, and find physical evidence for or against what the encyclopaedia says.

7.3   Comparing the Process of Doing Science to the Process of Learning

As we were observing and working alongside the family cultivating paddy, we were noticing that in doing these things, not only was the family doing science (at least to some extent), but also, the family was learning, teaching and developing paddy cultivation. Their process of learning and teaching was actually not really different from the process of doing science.

As we have discussed (Sect. 4.2.3), the family is learning by doing. But this is not just watching and copying what someone else does. It is rather a very mixed up combination of observing, questioning, trying out for oneself and creating ways to do things. Different family members do things differently (Sect. 3.2.4). For example, being left-handed, Bhimraj could not exactly copy from the other right-handed people, and when right-handed people learned from him they could not just copy him.

Some educationists think scientific observation needs to be explicitly taught (Catherine Eberbach and Kevin Crowley 2009). But the family had not been formally taught in school how to closely and scientifically observe. Neither had they first read or been formally taught a lot of 'background knowledge'. As we discussed (Sect. 7.2.5), their authority was the stuff itself. – the paddy, other plants, animals, and the environment. They keep observing the stuff and getting to understand it intimately.

However, the schools around Rudravali are mainly concerned with textbook teaching, without giving students opportunities to observe or manipulate real stuff. This is in contrast to how the family was learning in the field: by working, figuring out how to do things with their own hands, observing specific processes being carried out by themselves and each other, and becoming confronted by conflicts between observations, and between observations and beliefs. These conflicts gave rise to implicit or explicit questioning. The process of observing was used to inductively reason, investigate, make generalisations, and ask further questions. All of these aspects were interdependent with each other.

Besides being in the field directly interacting with what is being observed, the farmers also had to have time to observe, patience to observe closely over extended periods of time, and they had to care about what they were observing, or have some real need to observe. With developing capitalism, things get speeded up, alienation between people and nature increases, and people may lose this time and opportunity to observe. If this happens, they may also do less science and less learning.

Based on our interactions with the family, we concluded that much of the process of learning to do something requires trial and error on the part of the learner, rather than detailed instruction (Sects. 3.2.3, 3.2.5, and 4.2.3.2). People have to experiment and figure out things for themselves, rather than being told exactly how to do something. This is the way we (and the others) were learning. Besides just copying what someone else is doing, learners combine their developing understanding of how and why things work with trial and error, observing, questioning, considering suggestions and hypotheses, and trying them out to see how and why they work. In other words, people do science in order to learn how to do things like paddy cultivation. It is a type of constructivism in which learners need to create their own methods of doing things or adapt the techniques they see others doing. This pedagogy gives space for creativity and variation, even if this is not the intention. We wonder whether it may actually be the 'traditional' way of learning.

As teachers, the researchers often find it difficult to let learners try doing things for themselves. When we watch children doing things, and they are having trouble doing what we think they should do, it is sometimes difficult to stop ourselves from immediately interfering, trying to help, or just taking over and doing it for them. For example, it requires both patience and will-power to stop oneself from 'helping' when watching a very young child trying to fit a piece of a puzzle in an obviously backwards position. But turning the piece into the correct orientation probably does not help the child, although it may help get the puzzle done more quickly. In such a situation, young children do not seem to get as easily frustrated as an adult onlooker does. It requires patience to let learners make mistakes and figure things out for themselves.

Similarly, we would have thought that the farmers would have felt frustrated when they saw that the researchers were having so much trouble making rope (Sect. 4.2.3.2). But they did not express any frustration. They just continued their own work, making their own rope. Perhaps one reason why the farmers let the researchers learn by doing was that they were not just teachers – they were primarily busy in working, doing the same things that the learners were doing (or learning to do).

The reason that teachers get frustrated and do not let students figure things out for themselves, may be that teachers are not doing the work (or making the puzzles) themselves. Their objective is not to do the work, but to teach students to do it. But they seemingly forget this when they just take over and do the work for the students.

7.3.1   Work with Hands vs Using Language

We saw that the language the family used is very different from the language used in textbooks. It is even more different from the way professional scientists communicate with each other orally and through their written reports.

The development of science required an integration of work with the hands and work with the mind: an integration of observing and manipulating actual stuff with both oral and written questioning, description and comparison of observations, and analysis and communication (Sects. 6.3.3.2 and 6.3.3.3). Technical jargon developed as scientific writing developed. Professional scientists need to use complicated terminology in order to describe things that the readers do not have in their hands for direct observation. To some extent, written description and jargon developed in order to communicate observations and appearances that were difficult to communicate without easy ways to incorporate illustrations or photographs in reports. Nowadays, even when showing evidence through diagrams and photographs, scientists use arrows, labels, and descriptive figure captions in publications, because they are not present in person to point, explain, or discuss and the viewer is not able to directly question the presenter.

Scientific discussions and writing became very different from the way ordinary people speak. Scientists needed jargon that was more universal than the local terminology that was used to describe and identify plants, animals, and other things. This jargon has developed as a dialectical relation between being a universal language and a highly specialized 'secret code' that is not understood by ordinary people. Even though this may not be intentional, scientific terminology functions to exclude non-scientists from understanding or participating. It helps to make science a privileged domain, and it even makes each speciality inaccessible to scientists in other specialties. This helps secure science its position of high authority and power.

We have discussed the advantages and disadvantages of oral vs written communication (Sect. 7.2.1.3). In doing science, written communication has its advantages, but it is also a more alienated form of communication than direct discussion and interaction between small groups of people handling the stuff that is the subject of investigation. If there is less hierarchal division between the people in a small discussion group, different people can observe the actual stuff, raise questions, and become part of a creative collaborative process of investigation. It may not just be for reasons of convenience that scientists do not usually bring actual stuff (e.g. a heap of paddy plants) to their conferences to show each other, or even make their raw data available to each other. It may also be that by doing so they may risk their legitimacy, authority and competitive advantage. Without having the actual stuff or the raw data in one's hands, it is difficult to refute a claim written in highly erudite language.

But more significantly, professional science has become more commodified as it has developed, and nowadays it is usually closely connected to commercial interests and the development of technology that becomes commodified. This is especially true of agricultural science, with its trade secrets. Collaboration, communication, and social advancement have suffered as agricultural science has become more privatised and more directed by corporations who must keep profit as the bottom line.

The family is reluctant to buy a new type of seed until they have seen the results in neighbouring fields. They distrust advertisements, seed companies, and government officials who tell them to try new seeds. But this distrust – and questioning – are aspects of doing science. We saw that the farmers know that the quality of a crop depends on many factors, and seeds from the same the bag will inevitably yield variation. When they buy seeds in the market, they can see what the seeds look like, but not how they will grow and what the other characteristics of the plants will look like. Seeing and hearing testimonials on TV and internet are more believable than just words, but there is also more danger that farmers will become victims by believing fraudulent claims. Thus, compared to using the seeds they produce themselves, buying seeds is actually less scientific because it is based more on faith in authority (and faith in an authority that has ulterior motives – making profit).

Increased communication and access to media is dialectical – with increased communication, deception also increases. First-hand observation and experience can include questioning and direct investigation. One cannot question a farmer on TV or handle the paddy for closer inspection. Only a select few professionals attend scientific meetings in which they can directly question other scientists and have discussions with them. If the scientists work for private companies who are in competition with each other for increasing profits, communication is necessarily limited. When communication is purposely limited in order to maximise profits, research is needlessly duplicated or obstructed and learning is limited.

7.3.2 Controversies Regarding 'Traditional Knowledge' and Science

Science education researchers have been having an ongoing debate between 'multiculturalism' and 'universalism' (Michael Matthews 2015). Some argue that science is universal and science education should not be concerned with differentiating between different cultures. Others say that science is only one 'way of knowing', and various other equally valid ways found in other cultures should be included in school education.

Of course, we all agree that different cultures at different times produce different technologies, tools, machinery, cultural artefacts, and products such as food, clothing, and shelter. These are specific to particular local environments and historical times. We also see that there are also certain commonalities between some technologies of communities who are widely separated in space and/or time (e.g. similar plough designs have been produced independently in different parts of the world). However, besides these physical objects and technologies, different communities also produce different theories, stories, and mythologies. Some of these may also be important or useful in the daily work on farms and in the development of materials and methods of cultivation. These understandings, narratives, concepts, and explanations of physical phenomena are sometimes called 'bodies of knowledge'.

Many researchers in the field of 'ethnobiology', at least until more recently, seem to assume that science is a 'body of knowledge' or a 'way of knowing' (Geilsa Baptista and Charbel El-Hani 2009; Rebecca Zarger 2011). We agree with their plea that indigenous culture should not be excluded from classrooms (Geilsa Baptista 2018). However, they also usually consider 'indigenous knowledge systems' to be the definitions and explanations of phenomena that are given by members of a particular community. They distinguish between traditional knowledge and 'modern western science' by comparing the end results of the processes of production of 'knowledge', as well as the technologies that emerge from this knowledge. This contrasts with our approach, in which we are concerned with understanding the *processes* of producing and learning to produce and reproduce the 'knowledge', theories (explanations of phenomena), techniques, and technologies.

Even when traditional or indigenous cultures/systems are described by researchers as 'ways of knowing', there are very few efforts to look at the different ways/processes/practices of the indigenous peoples and how those practices do or do not change with different physical and social contexts and with time. In particular, the historic, political, economic aspects are seldom mentioned. Therefore the cultures and their ways of learning cannot be understood.

Some researchers consider traditional knowledge to also include practices, but there is little detailed analysis of the practices, other than claims that they are holistic, without separations between rituals, religious practice, and livelihood. There are only a few researchers who study the processes that these communities engage in order to come up with and pass along the definitions and explanations. We have listed a few of the exceptions – researchers who discuss characteristics of the processes – in Table 2.1. Where we differ from most of these researchers is that we have found that in the family we are working with, these practices are much more complex than they first appear to be, engulfed in conflict, and interdependent with the political economy. As we have shown, and summarised in Table 4.1, we see that the family is continuously confronted by dialectical conflicts and their practices are neither traditional nor untraditional.

Most researchers do not discuss how what they call 'traditional knowledge' is produced. Perhaps many people would not object to our claim that people produce it by observing, experimenting, and doing other aspects of science. However, some may think that it comes from some sort of intuition or inner-knowledge that is passed on and believed on the basis of authority without questioning. For example, Oscar Kawagley et al. (1998) say that Yupiaq villagers do 'indigenous science', making detailed "observation of the natural world coupled with direct experimentation in the natural setting". However, they say that they also obtain knowledge by "observing one's inner spirit". We would not include "observing one's inner spirit" as one of the aspects of doing science (unless it just means using one's mind to analyse and reason).

Even if researchers do not discuss the process by which the so-called traditional or indigenous knowledge is produced, the implicit assumption is often that it is not produced by doing science. The assumption is often that cultural tradition is passed along by the testimony of authority rather than learning by doing, questioning, and investigating.

Thus, in the debate between universalism and multiculturalism, what is usually being debated is whether different cultures produce different 'bodies of knowledge', and whether science is just one amongst many kinds of 'bodies of knowledge' or 'ways of knowing'. If science is seen as a in this way, it may make sense to think that it is not universal. For example, if 'stories' about the origin of paddy – or the origin of life – are considered to be bodies of knowledge, whether they are the result of 'modern western science' or indigenous cultures, then it is obvious that there is a large diversity of accounts, with even various 'modern western science' accounts being far from universal. But if science is the process of questioning and investigating what the origin is, then we might find some universality in this process.

Others consider science as a process of doing – but they may consider the doing of science to be mainly an act of reasoning. In that case, they may see different kinds of reasoning and different systems of logic as indications of different multicultural sciences. Others insist that there is only one, universal kind of reasoning in all science, and ways of knowing that do not use that method are not science.

However, when analysing reasoning, we think it is inappropriate to disregard or de-emphasize the associated and the entire process of various aspects associated with reasoning. As we have discussed, we have found that it is inappropriate to separate work with the hands from work with the mind (Sects. 4.4, 5.3, 6.3.2, and 6.3.3). It is also necessary to consider the context and the reasons why science is or is not being done to a greater or lesser extent at particular times and in particular contexts.

Some say that it is hegemonic to refuse to see 'indigenous knowledge' as being just as valid as 'modern western science'. But on the other hand, maybe it's hegemonic to think that doing science is only western. A number of scholars have claimed that indigenous people have for long been scientists and inventors, and they claim that science is multicultural (Bryan Brayboy and Angelina Castagno 2008; Gloria Snively and John Corsiglia 2001). They argue for a culturally responsive science education in which a range of kinds of science, including a range of kinds of indigenous science, are included – because they all have something to learn from each other.

There are others like Masakata Ogawa (1995) who argue for a multiscience perspective rather than a multicultural perspective. We agree with Ogawa that the science studied in classrooms should include materials and contexts that are of concern to local communities. However, we do not agree with his definition of science as a 'rational perception of reality'. Our problem with such an understanding of science, as described earlier, is that it places the mind as central and separate from hand/

work/practice. Also, how can science education be understood or changed without taking into consideration the political, economic system in which it occurs, and the effects of colonialism, imperialism, and capitalism?

Some researchers argue that it is better to use the word 'knowledge' rather than 'science' for indigenous ways of knowing, in order to avoid being committed to the "scientistic myth about science as an unproblematically truthful, superior form of knowledge" Charbel El-Hani and Fábio de Ferreira Bandeira (2008). However, since the word 'knowledge' commonly refers to true, justified beliefs, it suffers from the same problem.

In her book, "Cognition in Practice", Jean Lave (1988) critiqued the simplistic definitions and distinctions between so-called 'everyday' (concrete, subjective, intuitive) and 'scientific' (abstract, objective, rational) thought (p. 79–80). She pointed out that there was an apparent shift amongst anthropologists and psychologists away from this binary, towards an increase in the number of categories: distinguishing between occupations and social classes. She discusses how the categories have become blurred, and based on empirical studies, a number of researchers have characterised science as an everyday practice. However, she says, this may not have changed the underlying hegemony of more scientific, rational thinking as somehow being higher than other, 'simplistic' ways of thinking. Nor has it necessarily dissuaded people from thinking that scientists consider science to be objective. Jean Lave concludes that, "...the concept of rationality has no general scientific power (being ideological) to account for more and less powerful forms of cognition, the efficacy of schooling, or anything else" (Jean Lave 1988, 174).

Some people have claimed that cognitive processes are universal, and that cultural differences are just matters of content, which depend on context (Jean Lave 1988, 85). This may make culture appear to be just a series of discrete 'facts' – a 'body of knowledge' – which is questionable. It is problematic in that it makes culture sound less dynamic than it actually is. Other objections to this universality, and to the universality of science, are often objections that what is occurring is not positivistic or objective.

Most of the researchers on various sides of this debate are focusing on the knowledge itself, or on the reasoning, rather than the process of learning or producing the knowledge with hands and minds. As we have explained, we are concerned with science as a process.

If, as we claim, science is a way of doing, and if a broad enough definition of doing science is used, then it may be difficult to show that it is not universal. For example, if the doing of science involves engaging in a variable network of aspects such as questioning, investigating, observing, analysing, comparing, and testing, then these aspects might be unavoidable and universal as part of the processes of cultivating, cooking, making baskets, or engaging in any everyday activity. Who can argue that questioning and observing are not universal? Of course the particular questions, observations, results, etc., are specific to the context, but the process of questioning and investigating is carried on everywhere. Disagreements arise only if people think that the ways that questioning, observing, analysing, etc. are done in professional science are very different from how they are done in cultivation or in everyday life. We do not think this is necessarily true. We have presented evidence that there are no fundamental differences between the way each of these aspects are done in professional science and in the process of cultivation we have seen and experienced on this farm. Perhaps people do things very differently on other farms, which we have not studied. The extent to which people do science may depend on the context, the culture, and the point in history. And the number and order of aspects of science that people engage in may also depend on the context. But there may be something universal about the aspects themselves. People might use very different techniques to do cultivation, and many different types of food or baskets might be made. But still, in the process of figuring out how to do these things, we suspect that people might, at least to some extent, be engaging in a universal process of doing science.

However, we are not basing our definition of universality on a mechanical, cartesian logical structure of knowledge. Our definition of 'universal' is related to our dialectical understanding of 'essentialism' (Sect. 7.2.7.4). We do not deny that essences exist. But we reject an idealist definition of 'essence' in which inherent diversity is reduced to a single crucial characteristic. We recognise inherent diversity as well as similarity as being an aspect of the essence.

Thus, the universality that we see is not in a 'body of knowledge' or the technology or the results of the process of doing science. The universality is in how the process is done: through the recognition of dialectical conflict that gives rise to questioning and investigating. The particular questions depend on the context, who is doing the questioning, why they are doing it, and for whom they are doing it. Therefore, different groups of people in different contexts will produce different results. This is how the diversity arises. But they may all essentially be doing science, even though in different contexts, different groups of people will engage to various extents in different aspects from the 'science toolbox'.

However, we question the sort of universalism that William Stanley and Nancy Brickhouse (2001) have critiqued in their defence of multiculturalism. According to them, universalism includes a (realist) belief that physical reality exists and has a structure that is invariant over time and place and can be known and understood increasingly more accurately by doing 'modern western science', which overcomes individual biases.

We ask whether the process of doing science may have a sort of universality that is different from this. People in different cultures may all be engaging in various aspects of doing science – questioning, observing, and searching for understandings. Broadly speaking, they all may be engaging in fairly similar processes of doing science, even as they come up with very different concepts and technologies. We see nature/society as being very dynamic and changeable. We think science can be done even to question whether or to what extent the structure of reality is invariant. Science can be done even to question the nature of science, as we have done.

We also do not find evidence that individual biases are necessarily overcome as science progresses. Biases may change, but how can the doing of science escape subjectivity? We have discussed how we see the doing of science as being dialectical and materialist, but at the same time not objective or positivist, or in line with logical empiricism. But it still may be universal. The universality we see is this dialectical nature of the process, rather than a mechanical, cartesian logical structure of knowledge, as suggested in some definitions of universality.

In sum, the science that we saw being practiced on the farm differs from what many other educationists and ethnographers have described in four main ways. First, we see the family's science as a process or practice, and many others are instead referring to science as more of a thing than a practice. Second, many people see science as a way of thinking and reasoning, where as we see it as a way of doing/thinking, in which the hand and mind are interconnected. Third, we have not found researchers of ethnography or education who see the doing of science as being both dialectical and materialist. This means that we allow that universal truths exist, however difficult it may be to find them, but we think that truth can be ascertained only with reference to physical reality, as it changes throughout history. As we have also discussed, if we use a dialectical understanding of the process of doing science, we see that science is necessarily socially mediated – human praxis and theory (consciousness) are dialectically related. Relations to nature cannot really be considered separately from human relations. They both simultaneously cause and affect each other. Since we do not see any kind of process of doing science as being objective (i.e. not socially mediated in this sense), we do not see 'objectivity' or positivism as a way of distinguishing what professional scientists do from what ordinary people do. We think that neither are positivist, and both may be science, to some extent. Fourth, we have found that we need to consider the political economy of the family and the society in order to understand their science and their ways of learning.

Our small case study does not provide evidence that allows us to make generalisations about whether science is universal or multicultural. However, it does provide evidence that many aspects of what the family is doing on the farm are similar to what we have seen professional 'modern western scientists' do.

From a postmodern relativist perspective, the dominant culture (including 'modern western science') reinforces conceptions of truth and reality which are different from the conceptions of truth and reality of cultural minorities. We claim rather, that dominant powers oppress others by limiting them access to the doing of science. By doing more science, minorities would be more apt to question and distinguish truth, rather than faithfully believing the dominant authorities. Oppressed people, such as the farming family in Rudravali, may be wanting and needing to do non-capitalist science in order to figure out ways to improve their own lives and reduce their drudgery. But doing science to a greater extent means experimenting to a greater extent, and experimenting with paddy cultivation is very risky when the family is on the edge of survival. From their point of view, their problem is not that they are prevented from believing in or doing traditional ways, but that they are forced to keep doing 'traditional' old ways of paddy cultivation even though they realise that they need to change. The dominant powers are preventing them from developing in ways that they need and want, and from being scientists who develop their own ways of cultivating. If they do manage to develop, they are forced to develop in a capitalist system in which profit, rather than their own well-being, is the bottom line.

We would rather use the word 'science' for what we saw the farming family doing, while explaining that 'science' is actually not a collection of facts, but a process of questioning and investigating. We see a need to avoid assuming that a fixed 'body of knowledge' even exists, whether it is called science or knowledge. The reason to use the word 'science' is not just to value what indigenous people do (as if it is better than being non-scientific), but to emphasize the tentative nature of all forms of science, and to emphasize that they are processes, not things. We are not using the word 'science' in order to claim that indigenous people are inventors or 'knowledge producers'. We are claiming that the way they are working and learning is that they are engaging in the process of doing science, at least to some extent.

Another reason some people object to using the term 'science' to refer to what indigenous people do, is that there certainly are some differences between what they do and what professional scientists do, and why they do it. We agree that we have to recognise the differences. One of the main differences is whether people are doing science more for its use value or its exchange value. Rather than calling both as science, it may be better to use different terms in order to emphasize these differences. We suggest using the terms 'HDM science', and 'capitalist science'.

Another reason not to use the term 'science' in a more universal sense, is that it has negative connotations. Some people blame science itself for the horrendously destructive negative effects of technological development that have been produced through science. It is easy to become totally disenchanted with science and development when we are displaced by a large dam, our farm is converted into an open-pit mine, our small-scale industries are replaced by high-tech multi-national corporations, we become unemployed due to automation, our taxes are used for making ever-more destructive weapons, we get frustrated when our new technology falls apart due to planned obsolescence, our air is polluted by automobiles and chemical industries, and we become increasingly alienated by science and technology. The teaching and learning of 'science' in school has also taken on very negative connotations. Too often, it is mainly or solely concerned with memorising answers to questions asked by authorities. And the main objective is to sort students and sift most of them out of opportunities for further education and better jobs, and to keep capitalism spiralling on.

This is why we think it is important to understand why science is being done and for whom it is being done. Doing so allows us to distinguish between capitalist science and people's science. It also allows us to avoid naturalising both capitalist science and capitalism, as if capitalism is natural and inevitable and capitalist science is the only possible kind of science. We need to recognise that science need not be capitalist (and throughout most of history, actual science has not been capitalist). Development need not be capitalist. Surely people can figure out how to have development that is not anti-people. Socialist, communist, and other non-capitalist systems are possible in which new forms of people's science will flourish. We hope that HDM science will be what will flourish in the future.

7.3.3 Connections to Other Research on How Farmers Learn

As we discussed in Sect. 2.3, Bhimraj and Smita seem to think that farming is something to be done, not something to be 'learned'. A similar problem with the use of the word for 'learning' was reported by Rebecca Zarger (2011, 377) in the Q'eqchi' language in Belize. She found that 'tz'olok' (to learn) is used only with regard to formal education in schools, and the word 'k'anhelak' (to work), is associated with informal learning about nature and the environment. She writes: 'Maya parents talked about their children "learning how to work" as synonymous with learning about the environment. Children learned on the way to the family farm, by helping adults with chores, and by completing tasks specific to their age and gender. That is, they learned farming by doing, by working, not by sitting in a classroom'.

7.3.3.1 Apprenticeship Learning

Some might consider the kind of learning that occurred on the farm in Rudravali to be 'apprenticeship learning', which has been the subject of considerable research, although not necessarily with regard to farming.

Jean Lave claims that, although there are many different kinds of apprenticeship, in the case of the tailors she studied in Liberia, apprenticeship was not considered to be a way of teaching and learning, but rather a way of living and a way of working (Jean Lave 2011). She says that she began her research by thinking that there was a binary distinction between formal and informal education. Accordingly, formal education involves learning out of context, through verbal instruction and general understanding of abstract principals, which results in a context-free understanding and "general learning transfer". Informal education in a traditional society involves learning in everyday activities, through demonstration and observation and concrete, non-verbal comprehension of specific activities, which results in context-bound understanding and no general learning transfer. This description of informal learning is similar to some of the superficial appearances that we initially noticed in the farm in Rudravali, as summarised in Table 4.1.

However, Jean Lave writes that as her research continued, her evidence contradicted this sort of binary contrast between formal and informal learning. She also found that the formal education that some tailors obtained did not result in any significant superiority in "learning transfer", compared to those without formal education. She came to realise that even formal education – and mathematical practice – occurs in a socially situated context. Also, she eventually realised that she could no longer ignore the "conflictual, political-economic relations shaping master-apprenticeship relations and practices of learning, making, and doing." She describes her own research as a critical ethnographer as her own apprenticeship learning: "... over and over, a detailed descriptive account led in unexpected directions of inquiry to changing conclusions, to ideas for next segments, and to questions about the theoretical limitations of the enterprise" (ibid, p. 149). She developed 'social practice theory' as part of the historical materialist social-theoretical tradition of investigating how the aspects of social life she was studying were interdependently changing processes and were historically grounded.

In some ways this is similar to the way we have been developing our framework, and to the framework we have developed. While she also finds that knowledge is dynamic, we go further and question whether the term 'knowledge' is even appropriate and whether a 'body of knowledge' even exists (Sect. 7.2.2). Our main difference with her is in our stress on dialectical aspects and historical development – in how inherent conflicts give rise to questioning and change – which we recognise as being aspects of doing science. In her book that summarises her years of study of apprenticeship (Jean Lave 2011), Jean Lave does not discuss the nature of science or mention any similarity between what she calls a "historical-materialist problematic" and the doing of science. Instead, she sees various "relativistic" alternative forms of education, mathematics, and "learning transfer". She criticises western research strategies that are based on comparing indigenous practices to Western universalist categories. Some might say that our claims regarding the nature of science are problematic for this reason. However, as we will discuss, we do not think that such comparisons are necessarily problematic (or that they are even avoidable), and we think that even Jean Lave is doing the same thing – comparing – when she defines various alternative ways of learning. But it would be problematic if she – or we – did not recognise the reasons for hegemonic power relations between different practices. In order to understand the possible hegemony, we need to analyse and compare the reasons people learn or practice 'science' in different contexts.

7.3.4 Research on Farmers' Skill and Decision Making

There has been some research by anthropologists in the dominant countries concerning how farmers in developing countries make decisions – e.g. how they decide what to plant and how to plant it (Peggy Barlett 1980). However, this research may not be aimed at just understanding the pedagogy of how farmers learn and teach or whether they do some sort of science. It may be aimed at understanding the psychology and mathematics of decision making as individual (or sometimes cultural) behaviour, in order to 'modernise' (Glen Davis Stone 2007). Most of this sort of research is done within a mainstream, neoliberal economic framework, sometimes corporate sponsored. It is often done by and from the perspective of outsiders.

Officials may say that farmers need to 'be educated' so that they learn the language of science. Moreover, they may say that farmers need to overcome their old, ignorant ways, trust the authorities, and take up new seeds and modern methods. Companies actually need to convince farmers to buy their seeds and equipment, mainly on the basis of verbal communication and advertising, not on the basis of their own scientific analysis or experimentation.

The apparent aim of some of the research is to learn how to teach the 'natives' modern western science (capitalist science). And 'modern western science' is usually taken to be just a 'body of knowledge', not a method of questioning and investigating. Even when researchers seem to be advocating multi-culturalism, and condemning ethnocentricity, they do not necessarily question the aim of teaching 'modern western science' to indigenous communities:

> One wonders in retrospect how much more successful curriculum science projects which were supported by the international aid agencies would have been, if the agencies had put substantial funds into preliminary culture-based research, before embarking on the adoption-adaptation process. Research studies which focus on investigating the fundamental scientific knowledge and cultural knowledge possessed by indigenous societies should be an area given a high priority by international funding agencies. ... [This] would probably have been of much greater value ... than many of the efforts of insensitive experts trying to hammer an American or British curriculum package into a rough fit for an African or Asian classroom.
>
> The use of the cultural base of a society, both scientifically and linguistically, as the starting point and nucleus of a science education programme, implies that something is known about both aspects. Unfortunately, in many places, such as in African countries and in Papua New Guinea, very little is known about either. (Max Maddock 1981)

Although this may appear to be an effort to appreciate indigenous culture, is this sort of research really helpful? Surely the 'natives' must know about their own culture! And, even without formal education, they must have opinions and questions about their own needs and desires for particular forms of "aid". Who is defining the success and value of curriculum science projects? And for whom are they successful? It seems that in many cases, even if the stated starting point is 'indigenous knowledge', the implicit ending point is the western 'body of knowledge' and the promotion of capitalist development.

Many researchers hardly acknowledge the possibility of any course of action other than capitalist development for indigenous communities. Their underlying aim is to not only understand individual decision making, but to influence decision making, e.g. to get farmers to adopt particular seeds, fertilisers, pesticides, and/or grow different crops in different ways. And in many of the studies, the researchers do not question whether these materials and methods should be adopted. Their aims are focussed on how to 'teach' (convince) farmers that they should be adopted, in order to guide agricultural development programmes.

There is separate research that evaluates whether or not the seeds, materials, and methods should or should not be adopted, their efficacy, functions, and larger social, economic, political, and environmental effects. But these studies are often not integrated with research on education.

A 'lack of skill' is a convenient excuse for poverty and the lack of development, and for corporate profits not being high enough. According to companies such as Monsanto, their innovations and new seeds are improvements, and farmers adopt them if and when they are knowledgeable and 'skilled'. For example, scientific researchers have studied the modes of adoption of new seeds, and found that the farmers' adoption of new seeds follows an S-shaped curve distribution which consists of the phases: initial knowledge, persuasion, decision, implementation, and confirmation. From this they have categorised different types of farmers as "innovators" (or "winners"), and "laggards" (or "losers"), the latter being those who do not adopt the new innovations. This study is based on the assumption that the innovation is good. However, Glenn Davis Stone (2007) has criticised this study, and has shown that, for example in Warangal District, Bt cotton seed was adopted as a fad, due to agricultural deskilling, social bias, and as a result of the farmers' failure to experiment and evaluate. The farmers had to pay the price.

Obviously, it would be advantageous to seed companies if farmers' decisions to adopt new seeds are based on 'cultural' practices and beliefs that can be used or manipulated by the companies. Therefore, it may also be to their advantage if farmers are educated to believe that they can acquire knowledge by transmission from authorities, rather than through a collective process of questioning and investigating. They may be more apt to 'learn' to use new seeds by doing what authorities 'educate' them to do. An educational system that stresses the transmission of so-called 'knowledge' happens to be what we have at present. Other types of education may not be so convenient for the seed companies. It seems that typical formal 'education' (i.e. deskilling) will be profitable for capital, not for farmers.

In other contexts, researchers have made efforts to understand farmers' practices and create Participatory Plant Breeding approaches. For example, to understand and collaborate with farmers' seed selection in maize cultivation in parts of Mexico, Daniela Soleri and David Cleveland (2001) have done research in order to bridge "the dichotomy between scientific and nonscientific practice" and recognise elements of understanding and theory that are common to both. According to these authors, "The greatest obstacle to this has typically been the hierarchy of knowledge implicit in many conventional approaches to agricultural research and extension, particularly in the context of low resource, small-scale agriculture" (ibid, 125).

One of the main aims of learning science, and especially learning mathematics, has been thought to be that it will change the attitudes and abstract thinking and reasoning capabilities of the students. Max Maddock (1981) wrote that the UNESCO Secretary General Adiseshiah in1965 had said: ". .. the aim of science teaching and the whole of education generally should be to introduce students to that innovative mindedness and capacity for inventiveness and experimentation so urgently needed in the world – in other words, an attitude development aim."

The implication may be that people in traditional societies are not developed because they lack innovation due to a deficit in their thinking, their attitudes, and their 'capabilities', and at least the first two of these can be rectified through education. However, we doubt whether a lack of agricultural development is due to an attitude or individual ability. As we have explained, it is more likely to be due to farmers not getting the true value for their crops.

7.3.5   Research on Changing Traditions in Farming

Some researchers have studied what they call the cultural transformations or hybrid cultures of a new class of 'post-colonial farmers', who use landless labour to cultivate the land they own, using HYV seeds, artificial fertilisers and pesticides and mechanisation. We have discussed how this category of farming has not been nearly as widespread as was expected. Small family farms have persisted and are the most prominent type of farm in many areas (including Raigad District). Even people who are primarily wage labourers may own small plots of land. Nevertheless, it might be interesting to contrast the family in Rudravali to different kinds of farmers or peasants in other areas.

Akhil Gupta (1998), has done a case study in Alipur, a village on the outskirts of Meerut, in Uttar Pradesh. He has focussed on a few so-called 'farmers', although most of them do not actually do the day-to-day labour in the fields. All of them are male and almost all are upper caste and relatively well-off, owning fairly substantial amounts of land.

He reports on farmers understandings of cultivation – mainly discussing the 'knowledge', rather than the way they form their understandings. It is not known whether, or to what extent, their understandings are based on actual experience and observations.

For example, he writes that Suresh, the village priest, gave the following explanation of why it is better not to plough the field too many times before planting maize: "[If we plough many times] the soil will drink more water. If it drinks water, then heat will come out of the ground. Which will appear to make the sapling yellowish. Thus the plant keeps sitting down there instead of shooting upward. It doesn't run upward. It just sinks into the earth and remains there as a twig. Underneath, its roots rot from the heat [bhabhkaa]."

He certainly did not use the same kind of language that professional scientists use. But his explanation does seem to us to be based on observation and experience, making it relatively 'scientific'. He is saying that more ploughing loosens more soil, and then when it rains, it will absorb more water. Before the rain, the dry soil will be very hot. When it rains, perhaps the sudden change from being hot to being in water will damage the roots of the young plants, turn the plants yellow, and prevent them from growing. With fewer ploughings, rain would run off the soil rather than being absorbed, and this problem would be avoided.

According to Akhil Gupta, this shows how "Suresh brought an 'indigenously' developed knowledge of agronomy to bear on an agricultural practice conducted with chemical fertilizers, electric tube wells, and 'scientifically' bred hybrid seeds." This is what he calls a hybrid culture.

This reminds us of how the farmers in Rudravali explained the effects of rabbing – using colloquial language, but having understandings that were based on doing science themselves. However, Akhil Gupta claims that:

> Farmers in Alipur often used discourses regarding the health of plants. which relied on a
> theory about a proper balance of humors – that is, elements that were hot, cold, dry, and

moist. The balance of humors depended on its intrinsic properties but also on the properties of the soil, fertilizers, water, and wind. with which it came into contact. ... 'Scientific' theories of agronomy were freely interspersed with humoral ones... (Gupta 1998)

He seemed to think that there was something 'postcolonial' about the "unexpected intersections" of the indigenous and the modern. However, it is not easy to understand whether or when a farmer is talking about mystical 'humors' or simple qualities that can be observed. If, as in our case, the farmer actually bends down and shows you how hot the soil is, it is clearly not a mystical, humorous heat.

As Meera Nanda points out in her critique, Akhil Gupta ends up celebrating and calling for a preservation of traditional knowledge, in a hybrid form with modern technologies, evidently also including highly problematic aspects which are deeply hierarchal, casteist, and patriarchal. She objects to his postmodern relativism, in which "the need for questioning contradictions and removing falsified conceptions is considered as oppressive and imperialistic" (Meera Nanda 2003, 271).

We agree with her assessment. Contradictions are not to be revelled in. Contradictions, particularly the materialist dialectical contradictions we have been discussing, give rise to confronting, questioning, investigating, and overcoming the contradictions. In our study, we have seen that the practitioners themselves may do this, even without being explicitly taught by intellectuals. But when farmers (or intellectuals) are oppressive, even outsiders have a right to intervene.

There are some reports that suggest that there are similarities between methods of forming traditional knowledge and the scientific method. For example, David Brokensha and Bernard Riley write about their case study in Kenya: "Mbeere and other folk-belief systems contain much that is based on extremely accurate, detailed and thoughtful observations, made over many generations" (David Brokensha and Bernard Riley 1980).

But in most of the research that we have seen, what is being studied is the content of beliefs, not how beliefs are formed, passed on, or learned. And these studies reveal many cases where people do have beliefs that are very different from what might be derived by doing science. For example, Ann Grodzins Gold (1998) interviewed villagers in Rajasthan, asking them to speculate on the reasons for natural phenomena such as (ostensible) changes in forestation, climate, drought, and rain patterns. For example, villagers told the author that these days there is less rain because of deforestation – and they explained how and why deforestation had occurred. When the author asked why deforestation caused less rain, most people said that the trees on surrounding hills used to cause wind to 'pull' rain to the villages. This appears to be a hypothesis based on physical causes. But the author also reports, "The other theory, less widely held and sometimes offered with less conviction, has to do with the decay of social life, religion, and love among human beings, all of which are understood to displease God." Interestingly, it was a Brahmin who claimed that the decrease in rain is due to the gods – because people do not obey caste proscriptions as they used to. Perhaps he has more use for gods, because he is upper-caste. Whether he really believes the physicality of the mythology he espouses is another question.

This contrasts with how the family in Rudravali told us that rain is unpredictable and the gods do not determine the outcome of their hard work doing cultivation. The context is also quite different from the context of our study in Rudravali, where our analysis was of our observations and interactions with the family of cultivators, while working in the field. Our understanding of how they learned and did science was based on both observations of what they did as well as what they told us they did.

It is also not easy to interpret what people really mean when they explain physical processes as being caused by immorality or divine anger. It may not always be very irrational or anti-science. For example, deforestation may be caused by people who are cutting down trees and the act of cutting down trees may be immoral. Saying that the gods are punishing immorality may be just a metaphorical way of saying this.

There are interdependencies between physical processes and mental conceptions. As we have discussed, people may get some peace of mind by doing rituals. This peace of mind may help people do hard work in the field, which will then help in the production of a better crop. We would still call this a materialist explanation for crop productivity because the hard work is what is fundamental rather than the ritual. Hard work can (possibly) be done without doing rituals, and a good crop will be produced. But it is very unlikely that rituals without hard work will produce a good crop.

References

Ambedkar, B. R. 1944. *Annihilation of Caste: With a Reply to Mahatma Gandhi*. 3rd ed. reprinted 2011. Nagpur: Samata Prakashan.
———. 1957. *The Buddha and His Dhamma. , 2017 edition*. Delhi: Kalpaz Publications.
Anderson, Siwan. 2011. Caste as an impediment to trade. *American Economic Journal: Applied Economics* 3 (1): 239–263.
Athreya, Venkatesh, Kumar Deepak, R. Ramakumar, and Biplab Sarkar. 2017. Small farmers and small farming: A definition. In *How Do Small Farmers Fare? Evidence from Village Studies in India*, ed. Madhura Swaminathan and Sandipan Baksi, 1–24. New Delhi: Tulika Books.
Baptista, Geilsa Costa Santos. 2018. Tables of contextual cognition: A proposal for intercultural research in science education. *Cultural Studies of Science Education* 13 (3): 845–863.
Baptista, Geilsa Costa Santos, and Charbel Niño El-Hani. 2009. The contribution of ethnobiology to the construction of a dialogue between ways of knowing: A case study in a Brazilian public high school. *Science & Education* 18 (3–4): 503–520.
Barlett, Peggy F., ed. 1980. *Agricultural Decision Making: Anthropological Contributions to Rural Development*. New York: Academic Press.
Bellwood-Howard, Imogen. 2012. Reflections on teaching and learning scientific methods in the Ghanaian field and the British classroom. *School Science Review* 94: 85–89.
Bernal, J.D. 1954. *Science in History. Volume 1: The Emergence of Science*. New York: Cameron Associates.
Brayboy, Bryan McKinley Jones, and Angelina E. Castagno. 2008. How might native science inform 'informal science learning'? *Cultural Studies of Science Education* 3 (3): 731–750.

Brokensha, David, and Bernard Riley. 1980. Mbeere knowledge of their vegetation and its relevance for development: A case-study from Kenya. In *Indigenous Knowledge Systems and Development*, ed. David Brokensha, DM Warren, and Oswald Werner, 113–29.

Chakravarti, Uma. 2012. Women, men and beasts: The Jataka as popular tradition. In *Everyday Lives, Everyday Histories: Beyond the Kings and Brahmanas of 'ancient' India*, 198–221. New Delhi: Tulika Books.

Cornforth, Maurice. 1975. *Materialism and the Dialectical Method*. 4th printing. New York: International Publ.

Duckworth, Eleanor. 2012. When teachers listen and learners explain. TEDxPioneerValley – EleanorDuckworth.. https://www.youtube.com/watch?v=1sfgenKusQk. Accessed 30 Aug 2016.

Eberbach, Catherine, and Kevin Crowley. 2009. From everyday to scientific observation: How children learn to observe the biologist's world. *Review of Educational Research* 79 (1): 39–68.

Education Commission. 1966. *Education and National Development: Report of the Education Commission 1964–66, Government of India, Ministry of Education, March 1971 reprint*. New Delhi: National Council of Educational Research and Training.

El-Hani, Charbel Niño, and Fábio Pedro Souza de Ferreira Bandeira. 2008. Valuing indigenous knowledge: To call it 'science' will not help. *Cultural Studies of Science Education* 3 (3): 751–779.

Engels, Frederick. 1886. Ludwig Feuerbach and the end of classical German philosophy. In *Collected Works*, ed. Karl Marx and Frederick Engels, vol. 26, 353–398. New York: International Publishers.

Engels, Friedrich. 1876. *Dialectics of Nature, Trans. from German by Clemens Dutt, 1934, 6th Printing 1974*. Moscow: Progress Publishers.

Foster, John Bellamy, Brett Clark, and Richard York. 2010. *The Ecological Rift: Capitalism's War on the Earth*. New York: Monthly Review Press.

Freire, Paulo. 1970. *Pedagogy of the Oppressed*. New York: Seabury Press.

Gandhi, Mohandas Karamchand. 1927. *An Autobiography or the Story of My Experiments with Truth*. Ahmedabad: Navajivan Publ. House.

Gold, Ann Grodzins. 1998. Sin and rain: Moral ecology in rural north India. In *Purifying the Earthly Body of God : Religion and Ecology in Hindu India, SUNY Series in Religious Studies*, ed. Lance E. Nelson, 165–195. New York: State University of New York Press.

Gough, Annette. 2006. A long, winding (and rocky) road to environmental education for sustainability in 2006. *Australian Journal of Environmental Education* 22 (1): 71–76.

Gough, Noel. 2019. Playing within/against entombed scholarship: Episodes in an academic life. In *Critical Voices in Science Education Research: Narratives of Hope and Struggle*, ed. Jesse Bazzul and Christina Siry, 201–212. New York/Berlin/Heidelberg: Springer.

Gupta, Akhil. 1998. *Postcolonial Developments: Agriculture in the Making of Modern India*. Durham: Duke University Press.

Haldane, J. B. S. 1940. Why I Am a Materialist.. *Rationalist Annual*.

Harvey, David. 2010. A Companion to Marx's Capital. In *London*. New York: Verso.

———. 2014. *Seventeen Contradictions and the End of Capitalism*. London: Profile Books.

Haydock, Karen. 2015. Stated and unstated aims of NCERT social science textbooks. *Economic & Political Weekly* L (17): 109–119.

———. 2017. A Marxist approach to understanding ecology. *Economic and Political Weekly* 52 (37).

Haydock, Karen, and Himanshu Srivastava. 2019. Environmental philosophies underlying the teaching of environmental education: A case study in India. *Environmental Education Research* 25 (7): 1038–1065.

Ilaiah, Kancha. 2010. *The Weapon of the Other: Dalitbahujan Writings and the Remaking of Indian Nationalist Thought*. Delhi: Longman.

Jaiswal, Suvira. 2018. *The Making of Brahmanic Hegemony*. Chennai: Tulika Books.

Joseph, Tony. 2018. *Early Indians: The Story of Our Ancestors and Where We Came From*. New Delhi: Juggernaut.

Kawagley, A.O., D. Norris-Tull, and R.A. Norris-Tull. 1998. The indigenous worldview of Yupiaq culture: Its scientific nature and relevance to the practice and teaching of science. *Journal of Research in Science Teaching* 35 (2): 133–144.

Keller, Evelyn Fox. 1983. *A Feeling for the Organism: The life and work of Barbara McClintock*. San Francisco: W.H. Freeman.

Kirschner, Paul A., John Sweller, and Richard E. Clark. 2006. Why minimal guidance during instruction does not work: An analysis of the failure of constructivist, discovery, problem-based, experiential, and inquiry-based teaching. *Educational Psychologist* 41 (2): 75–86.

Kumar, Krishna. 1996. Agricultural modernisation and education: Contours of a point of departure. *Economic and Political Weekly* 31 (35/37): 2367–2369.

Lave, Jean. 1988. *Cognition in Practice: Mind, Mathematics, and Culture in Everyday Life*. Cambridge/New York: Cambridge University Press.

———. 2011. *Apprenticeship in Critical Ethnographic Practice*. Chicago: University of Chicago Press.

Levins, Richard, and Richard C. Lewontin. 1985. *The Dialectical Biologist*. Cambridge, MA: Harvard Univ. Press.

Maddock, M.N. 1981. Science Education: An anthropological viewpoint. *Studies in Science Education* 8 (1): 1–26.

Marx, Karl. 1867. *Capital: A Critique of Political Economy. Volume I, (in German)*, 1887 English Edition, Trans. Samuel Moore and Edward Aveling, edited by Frederick Engels. Progress Publishers, Moscow, USSR.

Marx, Karl, and Friedrich Engels. 1846. *A Critique of the German Ideology* (First Published in 1932, English Version, 1968). Moscow: Progress Publishers.

———. 1894. *Capital: A Critique of Political Economy. Volume Three: The Process of Capitalist Production as a Whole*. New York: International Publishers.

Matthews, Michael R. 2015. *Science Teaching: The Contribution of History and Philosophy of Science, 20th Anniversary Revised and Expanded Edition*. 2nd ed. New York: Routledge, Taylor & Francis Group.

Nanda, Meera. 2003. *Prophets Facing Backward: Postmodern Critiques of Science and Hindu Nationalism in India*. New Brunswick: Rutgers University Press.

Nehru, Jawaharlal. 1946. *The Discovery of India*. New Delhi: Penguin Books.

Ogawa, Masakata. 1995. Science education in a multiscience perspective. *Science Education* 79 (5): 583–593.

Quigley, Cassie. 2009. Globalization and science education: The implications for indigenous knowledge systems. *International Education Studies* 2 (1): 76–88.

Ramdas, Sagari, et al. 2004. Overcoming gender barriers: Local knowledge systems and animal health healing, Chapter 2. In *Livelihood and Gender: Equity in Community Resource Management*, ed. Sumi Krishna, 67–91. New Delhi: Sage.

Rampal, Anita. 1992. A possible 'orality' for science? *Interchange* 23 (3): 227–244.

Randhawa, Gurinder Jit. 2006. *Document on Biology of Rice (Oryza Sativa L.) in India*. New Delhi: National Bureau of Plant Genetic Resources Indian Council of Agricultural Research.

Roth, Wolff-Michael. 2007a. Busting boundaries: Rethinking language from an epistemology of difference. *Cultural Studies of Science Education* 2 (1): 290–300.

———. 2007b. Toward a dialectical notion and praxis of scientific literacy. *Journal of Curriculum Studies* 39 (4): 377–398.

Scharfe, Hartmut. 2002. *Education in Ancient India*. Leiden/Boston: Brill.

Singh, Gurinder, Rafikh Shaikh, and Karen Haydock. 2019. Understanding Student Questioning. *Cultural Studies of Science Education* 14 (3): 643–679.

Sinha, Savita, and Mohammad Husain Husain. 1994. *Lands and Peoples: Geography Textbook for Class VII*. New Delhi: National Council of Educational Research and Training.

Snively, Gloria, and John Corsiglia. 2001. Discovering indigenous science: Implications for science education. *Science Education* 85 (1): 6–34.

Soleri, Daniela, and David A. Cleveland. 2001. Farmers' genetic perceptions regarding their crop populations: An example with maize in the central valleys of Oaxaca, Mexico. *Economic Botany* 55 (1): 106–128.

Stanley, William B., and Nancy W. Brickhouse. 2001. Teaching sciences: The multicultural question revisited. *Science Education* 85 (1): 35–49.

Stetsenko, Anna, and Igor Arievitch. 2002. Teaching, learning and development: A post-Vygotskian perspective. In *Learning for Life in the Twentyfirst Century: Sociocultural Perspectives on the Future of Education*, ed. G. Wells and G. Claxton, 84–87. London: Blackwell.

Stone, Glenn Davis. 2007. Agricultural deskilling and the spread of genetically modified cotton in Warangal. *Current Anthropology* 48 (1): 67–103.

The Real Process of Science. 2017. http://undsci.berkeley.edu/article/0_0_0/howscienceworks_02. Accessed 7 Sept 2017.

UNESCO. 1977. The International Workshop on Environmental Education, Belgrade, October, 1975. Final Report ED 76/WS/95. Paris: UNESCO.

Varadarajan, Lotika. 1979. Oral testimony as historical source material for traditional and modern India. *Economic and Political Weekly* 14 (24): 1009–1014.

Vogl, Christian R., et al. 2017. Keeping the actors in the organic system learning: The role of organic farmers' experiments, Ch. 6. In *Sustainable Development of Organic Agriculture: Historical Perspectives*, 99–114. Canada: Apple Academic Press.

Vygotskiĭ, L. S., and Alex Kozulin. 1986. *Thought and Language*. Trans. newly rev. and edited. Cambridge, MA: MIT Press.

Weber, Max. 1958. *The Religion of India: The Sociology of Hinduism and Buddhism,* Translated and Edited by Hans H. Gerth and Don Martindale, Original German Edition Published in 1916. Glencoe: The Free Press.

Wivagg, Dan, and Douglas Allchin. 2002. The Dogma of 'the' scientific method. *The American Biology Teacher* 64 (9): 645–646.

Zarger, Rebecca K. 2011. Learning ethnobiology: Creating knowledge and skills about the living world. In *Ethnobiology*, 371–387. Oxford: Wiley-Blackwell.

Zinn, Howard. 2002. *A People's History of the United States: From 1492 to the Present*. Harlow: Longman.

Chapter 8
What Is Actually Happening in Formal Education?

8.1 Is Formal Education Useful for the Farming Family?

The researchers felt very ignorant and unskilled as they watched and tried to learn how to do what the family was doing, and tried to find out how they learned. During the ploughing of the nursery plot, Swapnaja asked the family how they had learned to farm:

06:16	Swapnaja:	तुमचे वय काय आहे? (What is your age?)
06:18	Bhimraj:	चौतीस पसूतीस. (34–35.)
06:21	Swapnaja:	कोणत्या वयात आपण शेती शिकिलात? (At what age you have learnt farming?)
06:25	Smita:	दहावीला शाळा सोडल्यानंतर माझा अंदाज आहे. (I guess after leaving school at 10th standard.)
06:46	Ankita:	त्याच्याआधी लहानपणी पण येत असतील ना, शेतावर? (In childhood you might have gone on farm, nah?)
06:51	Smita:	लहानपणी म्हणजे आमचे बबलु-प्रतिकि येत असतात तसं. (Childhood means, the way our Bablu-Pratik come, that way.)

Of course, it would be difficult for anyone to answer the question, "When did you learn farming?" Bhimraj was busy when he was asked, and Smita answered for him. At first, she said Bhimraj learned after Class X – when he finished his formal education. Probably this was when he became more seriously involved full time in doing farming. As we mentioned before (Sect. 2.3), Smita and Bhimraj thought of farming as something that you do, not something that you learn. But Smita also agreed, when asked a leading question, that Bhimraj was learning when he was younger, just like his sons are.

They were often vague about dates and the measurement of the years. This may appear to be a sign of illiteracy and a lack of scientific temper. Or it may simply be

K. Haydock et al., *Learning and Sustaining Agricultural Practices*, International Explorations in Outdoor and Environmental Education 7, https://doi.org/10.1007/978-3-030-64065-1_8

that precision for these things are actually not important. Knowing the date of one's birthday is one of the things that comes with a formal education: it is usually printed on the school certificates that are used for various bureaucratic and political purposes, and without such things it is difficult to prove one's existence. But this use of birth dates may be signifying more about the level of development of the system than of the scientific temper of a human being.

8.1.1   What Did You Learn About Farming in School?

Much later, during harvesting, there was a discussion about education:

02:20	Ankita:	तुम्ही कितिवी पर्यंत शाळेत गेलात? (Up to which class did you go to school?)
02:21	Smita:	दहावी. (10th.)
02:22	Ankita:	दहावी. (10th.)
02:22	Karen:	So anything she learned in school helps her in this work?
02:26	Ankita:	शाळेत शिकलेल्या गोष्टींची तुम्हाला काय मदत होते इकडे? (How did school help your work?)
02:30	Smita:	इथे? (Here?)
02:31	Ankita:	हां म्हणजे काही काम करताना कशाशी-म्हणजे शेतातलं, पुन्हा इथे काही विकायचं असेल तर, किंवा काय, कुठेही. मदत काय होते? (Yes. Meaning, whatever work you do, some work, in the field, that is, like if you are trying to sell something? Or anything? What was the help?) [Smita looks at and fiddles with a stalk of paddy, with a smile and a furrow between her eyebrows as Ankita finishes asking the question.]
02:43	Smita:	[raises her eyebrows and beginning to speak] अशी- (That way -) [pauses, as she hears Dnyaneshwar]
02:46	Dnyaneshwar:	[Smiling, teasingly] जेवण घेऊन जायला!, शेतावरती (To carry food!, in the farm.) [Smita laughs, shaking her head in the negative.]
02:51	Smita:	नाही. शे-म्हणजे शाळेचा काय उपयोग होतो का शिक्षणाचा काय उपयोग होतो? (No. What is the use of school, what is the use of education?)
02:55	Bhimraj:	शिक्षण म्हणजे कसंय काय कुठं जायचं झालं तरी पण ते तेवढं- (Education means that even if we want to go,)
02:58	Smita:	नाही नाही शेतीच्या कामासाठी काही उपयोग नाही. (No, no, nowhere in farming, is there any use.)
03:01	Ankita:	नाही पूर्णच सांगा तुमच्या आयुष्यात तुम्हाला काय उपयोग होतो-कुठे? (No. Tell exactly – how does it help you in your life – where?)

03:04	Bhimraj:	शक्षिणामुळे काय गाडी-बडी घेऊन कुठे पुरवास झाला, तर ते एसटी-बसिटी वैगरे याला. म्हणजे – (With education you can travel anywhere by vehicle, meaning by S.T [State Transport bus]-)
03:09	Ankita:	ते कळतं ना- (That is known, no-)
03:10	Bhimraj:	कोणत्या ठिकाणी जाणार ते – (-meaning which place it goes -)
03:10	Smita:	शक्षिणामुळे समजतं ना आपल्याला कोणती खतं शेतीसाठी वापरली जातात. [...] शक्षिणामुळे सांगतात ना आपल्याला! (From education we understand which fertilisers are used for farming [... inaudible] From education we understand.)

Thus, when asked whether anything they learned in school helps in doing the cultivation work, Dnyaneshwar jokes, "To carry food!", probably referring to the harvest they were about to cart on their heads to the house. But Smita seems to be either puzzled or thoughtfully considering the question, and then she laughs at Dnyaneshwar's reply (Fig. 8.1). She asks rhetorically, with a smile, "What is the use of school, what is the use of education?". Then, with a serious expression she says that education is of no use anywhere in farming. However, Bhimraj then suggests that going to school did help them in being able to travel around: they can read the name of the destination printed on the front of a bus. (Karen smiled, suddenly

Fig. 8.1 Smita upon hearing the question, "Was your education any help in farming?"

remembering how she finally began to learn the Punjabi script when she needed to travel on buses in Punjab.)

But after this, with further probing and a few leading questions from Ankita, they do mention a few other things they learned in school: something about manure and chemical fertilisers, and the production of gobar gas. When prompted by Ankita asking about mathematics, they say that it helps in doing the accounts for the farm – especially when selling the rice or when hiring labourers.

When asked if anything they learned on the farm helped them at school, Bhimraj laughs and replies, "नाही! शाळेमध्ये फक्त एक-झाडं लावण्यासाठी!" (No. Only in planting trees in the school!). He was referring to the occasional government tree-plantation drives in which students are expected to help plant trees on school grounds.

Thus, it is clear that other than general literacy, formal education did not help much in learning how to do work on the farm. No one mentioned that learning botany or other types of biology might have helped understand plants, which might have helped in doing cultivation. Most likely, the long lists of 'facts' which they had tried to memorise for exams were forgotten because they were not meaningful, important, or necessary for anything except the exams. Besides something like the tree plantation, they did not have any practical hands-on experience in school. Even the implementation of the school 'laboratory' exercises in 9th standard hardly included hands-on work.

However, this is not to say that education is not important.

8.1.2 Do You Want Your Children to Go to School?

Everyone in the farming family agrees that school education is necessary and desirable.

Smita explained why education is necessary: "म्हणून शिक्षणच महत्वाचं आहे. शिक्षण केल्यानंतर पैसा मळितो. मी तर अशी समजते." (So education is important. Money is gotten after education. This is what I understand.) The entire family thought that if the children become educated, they may get jobs outside of agriculture that would allow them to earn more and have a better life.

Smita and Bhimraj are very clear that they want their children to go to school, study hard, and get a good education. They will pay for the best schools for them, even if it is very difficult to find enough money.

A conversation about the use of formal education occurred when we visited the family in 2018, three years after our initial visits:

10:29 Gurinder: आप वो – अ – आपको लगता हैं कि बिच्चों को पढ़ाना चाहिए?
                   (You that – uhh – do you think that the kids need
                   to be educated?)

10:32 Smita: हां, थोडासा. अभी हम इतनी मेहनत करके क्या मलिता
हैं खाने के लिए? इसी लिए! ऐश-आराम तो कुछ नहीं! सिर्फ
दिन-रात ये मेहनत करते – हां, बच्चों को भी नहीं ना ऐसा थोड़ा –
ऐसा. (Yes, a little bit. Now we work hard but what do we get
for eating? That is why! There is not even any rest! He only
works hard all day and night – Yes, Even kids should a little
bit – This is the way it is.)

10:46 Gurinder: तो आपको क्या लगता हैं कि बि- पढ़- पढ़ाने के बाद – मतलब के क्या
हो जाएगा? (So what do you think that edu- after educating –
meaning, what will happen?)

10:54 Bhimraj   ... पढ़ाई अच्छी हुए तो इनकी life अच्छी होएगी.
(They [our sons] will have a good life if they do well
in studies.)

10:57 Gurinder: हम्म. हम्म. हम्म. (Hmm. Hmm. Hmm.)

10:58 Bhimraj   पढ़ाई अच्छी नहीं हैं तो life नहीं अच्छे (Without doing well in
studies, life will not be good.)

11:00 Gurinder: हां. हां. (Yes. Yes.)

11:00 Bhimraj: ऐसा हैं. (This is how it is.)

11:02 Gurinder: ख-ख- खेती में लाइफ अच्छी नहीं कर सकते? क्या कर सकते हैं?
(A f- f- farming life cannot be good? Or can be good?)

11:06 Smita: नहीं, कर सकते हैं- (No. It can be.)

11:07 Bhimraj: खेती में तो कर सकते हैं- (Even in farming it can be.)

11:09 Smita: [continuing to speak, over Bhimraj] लेकनि थोडा पैसा- वैसा तो
चाहिए ना! अभी हम मेहनत मेहनत करके करते, हैं ना साहबि? अभी
वो उनका कधिर नौकरी वैगरे मलि जायेगी, तो पैसा लगाकर खेती हो
सकती हैं! ऐसा नहीं, ना? खेती पैसा मेहनत कर भी हो सकती. अगर
हम नही वैसा करते – हमारे पास पैसा नही-नौकरी कधिर हैं इनके
पास तो. हम खुद मेहनत- बहुत मेहनत करते हैं. तो नौकरी कधिर
जायेंगे? नौकरी मलि जायेगी, अच्छी पगार मलि जायेगी- और खेती
तोकरेंगे, उनका उधर भी काम हैं, ये ऐसा हैं उनका. (But a little
money – money is just needed, nah! Now we just do work and
do work, isn't it, sahib? Now if they [the children] get a job, etc
somewhere, then by investing some money we can do farming.
Not like this, nah? Farm work can also be done after investing
money. Now we do not do like that. We don't do like that – we
don't have money- where does he [Bhimraj] get a job? We
ourselves work hard – extremely hard. So where to go for a
job? If a job is found, a good pay check will be found – and
farming will be also be done, over there is also work for them,
this is how they have it.)

Here Bhimraj is saying that doing well in school is required for a good life. Smita
says that with education, their children will get jobs somewhere outside of agriculture, so that that they will earn money.

Smita realised that more money is needed in order to do good farming. She did not think that farming should be done the way they were doing it. Farming could be a good life only if they had more money to invest in it.

Gurinder and Abhijit suggested that there might be some other purposes in education, besides just enabling people to get better jobs. But Smita and Bhimraj insisted that education is needed in order to get good jobs, and that without education it is not possible to earn good wages.

Abhijit and Gurinder were distressed when Bhimraj and Smita said that they did not find farming to be a good life. They were discussing this with them, maybe hoping to convince them that it is not so bad in some ways. Or maybe Abhijit and Gurinder were realising that Pranay and Pratik may end up in farming, despite their education, so they were trying to be compassionate. Perhaps they also were hoping that education could somehow be useful in farming. or that it might not necessarily be so bad if their sons ended up in farming. Abhijit asked Smita whether they are telling their sons how to do farming by discussing what they are doing with them, but again the conversation veered towards the difficulties of the farming life:

14:09   Abhijit: तसं, तुम्ही, मी असं नाही म्हणत की शेतीसाठी तयार करता आहात, पण असं तुम्ही असं कधी सांगता का- की अरे, शेतीमध्ये असं करतात.?
(Like that, you – I am not saying that are you readying [them] for farming, but do you tell [them] like this – that, hey, this is what is done in farming.)

14:14   Smita: नाही, नाही. नाही, नाही, आम्ही नाही सांगत असं.
(No, no. No, no, we don't tell like this.)

14:18   Abhijit: का नाही? (Why not?)

14:19   Smita: नाही सांगत. सांगू का? म्हणजे, त्याच्यामध्ये मेहनत पण भरपूरच लागते. आणि main म्हणजे पैसा पाहिजे. आता आम्ही थोडं कमवून ठेवला पैसा तरच त्यांना तो उपयोगाला येईल की नाही? (Do not tell. Shall [I] tell you? Means, it requires extreme hard work, and mainly, money is needed. Now, if we make a bit of money, it will only help them, isn't it?

14:27   Abhijit: बरोबर. (Right.)

14:28   Smita: ते नुसते शेतीत गेले तर त्यांना काहीच life नाही. आमची सुद्धा काही life नाही. फक्त कमवायचं, खायचं. हेच life आहे की नाही? मग आम्ही काय म्हणतो, शेतीमध्ये दुर्लक्ष नका करू, त्यांना म्हणत पण नाही करा म्हणून. (If they go just into farming, they have no life at all. Even we have no life at all. Only earn, eat. This is all the life, isn't it? Then what we say, don't neglect farming, don't even say that do not do.)

14:39   Abhijit: अच्छा. (Achha.)

14:40   Smita: मूहणजे आम्ही काय मूहणतो, थोडं शक्तू द्या. दहावी होऊ द्या, बारावी
होऊ द्या. मग दहावी झाली की मग आम्ही त्यांना वचिरू त्यांना
कोणत्या side मध्ये जायचंय, त्यांना कशामध्ये इंटरेस्ट आहे, आम्ही
अगोदर आमच्या मुलांनाच वचिरू! जर ते बोललेच की "मम्मी
तुमच्यासारखीच आम्हाला शेती करायची आहे, तर मग चांगल्या
पद्धतीने करा. आमच्यासारखी नका करू. (Means, what do we say,
let [them] study a bit. Let tenth be done, twelfth be done.
Then after tenth is done then we will ask them which side they
want to go, in what they have an interest, we would ask our
child first! If they insist that "Mummy we want to do farming
just like you, then do it in a good manner. Not like us.)

So as we see, Smita and Bhimraj are saying that they do not try to teach farming
to their sons by discussing it with them. Again they are repeating that there is no life
in farming, and that is why their sons should not go into farming. Smita is saying
that they want their sons to decide for themselves what to do. But in case they
decide to do farming, they should not do it the way the family was now doing it.
They should first get enough money so that they can do it in a better way – with
modern technology, materials, and methods. But through farming they cannot earn
enough money for more investment. The only way to get more money is through a
job outside of agriculture, which their sons might be able to get if they get a good
education. This is what Smita and Bhimraj told us repeatedly.

They all agree that in farming, the physical labour is too hard. As Smita said,
"त्याच्यामध्ये मेहनत पण भरपूरच लागते" (It requires extreme hard work). And the
money is insufficient. Although they would like to, they cannot afford to adopt more
modern methods that would lessen their physical labour.

As we discussed above (Sect. 8.1.1), in 2015 we had asked the family whether
anything they learned in school helps them in farming. Their immediate answer was
in the negative, although with some probing they did mention a few things. Once
again, in 2018, Abhijit asked Smita and Bhimraj whether they learned anything
about farming in school, and they both said, "No. No." Even after Abhijit asked,
"Nothing at all?", Bhimraj said "Nothing." and Smita said, "No."

Furthermore, they do not want their children to go to school to learn agriculture.
Even now they discourage the boys from accompanying them to the fields. They say
that people know how to do farming by working in the fields, and it need not be
taught in schools. Children go to school in the hope that education will enable them
to get jobs outside of agriculture.

Regarding the nature and goals of the school education that they hoped their
sons were getting, Bhimraj said, "वो कैसा हैं, पढ़ाई होने के बाद में, वो, उनको सोच
आना मंगताय, की, कोनसा समझो – अभी Science है, ये सब. हैं ना? Arts है, -" (How it is,
after finishing studies, they need to have developed their [own] views [thinking],
that, which do they understand – now there is Science, all those. Isn't it? There
are Arts -). Nodding in agreement, Smita added, "Commerce," and Bhimraj continued, "Commerce है ना? वो-वो उनको कैसे, वो मेरेको ये होएगा, ये मेरेको अच्छा

जमेगा. ये पाठ मेरेको अच्छे से जमेगा. तो वो उसके हिसाब से वो करेगा. ऐसा हैं. (Commerce, isn't it? That- that's how is it, that would be for me, this would be good for me, this lesson [subject] I am very good at. Then he will do accordingly. This is how it is.)

Thus, they are saying that they would like their sons to be able to continue after Class X on their State Boards, where they could decide for themselves whether to study in Arts, Commerce, or Science streams. They said that they value a basic sort of education that would allow their children to go into various fields, rather than teach them specific skills or prepare them in just one kind of job. Bhimraj says that they need a basic education so that they will have the option of going into various occupations, from science to arts to commerce.

Smita and Bhimraj see farming as a career only if all else fails. But, as we discussed (Sect. 7.2.1.7), often all else does fail, and this is one of the main reasons why small family farming persists in India. Bhimraj said that if their sons do not get jobs as a direct result of their education, then they could try going into business. He said that only if that also fails should they go for agriculture. Unfortunately, there is a great chance that attempts at business will also fail.

They hope that their sons will not be trapped in doing manual labour. Although Smita said that the aim of education is to get jobs, she also mentioned that education is useful in helping people to teach their own children. She realised that, despite schools, the education of the children depends on the education of the parents.

Smita said later on, "अभी देखो, अभी पढ़ने से कुछ-पैसा कमा सकते हैं हम! पढ़ेंगे तो कधिर भी जाकर काम कर सकते हैं. अभी पढ़े-लिखे नहीं तो क्या – वो खेती में रहना पड़ेगा ना!" (Now look, now from studying we could have some money! If we study, then wherever we go, we can work. Now without reading and writing what happens – we have to stay in farming, nah?)

Bhimraj and Smita are alluding to the ideology that education can help one realise the apparent 'freedom of labour'.

However, in the case of both Smita and Bhimraj, education did not secure them other jobs. Bhimraj had many years ago gone to Nashik and gotten a low-paying contract job doing manual labour in a private gas agency. But then, when his father fell ill (due to overwork, Namdev said), Bhimraj had to come back to do the farmwork, and then he stayed on. When crops are being produced and animals need feeding and milking, not even one day can be taken off due to illness unless there are other people to help out.

On the one hand, Bhimraj appears to have the freedom to sell his own labour in the 'free market' in any way he wants. But actually, this 'freedom' is the freedom of capitalists to exploit labour power in a competition to increase their own profit. In the private sector in India there is no compulsion to even give workers minimum wages, or any kind of benefits. Employers are free to pay whatever they can get away with, or even to suddenly close down or relocate. Except for government jobs, workers have virtually no control over the length of the working day. In his work off the farm, Bhimraj was actually not free to decide where he would live, whether he would live with his family, how he would choose and perform his work, or what his working hours or wages would be.

8.2 Education, Skill, and Certification

Upon observation and analysis, we found that the family is already very skilled at doing farming, that to some extent they do science in the process, and that they learn and teach each other using very progressive methods of 'learning by doing', 'trial and error', and 'constructivist learning'. We questioned our initial aims of trying to teach farming in schools.

Although when the boys were young they accompanied the rest of the family in the field, they were not forced to work. But as the boys go into higher classes, their parents no longer let them go into the fields, worrying that their studies may suffer. Smita said that Pratik will not get time to go to the fields after Class X. She worried that he may prefer going to work in the field because it is easier than studying and sitting for exams. She said, "ऐसा हैं. हैं ना? ये काम easy हैं, कोई भी कर सकता हैं, हैं ना?"! (It's like this, isn't it? This work [farming] is easy, anybody can do it, isn't it?!)

One reason for their thinking that farming is easier than school work may be that they have more experience in farm work and are better at it – more skilled. But probably their belief is also conditioned by general social beliefs that work with the hands is of less value than work with the mind. This is because value is thought to depend on the wages that are received, and people do generally get lower wages for manual labour, especially in agriculture. This interpretation is reinforced by the following discussion. It took place in 2018, between Bhimraj and Smita and the researchers: Gurinder, Karen, and Abhijit:

38:35 Bhimraj: बताता हैं, बताता हैं. अभी सोचो आपुन-एक मनिटि-आपुन – अभी आपुन चार आदमी हैं, ठीक हैं? मैडम बड़ी हैं, हां? मैडम बड-बड़े हैं. तो आपुन चार आदमी कैसे बोल रहे हैं? मैडम बड़े हैं. (I'm telling you, telling you. Now think we- one minute- we- now we are four people, right? Madam is great [the biggest, oldest, most privileged], right? Madam is very great. So what we four people are saying? Madam is great.)

38:55 Gurinder: हम्म. (Hmm.)

38:55 Bhimraj: ठीक हैं? मैडम बड़ी है कि छोटी हैं, उनको-मैडम को पता हैं, लेकिन चार लोगों का दृष्टिकोण कैसा होगा, अच्छी मैडम हैं. (Okay? Madam is great or small, that Madam knows herself. But what will be the view of people, Madam is great.)

39:06 Gurinder: नही, नही, पर उनको नही पता, ना? उनको जैसे आप खेती में-जैसे आप- (No, no, but she does not know, nah? She, just like how you -just like how you in the farming-)

39:11 Bhimraj: उसके हिसाब से वो जो बोलेंगी ना, मैडम जो बोलेंगी
ना, उसके हिसाब से अपुन चलता हैं. (In sum, whatever is said, nah-
whatever Madam will say, nah, in sum we go like that.)

39:14 Gurinder: नही नही. पर आप- पर आप- पर मान लो- नही- ठीक हैं, हम
एक- क्या करते हैं, कल- नही नही- कल से- कल से जो हैं,
आप खेती नही करेंगे, सिर्फ मैडम जा के खेती करेंगी.
तो मुझे नही लगता ये काम याब होंगी. (No no. But you- but
you- but suppose- no, okay, we one- what do we do,
tomorrow- no no- from tomorrow- from tomorrow what
happens, you will not do farming, only Madam will go and do
farming. So I don't think she will know (remember)
how to work.)

39:30 Bhimraj: नही. (No.)

39:34 Gurinder: तो उनको फिर ज्यादा कैसे पता हैं, ना? उनको फिर
ज्यादा कैसे पता? (So then how come she [knows] more, nah?
How do you know hers is more?)

39:36 Bhimraj: हां. वो ज्ञान से पता हैं. (Yes. It is known she has knowledge.)

39:38 Gurinder: नही पर मान लिजिये, उनको भेज देते हैं खेत मैं, और
इनके साथ मैं भी चला जाता हु, और अभिजित भी चले जाते हैं,
तो हम तो मुझे नही लगता के हम कुछ कर पायेंगे! और हमे कुछ
समझ मैं भी नही आएगा. (No, but suppose, she is sent to the field,
and I also go with her, and Abhijit also goes, then we, I don't
think, that we would be able to do anything, or understand
anything at all!)

39:52 Bhimraj: नही. नही आएगा. (No. It won't. come.)

39:54 Gurinder: तो फिर इनको ज्यादा कैसे पता हो गया, ये मैं बोल रहा हु!
(So then how does she know more? That is what I am saying!)

39:57 Abhijit: म्हणजे तुम्ही ठरवलं की ह्या मोठ्या आहेत. तुम्ही काय
म्हणालात की आपल्या दृष्टिने ह्या मोठ्या आहेत. बरोबर?
पण, तुम्ही शेती actually तुम्ही करता. मग तुम्ही मोठे
झालात ना, त्या अंदाजाने बघायला गेलं तर. (Means, you
decided that she is big. You said that from our perspective
she is big. Correct? But, you actually do the farming.
Then you are big no, if we have to see it that way.)

40:09 Karen: ये ज्ञान नही हैं? ये ज्ञान हैं.
(Isn't this knowledge? This is knowledge.)

40:14 Abhijit: तुम्हाला भाजवळ कशी करायची, कवा कुठल्या जागेवर
भाजावळ करायची, tractor ने केल्यावर काय फरक पडतो, बैलाने
केल्यावर काय फरक पडतो, हे तुम्हाला माहिती आहे. ह्या काय
करताएत, समजा कुठे तरी बसून-त्यांनी पुस्तकं वाचली, त्याच्या
जोरावर ते तुम्हाला-तुम्हाला मचाशी नाही का म्हणाले 'बाहेरची
लोकं तुम्हाला सांगतात असं करा, तसं करा' ती लोकं-
(You know how to do rabbing, or in which place to do
rabbing, what difference it makes by doing with a tractor,
what difference does it make by using an oxen, you know that.
What she is doing is, assume she was somewhere and she
read few books, on that basis didn't she tell you,
'People from outside tell you, do this, do that' those people-

40:34 Karen: कतिाब लखि सकते है, पर कतिाब नही खा सकते ना?!
(Books can be written, books can't be eaten?!) ...

41:00 Bhimraj: नाही, नाही, म्हणजे कसंय, त्यांनी आता कृषीचा म्हणजे
course केलेला आहे आणि त्यांना तो पद भेटलेला आहे. तसं
आम्हाला-आपल्याला आहे ना, अपुन को है ना, पद नही है,
(No, no, means how it is, They have done a course about
agriculture and she has that degree/post. That way for us-for
us, for us no, there is no degree/post,)

41:15 Abhijit: पण, पुस्तकात वाचणं की जमीनीची अश्या रीतीने भाजावळ करतात,
आणि प्रत्यक्षात तशी मशागत करणं, ह्यामध्ये जमीन-अस्मानचा
फरक आहे. (But, to read in books that this is how rabbing is
done, and to do the rabbing in real is a vast difference.)

41:27 Bhimraj: हो. (Yes.)

41:27 Smita: हो. आहेच. (Yes. It is.) ...

41:46 Bhimraj: नाही नाही. पण ते कसंय, जरका तुम्हाला सांगू, आता तुम्हाला जे
करायचं आहे, बरोबर आहे? ते तुम्ही काय करणार, आम्हाला पैसे
देणार, आणि आमच्याकडनं करून घेणार. तुम्ही शेतात पाय टाकणार
नाही! असंय! (No no. But how it is, if I should tell you, now
whatever you have to do, alright? What you will do,
pay us money, and it will be done by us. You will not set
your foot in the field at all! See!)

As we see, Bhimraj began by comparing the four adult members of his family
(Prabhavati, Namdev, Smita, and Bhimraj), to the researchers who were sitting in
their house (Karen, Gurinder, and Abhijit). But he was recognising that Karen was
the 'leader' of the researchers, so he was speaking about her position as compared
to the position of the family.

When Bhimraj says that Karen is ' बड़े ' (big, great), he probably means that she
is of higher status: older, foreign, white skin, earning more, more highly educated,
and also more 'capable' than the family is. But, interestingly, he may actually have
some second thoughts, or doubt about her value, as indicated when he said, "Madam
is great or small – that Madam knows herself".

He refers to the books Karen makes, possibly alluding to the belief that this sort of work with the mind is of higher status than work with the hands. But, there seems to be an implicit question behind his argument when he says, "Now she only makes books to be printed. That's correct. But she has worked so hard for that." Maybe he is asking himself whether making books is really so important – perhaps in comparison to producing food that no one can live without. But then, he provides an answer by justifying that books are important because in order to write books one has to 'work hard' in the sense of going through so much of schooling. Both he and Smita seem to consider intellectual work with the mind to be more difficult than work with the hands.

But is the work in the field actually less valuable, or does it actually require less skill, than, say, the intellectual work of writing a book? This point came up later on in the same conversation. Karen had just gifted the family a copy of one of the children's books she had written and illustrated. Karen (who has a PhD) had been earning around Rs 70,000 per month (before her retirement) doing research and writing this book. She knew that the family had received Rs 35,000 for the paddy they had cultivated last season.

Karen asked the (rhetorical) question: Why did she earn around Rs 70,000 in one month, while the entire season's harvest of paddy (work by several people, over a period of 8 months) was sold for only Rs 35,000? Bhimraj answered, "नाही, आता त्यांनी एवढी मेहनत केलेली आहे, तेवढी-असं." (No, she has taken this much of efforts, that much. That's why.) Smita said, "त्यांनी शक्षिणाला पैसा दलिय. तसं आम्ही काय केलंय, आम्ही आमच्या शेतीतच टाकलं ना?" (She has invested more money in her education. Like that what have we done, we have invested it in farming only, no?).

We were confused. On the one hand, Smita and Bhimraj seemed to believe that they are not skilled. But on the other hand, what they were doing indicated that they are very skilled. And probably they actually (subconsciously) knew that they were skilled. Maybe this was one of those 'unknown knowns'. They did not know what they knew. They were living with an ideology that devalued their skill – measuring it on the basis of its (fake) 'monetary value'. Their illusion lies in their not knowing that they know. This ideology is being socially imposed on us all, leaving highly educated people believing that they 'know' what they do not know, and believing that uneducated people really are less skilled than educated people.

When the researchers objected to Bhimraj's saying that Madam is greater, he and Smita finally agreed that Karen and the other researchers do not have the skills to cultivate paddy. But still, Bhimraj argued (at 41:46) that if Karen had to do farming, she would be able to do it – by simply paying the family to get it done: "What you will do, pay us money, and it will be done by us. You will not set your foot in the field at all!"

This reminds us of Marx:

The extent of the power of money is the extent of my power. Money's properties are my – the possessor's – properties and essential powers. ... Do not I, who thanks to money am capable of all that the human heart longs for, possess all human capacities? Does not my money, therefore, transform all my incapacities into their contrary? (Karl Marx 1844, 60)

Smita and Bhimraj actually know that they work very hard, and that their work is very important because everyone depends on the food farmers produce. But sometimes they also talk as though they do not know it. They value themselves and their labour by the amount of money they get. They submit to the ideology that they are poor because they do not work hard enough and they do not do important work, because they do not have enough education and are not smart enough. This ideology is not just an illusion – the effects are real: money does have power, exams and work with the mind are difficult, and the family actually is poor. We might rather call this ideology a social illusion which is rooted in real contradictions. Thus, it cannot be changed just by trying to change the way of thinking, say through education. Its change also requires removing the real contradictions through social activity – changing the real conditions that determine, and are determined by the ideology. As long as Karen has the money and the power to hire workers, the 'illusion' will exist that she has the power to do paddy cultivation.

So what then is the actual significance of school education? As Smita said, education is an investment. This means that an educational certificate is a commodity, that can be bought and has an exchange value for obtaining a job. The purpose of education is reified in the certificate it produces. A certificate is just a piece of paper, but it seems to have a magical property of proving that one is 'knowledgeable' or 'capable', and conferring employment. In other words, it is a sort of fetish that we 'worship', even if we actually know that it does not really prove that we have learned anything, and even if we know that its magic does not always provide jobs.

An educational certificate is a commodity, that can be bought and has an exchange value for obtaining a job. It's true that it is just a fetish, but like any commodity fetish, the magic it may perform is real!

8.2.1 Deskilling Accompanies Capitalist Development

We have described how Namdev's bamboo basket making is highly skilled in comparison to the making of plastic baskets by workers in factories (Sect. 3.2.4.3).

One reason why this contrast between skill and unskill was so striking to Karen is her experience of deskilling in her own family. Her parents were visual artists whose formal education consisted of high school followed by art school (not university). But they were both highly skilled. The entire family – mainly her father – designed and built their large, modern house from scratch with their own hands over a period of more than 10 years (beginning in 1966), while the three children were between 10 and 25 years old. This included using the skills of drafting, ditch digging, stone quarrying, adobe construction, plumbing, electrical fitting, welding, painting, masonry, and carpentry. Before this, her father had helped his brother build a house and a boat. He had also worked on a sailboat and a farm (as a conscientious objector during the 2nd World War). He was largely self-taught, learning by doing. He did all this work himself partly for economic reasons (he did not want to become too indebted), but also because he wanted to be more free and independent – in order to get what he wanted and what he designed and planned himself. It seems crazy that he spent so much time doing all this. He was going against the modern trend towards specialising (the Renaissance man, Leonardo do Vinci, was one of his heroes). He refused to be an amateur or a dilettante in any of the things he did. He studied and did experiments and investigations in order to figure out how to do things. For example, he planned the orientation of the windows and roof overhang after measuring the direction of the rising and setting sun, and its inclination at different times of the day and year. He got an electricians' handbook, taught himself, and passed an exam to become a certified electrician so that he could legally do the wiring. Almost all the furniture, light fixtures, door handles and locks, were also designed and made by her father (including making extensive drawings at the design stage). All this was done while he had a full-time government job as a curator in a museum. Although the other family members also helped, and picked up many skills, clearly her father was much more skilled than any of them, in the sense that he was more experienced and spent more time producing more things. We found the things he made to be well-made, long-lasting, repairable, cleverly designed, functional, and beautiful. He always stressed the need to try to harmonise structure and function. He made things for our own use value, not for exchange value. Both of Karen's parents were also highly skilled at artwork, various crafts, cooking, cleaning, and sewing clothes, quilts, and upholstery. Karen's mother's art-work included book-making (her father was a professional bookbinder). Karen and her two brothers were encouraged to complete university education, and they all ended up with PhDs. All of them moved off to distant places, not finding jobs locally. None of them went on to build their own house, although they all did repairs, helped other people doing construction, and occasionally made furniture and other things. Now Karen's brother has his own workshop at home that is well-equipped with modern wood and metal working machinery, much of which did not exist 50 years ago, although he hardly has time to use it. Compared to how things were done 40–50 years ago, it is amazing how much easier it is to work with this equipment. But even though work is quicker, who has the time? The skill of Karen and her brothers, although less than their father, is an order of magnitude greater than the skill of their children, even though they grew up in houses where all sorts of tools were available and where their parents were sometimes using the tools to make things. Just being around people doing skilled work was not enough to pick up the skills. Even the artwork that the three generations have done has involved progres-sively less skill. Their skills have also become more specialised. Karen's daughter does computer graphics, jewellery making, visual art, and some sewing, but at her age, her mother did all these in addition to pottery, sculpting, welding, and carpentry.

It is easy to assume that the deskilling over the generations in this particular case is personal and idiosyncratic. However, we claim that actually it may be a general phenomenon: as formal education increases, skill decreases. Of course, it is difficult to get evidence for this, because 'skill' is almost impossible to quantify, let alone define (see Sect. 5.3).

As capitalist industrialisation develops, tools become more specialised, machines replace hand tools, there is more specialisation and division of labour, and the labour each person does becomes more uniform. This happens in the industrialisation of paddy cultivation, just as it does in other industries. In large corporate farms (e.g. in Japan), one worker lifts trays of sown seeds off of a rice seeding machine. Another worker places the trays onto a transplanting machine, and another worker drives the transplanting machine up and down the rows of the field, clicking icons on touchscreens to regulate the transplanting.

It is obvious that all of these types of labour require less skill than the skill that is required by the members of the farming family who we observed. As we saw, ploughing a field with a pair of bullocks requires much more skill than ploughing the same field with a tractor.

However, it is commonly thought that people who are unspecialised are unskilled, incompetent, and clumsy, and in need of training and vocational education. A person whose is trained at driving a tractor and whose only work is driving a tractor may be classified as skilled. But a person who has no formal training, ploughs with a pair of bullocks, and does hundreds of other kinds of work on the farm, will be called unskilled, no matter how much experience they have or how good the results are.

Industrialisation, including the industrialisation of agriculture, requires deskilling, not skilling. This is true even if we include the various types of '21st Century Skills', such abstract thinking and mental 'work' along with manual labour in our definition of 'skill'.

With industrialisation, corporations make increasingly larger profits. This is accomplished by increasing productivity, increasing intensity of labour, speeding up the labour process, decreasing the time required for production, increasing specialisation, and decreasing the time taken up by gaps in labour (e.g. one person performing all the steps needs to move from place to place, get different tools, take rests, etc). "On the other hand, constant labour of one uniform kind disturbs the intensity and flow of a man's animal spirits, which find recreation and delight in the mere change of activity" (Karl Marx 1867, 239). But then, with industrialisation, profit increases. The questions that are too often not asked are: Why should our entire way of life be governed by the quest for profit by capital? Why should efficiency and speed be so important? What about actually making life easier, and reducing labour time so that we all have time to relax and construct our own creative, purposeful lives?

If we ask what is the actual source of the profit, we find that one of the main sources is that, since each task requires less skill, less expenditure of time and money is required to train workers. People do not need to be taught skills, because hardly any skill is needed for most types of employment, and anyway there are very few opportunities for employment, and people are forced to keep jumping from one job to another. Why should the 'efficient de-skilling' mentioned in the National Education Policy (see Sect. 5.2) require much expenditure of money?

Marx wrote about the differentiation that occurs with the development of Capital. It mirrors the differentiation that was performed by the caste system. Tasks, and the workers performing the tasks, are differentiated into hierarchies:

> Since the collective labourer has functions, both simple and complex, both high and low, his members, the individual labour-powers, require different degrees of training, and must therefore have different values. Manufacture, therefore, develops a hierarchy of labour-

powers, to which there corresponds a scale of wages. If, on the one hand, the individual labourers are appropriated and annexed for life by a limited function; on the other hand, the various operations of the hierarchy [of labour-powers] are parcelled out among the labourers according to both their natural and their acquired capabilities. Every process of production, however, requires certain simple manipulations, which every man is capable of doing. They [these manipulations] too are now severed from their connexion with the most pregnant moments of activity, and ossified into exclusive functions of specially appointed labourers. Hence, Manufacture begets, in every handicraft that it seizes upon, a class of so-called unskilled labourers, a class which handicraft industry strictly excluded. If it develops a one-sided speciality into a perfection, at the expense of the whole of a man's working capacity, it also begins to make a speciality of the absence of all development. (Karl Marx 1867, 243–44)

With the development of capitalism, work with the hands gets more and more separated from work with the mind. We have seen how this happens in agriculture. It probably happens in every field. For example, with capitalist development, art becomes individualised and commodified and 'fine arts' become segregated from 'crafts' (or 'decorative arts') and 'artists' from 'artisans'. Nowadays a basket maker is considered to be an artisan, not an artist. A basket is not considered to be a piece of art, because it retains too much use value. In today's world, art becomes pure exchange value.

Are we saying that we should remain (or go back to) doing basket weaving and gruelling manual labour in paddy fields? No. Rather, we need to ask fundamental questions and investigate our historical development in order to understand our world and collectively work towards making a new world that is more just and egalitarian. This is what learning should be all about, whether we call it education, training, or skilling. Real learning is a process in which dialectical conflicts are explicitly revealed. It is a process of realising that we were under an illusion that we had a complete understanding, which we now see is incomplete.

8.2.2   Ability and Skill

'Skill' is commonly defined as the 'ability' to do things. Ability is a strange conception. A number of questions arise when we try to define and assess ability. First of all, what are the particular things that people need to do and why is ability defined in terms of those things? This question is related to the question of how the syllabus is defined in an education system. There is a conflict between the perceived need to teach practical skills, the need to teach abstract knowledge, and the need to appear to function as a sorting device. Formal education has always been related to preparation for future employment. But especially in India, formal education has not been seen to be very necessary for most kinds of physical labour – such as agricultural labour. Formal education has been designed more for white-collar employment – work with the mind rather than work with the hands.

Another problem is that ability seems to be a prediction about what people will do in the future based on what they did in the past. However, the prediction may not

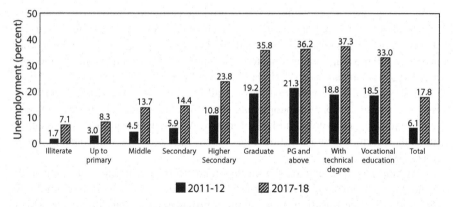

Fig. 8.2 Youth unemployment rates by level of education in India, 2012–2018

take into consideration whether or not people had the opportunity, need, or desire to do the things for which they are being assessed. For example, peasants who are not using modern farming methods may be assessed as not having modern skills, or not knowing how to do modern agriculture. There may also be an assumption that what is holding them back from agricultural development is this lack of skills. Their 'ability' may be defined in terms of their knowledge, inherent intelligence, access to education, or capacity to assimilate a body of knowledge, rather than simply the amount of money they have. Even if economic constraints are acknowledged, the root of the problem may still be blamed on a lack of 'ability': a lack of individual 'thriftiness', not knowing how to make the right choices, wasting money, or not having 'business acumen'. Some say that they might learn these things in school or in some sort of skill-teaching programme, if they work hard enough – and if they have the ability to learn such things. We think this is just blaming the victim.

Yet another problem is that ability is meaningful only in relation to disability – if everyone had the same ability to do a particular skill, then that ability would not be a useful concept because it cannot be used to differentiate between people's abilities. Skills are assessed in order to sort people and select a few for further education and jobs. But only a few are usually selected. Most are rejected: there are not enough positions for all. The assessment is done not just to find experienced people, but also to limit people – and to rationalise why they will not be allowed to do particular things.

8.2.3 Certifying Skill

In order to understand the meaning of 'skill', we have to ask who is defining it and why are they defining it. Many people have pointed out how skill has been defined in order to impose and sustain divisions of labour according to class, caste, gender, and ability (see Nicholas Whitfield and Thomas Schlich 2015).

Nowadays people are being asked to produce certificates in order to qualify for various kinds of work. For example, a woman at the grotesquely gigantic 'Statue of Unity' construction site in Gujarat said, "Every morning I go there to ask for work, and they tell me that I am uneducated." But she asks, "How much education do you need to carry soil and bricks?" (Lyla Bavadam 2018). The effort to proclaim 'Unity' is supposed to tell everyone that they are all the same (हम सब एक हैं). A certificate says otherwise, it separates people into different groups, and it provides a material excuse for the lack of jobs and lack of money. As a puny unemployed person gazes up at the enormous statue, it should be obvious that 'unity' actually means extreme inequality.

We claim that the main reason for the grand efforts to ostensibly 'develop skills' is not to teach skills, but actually to establish Certification for work that, until now, did not require certification. In so doing, it provides a rationalisation as to why so many people do not get jobs with adequate wages: they are not qualified. By design, only a small percentage of students will get certified at the top ranks. Thus, 'skill development' will deskill, increase class division, and benefit capital.

This reminds us of what we have heard others say about education. When we asked a social worker who was providing girls with sewing skills whether there were jobs available for them after they finished the course, she answered, "No, but they will have certificates! They will no longer be empty-handed!" It sounds like a new use for hands: rather than using hands for manual labour, hands can be used for holding certificates.

The government claims that people are unskilled, and that is why they do not find employment. At the same time, in 2019, the government has been trying to deny and prevent government agencies from publishing statistics – statistics that show that there is a huge unemployment problem in the country.

Santosh Mehrotra and Jajati Parida (2019) have analysed the relationship between unemployment and education, using data from National Sample Surveys and Periodic Labour Force Surveys produced by the Government of India. They report that "total employment (Workforce) is declining, and open unemployed and disheartened Not-in-Labour Force-Education-Training (NLET) youth (a reserve army) are rising massively." Comparing 2011–12 to 2017–18, total employment declined by 0.9 crore people. This included reductions in manufacturing as well as agricultural employment. It is interesting that the rates of unemployment increase with increasing education (Fig. 8.2). This may indicate that education does not necessarily increase one's chances of the kinds employment they were measuring. The data are open to many interpretations, but they are not incompatible with our claim that education plays an important role in sifting people out of employment. Of course the stated objectives are the reverse: that education helps select people for jobs. But since, for the top jobs, more people are sifted out than in, it is more accurate to state the objective as sifting out. Because unemployment at all levels is very high, education is not a guarantee for a job.

The government's skill programmes may free Industry from having to train their own workers, but more importantly, they may function as sifting certification programmes that select a few winners for jobs and keep the rest out. The government claims that teaching skills is a way to solve the agricultural crisis: by making agricultural workers eligible for non-agricultural employment. But it makes more people ineligible than it makes eligible. As Anil Sadgopal explains:

> A substantial majority of the skilled and knowledgeable workers are likely to "fail" the certification test, partly because of the uniformity of NOS [National Occupational Standards] and partly because these standards are formulated as per global parameters, which are irrelevant to their livelihood needs. This is how the Skill India mission will lead to deskilling India on a massive scale, making crores of skilled workers in the unorganised sector essentially "illegal" uncertified workers, who will be compelled to work for even lower wages than at present! (Anil Sadgopal 2016)

8.3   The Economy and Ideology of Formal Education

Education in India is fragmented and uncoordinated, with the Constitution relegating responsibility for education to the Concurrent List – which in effect ensures that neither the central nor the state governments take full responsibility or blame. At the same time, teachers, schools, colleges, and universities are suffering from a lack of autonomy and academic freedom, which extends to the details of syllabi, books, and pedagogy. This prevents individual as well as collective efforts to improve education – and also provides an excuse to give up before even trying.

While stating that it supports education, the Indian government spends an alarmingly small proportion of its funds on education compared to most other countries. The total allocation to education by the central as well as state governments, as a proportion of GDP, has not been satisfactorily progressing (Jandhyala Tilak 2008). In fact, the government continues to make substantial cuts in the education budget (Saumya Tewari 2015).

> The Economic Survey Economic Survey of the year 2017–18 shows that the states and the Union government together have been investing less than 3 per cent of the country's GDP in education or, in the survey's definition, in education, sports, arts and culture. (Education Desk 2019)

Education is becoming increasingly privatised and commodified, which is resulting in a decrease in training, permanent jobs, salaries, and support for teachers, and an administrative concern for profit rather than learning as the bottom line (Govinda and Josephine 2005; Manish Jain and Sadhana Saxena 2010; Geetha Nambissan and Stephen Ball 2010). When large corporations run schools and dictate policies, the effects of the interests of those corporations on education, and particularly on education related to agribusiness, is inevitable.

At first glance, it appears that the faults in the education system are mainly due to inadequacies of school infrastructure, textbooks, syllabi, and teaching methods. Better government funding is required in order to address all of these problems. It seems that agricultural education could be improved by building better buildings, hiring more teachers, having better teacher/student ratios, writing textbooks that have fewer errors (contradictions), and using more activity-based teaching methods.

Upon closer investigation, it is clear that problems in all these areas are there, and these solutions are needed. But this situation is baffling because it seems that many educationists and teachers realise that these problems exist and that these solutions are required. So why do the problems persist?

We claim that the real problem is understood only when we recognise the hidden contradictions and ask why the situation is as it is. The real problem is revealed only when we realise that ideologies are not what they appear to be. A process of reification is occurring in which the actual social relations are being disguised behind the physical manifestations such as school infrastructure, textbooks, syllabi, teaching methods, and the marks given on formal examinations. Because of the actual social relations, the government cannot allocate adequate funds for education, radically different textbooks cannot be written, teaching methods cannot be fundamentally changed, and examinations cannot be disposed of. The education system is the way it is because it exists in a capitalist system and it is required to be this way for the maintenance of the capitalist system. Most aspects are not explicitly designed to uphold capitalism. But because they do uphold the system, their existence is implicitly reinforced. The stated aims and objectives are according to an ideology that makes sense: e.g. students should learn about agriculture so that they will have opportunities to improve agricultural production in India and become more knowledgeable, more highly skilled, and wealthy. But this ideology is a facade – an illusion that hides what education actually does.

8.3.1   Capitalist Education Disconnects People from Science

As capitalism develops, agricultural science also develops, and education also develops. But, as we have discussed (Sect. 6.3), agricultural science develops in a way that removes the doing of science from the people who are working in the fields. A division of science occurs, a division between professional science and people's science – and science becomes more rather than less alienated from ordinary people. This is a general characteristic of capitalist development. As Harry Braverman said, "...the multiplication of technical specialties is the condition for dispossessing the mass of workers from the realms of science, knowledge, and skill." (Harry Braverman 1998, 295) Education develops in a way that increases the gap between a small number of educated intellectuals and professional scientists and the rest of the people, who actually become less educated – in schools, children learn to stop asking questions and stop trying to do science. Simultaneously, deskilling occurs.

Some educationists stress that we need to design effective learning environments based on the conclusions of cognitive science studies of how people learn (Sue Berryman 1991). For example, various types of 'apprenticeship learning' have been designed based on cognitive science studies of how apprentices learn.

If the desired outcome is just the monetary profit, it may be possible to use an apprenticeship learning algorithm to devise a plan for learning how to produce paddy with maximum profit, at least for a given set of inputs and technology. Even if we do not know how to assign different weights to trade off the various factors, we could use inverse reinforcement learning to learn a task. For example, if we want to learn how to reap paddy, the factors may be: how the left hand grabs a bunch of plants; how many plants to grab; how close to the ground to grab; how to swing and draw the sickle to cut the plants with the right hand (the angle of attack, the speed, etc). If we know that the desired result is the profit realised by the paddy harvest, this may be possible.

However, using money in order to quantify outcomes of education circumvents the difficulties of defining and quantifying actual learning.

8.3.2 Questioning

As we have discussed in previous work (Gurinder Singh 2020 and Gurinder Singh et al. 2019), in most classrooms, students are only exposed to 'transmissive' pedagogies in which they seldom talk or ask questions. The teacher does most of the talking.

As we saw on the farm, people would ask each other questions because they did not know the answers, and they needed to find out (Sect. 3.2.5.6). But in schools, it is peculiar that the 'learners' (students) do not ask questions, even though they are ostensibly there to find out answers. Instead, the teachers, textbook writers, and examination writers ask questions – questions for which the teachers and authors think they themselves already know the answers. In this sense, they are inauthentic questions. Students are trained only to give answers – but rather than giving the answers the students think are answers, they give the answers that they think the teachers know and want. This process of fake questioning and answering is not science. Neither does it help encourage authentic questioning that is part of the process of doing science. Neither does it promote science by providing learners with facts, or evidence, or an understanding they might need in order to engage in a more meaningful way with science. The questions are often not the questions that are most relevant for the students in their various contexts, and students do not have the power to ask more relevant questions. It mainly functions to encourage mindless (and short-lived) memorisation that is disconnected from understanding.

Textbooks appear to be lists of 'the facts'. Educationists complain when they find contradictions and untruths in textbooks. But how can a textbook avoid contradictions when contradictions are the nature of nature? And even if a textbook without contradictions could exist, what use would it be? Actual education is concerned with questioning, which arises from contradictions. If a textbook could be written that tells the truth, the whole truth, and nothing but the truth, then what would be the point of reading it? It would not lead to any change or any process of doing science, because doing science means questioning and investigating.

Thus, questioning and answering appear to be occurring, but the questions are not questions and the answers are not answers. We see this contradiction as being dialectical because the whole is a unity of opposition. This inherent opposition gives rise to the 'education' process that is occurring. However, the existence of the dialectical conflict is usually not recognised. It is disguised as something which it is not.

Actual questioning arises out of contradiction (Sect. 3.2.5.8). Actual learning/teaching arises out of actual questioning/answering. But in formal educational systems, actual education is not occurring because actual questioning/answering is not occurring. Teachers and textbooks are supposed to clearly state the 'body of knowledge' so that students do not become confused. Thus the purpose of schooling appears to be education, and yet it is inherently its opposite: anti-education. Students learn to stop asking questions, and stop doing science. They become mechanical copiers.

Real, meaningful school education would actually encourage students to question, investigate, and do science. But this is not compatible with capitalism and is dangerous to its existence. And that is why, even if educationists and policy makers may say that students need a real education in which they practice science, it is very difficult to implement.

There is also a disconnect between work with the mind and work with the body, both of which are necessary in order to do questioning and investigating. School work is mainly 'mental', based on prescribed textbooks and State Board Examinations. Reading and writing is over-emphasized and oral discussion between students is de-emphasized – or not even part of the main classroom discourse. Fixed text in the books defines what is supposed to be 'learned', where 'learning' is taken to mean remembering.

We showed an example of the sort of material students are asked to memorise in school in relation to plant science (Sect. 4.1.2). Even if students understood the concepts and remembered all the difficult word definitions, it is hard to see exactly how this would help them to do paddy cultivation.

But it clearly does function to present students with some text that is difficult to memorise and reproduce in examinations. It is useful in order to frame a difficult question that can be used to sift students out of further education and employment opportunities. And it makes people think that the reason they cannot get better employment is that cannot answer questions about such text because they are not smart enough or did not work hard enough at their studies.

Many educationists report that indigenous students face a conflict between different 'ways of knowing' or 'systems of knowledge' when they encounter the teaching of 'modern western science' in schools. We doubt the extent to which they are really confronting authentic ways of doing *any* sort of science in school. In Chapter 4 we discussed evidence that this has generally not been happening in Indian schools. More often, students are mainly confronting different 'bodies of knowledge', or lists of facts and word definitions. The differences and conflicts between this body of knowledge and some indigenous beliefs are obvious. Without actually doing science, it may be just a matter of faith or habit to accept one belief or another. Educational systems that try to impose the belief in any particular 'body of knowledge' is problematic. Education systems should not require students to memorise and regurgitate any 'body of knowledge'.

Rather, we suggest an educational system that encourages people to use various aspects of a scientific method in order to question, test, analyse and assess traditional beliefs, scientific beliefs, or any kinds of justified or unjustified beliefs. At the same time, we would not insist that people should question or disbelieve in religious beliefs, unless their beliefs are obviously leading them to act in ways that are harmful. If their beliefs are helpful, we do not think we should object. But there are some traditional as well as some scientific beliefs that are causing harm (often the harm is to minorities or oppressed classes or castes), and we would encourage people to question such beliefs. However, we do not see the benefit in trying to preserve traditional beliefs just for the sake of their preservation, or of preserving diversity for the sake of diversity. The beliefs, actions, needs, and desires of oppressed classes need to be heard and acceded to – not in order to have a diverse spectrum of beliefs, actions, needs and desires, but because oppression must be stopped. What education needs to focus on, is the process in which students, working together with teachers, explicitly analyse and choose which questions and beliefs to investigate, and ask why some questions are being chosen while others are not chosen, and for whom is this being done. In some cases, conflicts between indigenous and scientific beliefs could provide appropriate conflicts from which questioning may emerge. In other cases, indigenous beliefs are needlessly – or hegemonically – or maliciously – chosen for questioning. For example, why should a group of researchers who know very little about paddy cultivation tell the farming family to stop believing in a god or to stop putting an onion in the basket of seeds being sown, or to stop attaching mango leaves to the plough? Or why should they even encourage the family to question these beliefs? It would be neither useful nor ethically correct for researchers to do so. However, if the farming family was mistreating hired help on the basis of religion or caste, this needs to be opposed. (We did not see indications of this.)

Charbel Niño El-Hani and Ferreira Bandeira (2008) see understanding rather than changing 'beliefs' as the goal of science education, and this leads them to argue for a form of education that includes different cultural understandings, but without subsuming them in the definition of 'science'.

Our criticism of the use of the term 'science' in schools – and of the way science is taught – is not that it excludes or includes 'indigenous knowledge', but that it gives the wrong impression of science as being a fixed body of knowledge. We argue that even capitalist science is tentative, dynamic, and subjective (especially in the choice of questions and applications). Students are not even being taught capitalist science when classroom 'science' is just a matter of memorising (and perhaps understanding) a body of knowledge.

8.4   What Does Formal Education Actually Do for the Farming Family?

Contrary to what Smita and Bhimraj say, educationists state that schools are to provide all children with knowledge, skills, and opportunities that will help them in whatever they 'choose' to do in life – even agricultural labour. This educational ideology obscures the actual social relations, which are apparent to Smita and Bhimraj. They know very well that society is divided according to class, caste, creed, gender, ability, etc. They know through their direct experience of it. The education system reinforces this structure (see Samuel Bowles and Herbert Gintis (2002) on how this happens in USA). They know that school education is not very useful to them in their work in the field. At some level, they also know that school education may not do much to help get jobs outside of agriculture. But still, their only hope for their children lies with them getting jobs outside of agriculture, and with education enabling this to happen. They really cannot give up this hope, because it is still their best chance at escaping from the agricultural crisis.

As we have discussed, school textbooks and formal education divert people from actual learning. Schools actually teach children to stop asking questions, to stop doing science, and to become less skilled (less experienced) at doing work with their hands. Jobs are scarce and unemployment is high for both the uneducated and the educated. And unemployment is actually needed in order to maintain a reserve labour force and keep wages down and corporate profits up.

There is only a small chance that through education a few children of small farming families may eventually find some sort of employment which will enable them to escape the poverty and hard labour that agriculture presently entails. But this chance, however small it may be, is the advantage over the feudal relations that would otherwise bind them to the class, caste, and gender into which they are born.

Even if all students work hard, they cannot all become toppers – the very definition of 'topper' excludes this possibility. Even if everyone could become 'educated', they cannot all get good jobs because in this system there cannot be enough good jobs for everyone. There must be enough unemployment to keep wages down and profits up. But people continue to hope that they, or their children, will become the exception that does 'succeed'.

Is it even true, as Smita and Bhimraj say (under some duress brought about by our leading questions), that school education helps farmers because they can make use of literacy to do their accounting and read the destinations of buses? Actually, many children finish Class V, or even finish their entire education, without attaining a level of literacy that allows them to read the destinations of buses or calculate simple sums. While many people blame the teachers, teacher training, textbook design, and teaching methods, the actual reason for this situation is that the educational system is designed to do exactly what it is doing. It is designed to sort people and select a few toppers and a large number of non-toppers (failures). But even the sorting is only an apparent sorting. The toppers are actually pre-selected: except in a few rare cases (which come out on the front pages of newspapers), all toppers come from wealthy, high-caste families and are non-disabled (and more often than not, they also identify as cis-gendered males).

In the present system, it is not acceptable to have a school in which teachers and textbooks do not state the names of the parts of a hibiscus-type flower. But it is acceptable – and perhaps necessary – to have a school in which most children do not really remember or use these names or understand anything about the structure and functioning of flowers. It is not acceptable to have a school in which teachers and textbooks do not state that a particular set of agricultural techniques are the best (whether the latest techniques include tractors, genetic engineering, organic farming and/or 'traditional' methods). But it is acceptable – and desirable – to have a school in which students do not understand what the techniques are or remember which ones are better or worse, because they will not actually have any say in deciding which methods will be used.

Our work on the farm in Rudravali has revealed the actual meaning of 'learning' and 'education', which is very different from what we see in schools. Actual learning is done by doing science, in its broadest sense. In other words, we learn by questioning, investigating, and experimenting. This includes questioning the basic structure of our social / economic / political system. But the imposition of the capitalist system with its profit motive makes it very difficult to do this. It obscures our understanding of what we are doing. It obscures the dialectics of learning. The formal education system actually reduces our learning and reinforces our unconsciousness by separating the mind from the body, separating the concepts from the acts, and separating the ideal from the material. It reinforces an ideology that masks our understanding of our own activity. It almost makes us believe that our failures and our difficulties are due to our personal 'inabilities' or lack of hard work. But perhaps this is better than believing, and acting as if our problems are inevitable.

This shows how and why formal education does not help cultivators. Formal education in fact makes things worse by creating more layers of false consciousness. This makes liberation less likely. Meaningful education is necessary for systemic social change, but at present, it seems to be possible only through cracks in the existing system.

References

Bavadam, Lyla. 2018. Sardar as pawn. *Frontline* 35: 33–36.

Berryman, Sue E. 1991. *Designing Effective Learning Environments: Cognitive apprenticeship models*, IEE Brief 1, 1–4. New York: Institute on Education and the Economy, Teachers College, Columbia University.

Bowles, Samuel, and Herbert Gintis. 2002. Schooling in capitalist america revisited. *Sociology of Education* 75 (1): 18.

Braverman, Harry. 1998. *Labor and Monopoly Capital: The Degradation of Work in the Twentieth Century*. 25th anniversary ed. New York: Monthly Review Press.

Education Desk. 2019. Economic Survey 2018–19: Sharp fall in elementary school-going children; Key highlights of education sector. *The Indian Express* 4 July.

El-Hani, Charbel Niño, and Fábio Pedro Souza de Ferreira Bandeira. 2008. Valuing indigenous knowledge: To call It 'science' will not help. *Cultural Studies of Science Education* 3 (3): 751–779.

Govinda, R., and Y. Josephine. 2005. *Para-Teachers in India: A Review*. New Delhi: National Institute of Educational Planning and Administration.

Jain, Manish, and Sadhana Saxena. 2010. Politics of low cost schooling and low teacher salary. *Economic and Political Weekly* XLV (18): 79–80.

Marx, Karl. 1844. *Economic and Philosophic Manuscripts of 1844*. Ed. Dirk J. Struik, 1964. New York: International Publishers.

———. 1867. *Capital: A Critique of Political Economy. Volume I, (in German), 1887* English Edition, Trans. Samuel Moore and Edward Aveling, edited by Frederick Engels. Progress Publishers, Moscow, USSR.

Mehrotra, Santosh, and Jatati K. Parida. 2019. India's Employment Crisis: Rising Education Levels and Falling Non-Agricultural Job Growth'. In *Centre for Sustainable Employment, Working Paper 2019–04*. Bengaluru: Azim Premji University.

Nambissan, Geetha B., and Stephen J. Ball. 2010. Advocacy networks, choice and private schooling of the poor in India. *Global Networks* 10 (3): 1–20.

Sadgopal, Anil. 2016. 'Skill India' or deskilling India: An agenda of exclusion. *Economic and Political Weekly* LI (35): 33–37.

Singh, Gurinder, Rafikh Shaikh, and Karen Haydock. 2019. Understanding student questioning. *Cultural Studies of Science Education* 14 (3): 643–679.

Tewari, Saumya. 2015. Jaitley slashes education, health spending.. *India Spend*. http://www.indiaspend.com/budget-201-modis-moment-of-reckoning/jaitley-slashes-education-health-spending-67467.

Tilak, Jandhyala B.G. 2008. Higher education: A public good or a commodity for trade?: Commitment to higher education or commitment of higher education to trade. *Prospects* 38 (4): 449–466.

Whitfield, Nicholas, and Thomas Schlich. 2015. Skills through history. *Medical History* 59 (3): 349–360.

Chapter 9
Conclusions and Implications

We have been investigating how one family in Rudravali, Maharashtra learns and teaches paddy cultivation. We have been doing this by observing and interacting with the family as they learn and teach each other, and as they teach us how to do paddy cultivation.

The main questions we have been asking are: How does the family learn, teach, and confront problems they encounter in the course of their daily labour cultivating paddy? How do they pass on traditions and/or develop new ways of doing cultivation? Do they do science in order to solve problems they encounter? How and why might their approaches be different from the ways professional scientists do science? In other words, how do they cultivate cultivation?

It is not our intention to use this study of one family as the basis to generalise. Our aim is to question and try to understand what was happening on this one farm, with reference to historical, empirical, and theoretical research that others have done. We have been trying to understand how the current agricultural crisis affects the family, and what might be done to end the crisis, especially with regard to what kind of education this family needs and desires.

9.1 Dialectical Conflict and the Need for Change

One of our main findings is that what at first appeared to be a traditional process that has existed for hundreds of years, is actually not what it appeared to be. The process of paddy cultivation on this farm is not an unchanging, harmonious process that has gotten carried on unaltered from generation to generation. The process is full of material dialectical conflict. And this conflict gives rise to the learning and to the changing of paddy cultivation. Not only the materials and methods used in cultivation, but the paddy itself, as well as the environment also keep changing

© The Author(s), under exclusive license to Springer Nature
Switzerland AG 2021
K. Haydock et al., *Learning and Sustaining Agricultural Practices*,
International Explorations in Outdoor and Environmental Education 7,
https://doi.org/10.1007/978-3-030-64065-1_9

interdependently, through dialectical conflicts. But due to the nature of the dialectical conflicts, some kinds of change are stifled.

Upon observing and working with the family doing cultivation, it is clear that they are tremendously over-worked and they are not getting the true value of their labour. This is what we saw and also what they have explicitly told us many times. These dynamics are based in dialectical class conflict.

The family is actually very skilled at cultivation. A lack of skill or a lack of formal education or certification is not the reason that they are not getting the true value of their labour. That is not why they have not escaped poverty. That is not why the materials and methods they use in paddy cultivation have not changed more than they have. The problems are systemic, and are part of the present agricultural crisis, as we have discussed.

The family members told us that they need more money in order to eat and live adequately. They are no longer able to survive through subsistence farming, as was possible as recently as 30 years ago. Paddy cultivation is not what it used to be. Therefore the family keeps trying to find ways to earn outside of agriculture. They are doing whatever they can so that their children get an education, in the hope that they will get jobs outside of agriculture and escape a life like their parents. However, education is no guarantee of this. There is a huge shortage of jobs and high unemployment and underemployment throughout the country for people with all levels of education. These are systemic problems, not individual failures.

Through a review of the historical development of paddy cultivation we have seen that even before people domesticated it, rice kept evolving through a non-teleological process of natural selection. This happens because there is always variation between individual rice plants in any field or population of rice, and some of this variation is due to genetic differences interdependently interacting with the environment. Since some variants of plants tend to survive and reproduce more than others, the population of rice evolves. Thus, evolution by natural selection occurs unintentionally, through dialectical conflict. However, since rice began to be cultivated, people have purposely tried to change rice according to their various changing needs and desires. They have used a process of artificial selection, which we saw is still being carried out by the family in Rudravali. The process of cultivation has not been a balanced process in which the same kind of rice keeps getting produced year after year, century after century. There is no balance of nature. Even before people existed, fossil evidence shows that nature has always been evolving interdependently with inherent internal conflicts. This is why we say that nature (and nature/society) is historical dialectical materialist (HDM). By 'historical' we mean that it changes over time. 'Dialectical' means that it changes through inherent opposing forces. And by 'materialist' we mean that the changes are rooted in physical reality rather than in ideas, concepts, or some sort of spirit.

9.1.1 Dialectical Conflict Gives Rise to Learning and Doing Science

It has been claimed that traditional farmers have different ways of passing on methods, learning, and developing their ways of knowing than 'modern western scientists'. Researchers have suggested that they tend to be vague, approximate, pragmatic, and specific, act locally, respect elders and authority figures, use local languages and mainly oral communication, have 'collective agency', and have little knowledge of modern agricultural methods.

But in our interactions with the family, we have observed that the situation is much more complex than this. We have found that their paddy cultivation necessarily involves dialectical conflicts. These are material conflicts between ways of doing, not just conflicts between beliefs. We have seen a number of types of dialectical conflict: between using old and new technologies, between being pragmatic and abstract, between being specific and general, between acting locally and globally, between cooperative collaboration and respect for authority, between literacy and oracy, between doing rituals and hard work, and between availability and non-availability of technologies. These conflicts give rise to and define the sort of paddy cultivation that occurs and develops on the farm.

We have observed how small and large problems keep arising during the process of paddy cultivation, and the family engages in cooperative and collective processes of questioning and investigating in order to learn and figure out how to solve these problems. They learn by doing, using both hands and mind interdependently to investigate the problems. In other words, at least to some extent, they do science: they use variable, complex networks of aspects of science, such as questioning, experimenting, hypothesising, observing, analysing, testing, comparing, categorising, trial and error, reasoning, etc.

Through this process of learning and doing, they continuously develop new materials and methods, and the process of paddy cultivation keeps changing. However, we do not claim that the family's practice of paddy cultivation is either science or non-science. Rather, we have showed that it is fairly scientific in many respects, if we use a broad definition of science as a way of doing: a network of different aspects in various non-linear orders and combinations.

The science that the family does is different in several significant ways from the way capitalist science is done. There is emphasis on different aspects of the scientific method. The type and amount of communication is different from what is involved in capitalist science. It includes more oral communication and less access to reading and writing and communicating over long distances. They also do less experimentation that involves isolating variables and using controlled variables. In their way of doing science, the social and the natural are interconnected, unlike the more reductionist approach that is taken by many capitalist scientists. And on the farm there was a more intimate connection between work with the hands and work with the mind, with the doing of science directly connected to their immediate needs.

Another important difference is that most of the science on the farm is done for its use value. We observed the family doing science because they need to do it in order to solve their own problems, and produce food for their own use.

However, they also do another kind of science: according to one way Smita defined science, it is what they do in order to discover how to progress and surmount their economic problems. Rather than just being concerned with use value, this kind of science is concerned with exchange value. They have opened a small road-side shack in the hopes of surviving the agrarian crisis by engaging in petty trade on the side.

This contrasts with the way science is often done by professional scientists, who may be either working on rather abstract, basic questions that do not have obvious or direct bearing on their everyday lives, and/or whose science may be restricted by capitalist compulsions to keep profit as the bottom line.

We have discussed how the family's way of doing science is, in many ways, historical dialectical materialist science (HDM science). By this we mean that their collaborative questioning and investigating arises in a process of dialectical contradiction, and also, in this process they investigate the dialectical nature of nature/society. Rather than just finding answers that they pass on as 'facts' or 'the ways things are', they keep confronting new contradictions as they arise, and keep trying to figure out new ways of doing things.

The way the family learns and teaches is not really different from this process of doing HDM science. But it is very different from the way science education is usually done in schools, in which there is very little learning by doing. The family was not lecturing, discussing, or analysing in isolation from the stuff they were working with in the field, as happens with formal science education.

We have seen how dialectical conflict is inherent to nature/society, and how it gives rise to both explicit and implicit questioning. The questioning and investigation of questions is what occurs during the process of learning cultivation. Learning can occur with or without explicit recognition of conflict – we saw examples of both. However, we claim that meaningful and useful education should probably be concerned with trying to recognise dialectical conflicts and form explicit questions. In other words, this learning is a Hegelian dialectical process of consciousness coming to an awareness of itself and its world. This is what is called as 'negation', because it involves realising that there is a conflict: realising that we were under an illusion that we had a complete understanding, which we now see is incomplete. Maybe this also involves turning unknown knowns into known unknowns.

Note that in this process of learning cultivation, we do not necessarily come to know 'the correct answer' or the right method of cultivating. Rather, we become conscious of the conflicts and the questions. We try to resolve conflicts through questioning, learning, and doing science. But as some conflicts are resolved, new conflicts and new questions arise. Both questions and answers keep evolving, although in practice some answers are so probable that we need to act as if they are certainly correct. This is why it is important to explicitly and purposely choose questions, based on an analysis of which questions are possible, relevant, important, and for whom are they important, and why.

These aspects are usually not explicitly addressed in school education, either by students or teachers. There is very little hands-on work in the local schools. At most, students are asked to perform science practicals by following a rigid, prescribed procedure – a series of ordered steps – in order to produce prescribed results. Or they may just watch as teachers demonstrate the 'science experiments'. There is no real need to perform these activities because the 'questions' that they are designed to answer are inauthentic: their answers are already known beforehand. This is not really science. And it is not helping people do science in their everyday lives, or do cultivation. The farmers told us that except for the benefits of learning to read and write, their school education did not help them do cultivation.

But in the paddy field, we saw that questions and investigations arise by necessity, through the process of working with hands/minds. And perhaps the doing of science also required an awareness that the problem was not inevitable and that new answers were possible. Maybe this is what DD Kosambi meant when he said, "science is the cognition of necessity" (Kosambi 1952).

HDM science is done because it has to be done. It arises due to the recognition of dialectical conflict, and it may not have a clear-cut beginning and end, or result in a very well-defined answer. We saw this, for example, in the question of when to harvest the paddy. In the paddy field we learn that if we wait a few more days before harvesting a crop of paddy, it may be drier, and be less likely to be attacked by fungus, but it may rain in the meantime, and then the entire crop may be lost – or if left standing in the field too long, the grains may fall and be lost. These sorts of conflicts and questions are crucial, and learning to cultivate means recognising the conflicts between being too wet and being too dry, being too immature and being too mature, as well as learning to ask questions that arise from the recognition of these conflicts. Learning cultivation does not mean knowing exactly when to harvest. It is actually not possible to know exactly when to harvest, and each year new factors will complicate the decision. It is not possible to 'know the right answer'. But through the process of questioning and investigating, we get experience and may get closer to better answers and become better at cultivating.

Doing HDM science in the paddy field is in conflict with an idealist approach of passing on tradition through authoritarianism, indoctrination, faith, doing rituals and passing on mystical, irrational beliefs. Cultivation cannot rely too heavily on an idealist approach – if it does, there is a greater probability that it will fail to produce useful physical results. Just performing a ritual does not work. Just relying on some unexamined 'inner belief' or intuition does not work.

The farming family does also engage in some practices that are not very scientific. For example, they do various rituals, such as attaching a few mango leaves to the plough. However, they say that these rituals do not determine the outcome of their crop. The outcome of paddy production, they say, is determined by their hard work. Rituals are for their social well-being and self-respect.

9.1.2 How and Why is Doing Science on the Farm Inhibited?

We found that in a number of ways, the family is hindered from doing more science, and a more HDM kind of science, than they do manage to do. How and why are they hindered? We list the following ways, most of which are rooted in systemic oppression due to class and caste.

- They are being robbed of the true value of what they produce as commodities.
- Because of this, food insecurity and poverty are constant worries.
- Their work is too hard and physically exhausting, and they do not have enough free time.
- Their production for use value is decreasing, while production for exchange value increases, and thus the doing of HDM science decreases.
- They are subjected to too many risks, and experimentation would add more risks.
- They have inadequate access to widespread written and oral communication, which is important for doing science.
- They have a lack of access to education that encourages questioning and investigating.
- They have some access to schooling that teaches students to stop questioning and stop doing science.
- The hegemony of capitalist industrial science interferes with people's science.

Note that there are several things that are not on this list. They are not being hindered by a lack of knowledge of cultivation, a lack of knowledge of how to do science, a lack of skill, a lack of intellectual ability, or by a lack of motivation or need to do science. We did not find evidence that any of these deficiencies were holding them back, even though many people have claimed that such problems exist amongst peasants.

In our discussion of the history of agricultural science we have seen that science does not begin with the advent of capitalism – it gets transformed, and actually decreases, in the sense that fewer people practice science, and that HDM science for people is suppressed. Capitalism alienates science from people. A division of labour occurs and the profession of scientist emerges, which is an option only for a tiny minority. Science becomes a tool by which capital makes profit. Science gets commodified and used to produce technology that is for profit rather than for people's needs, which may not even be recognised. Rather than doing science for its use value, capitalist science is done for its exchange value.

We have discussed how and why agricultural science in India has been stifled, and why the small family farm persists. Our brief overview of the history of paddy cultivation shows how the structures of feudalism, colonialism, and capitalism have hindered the development of agricultural science and technology. At present, both capitalist and non-capitalist science are inhibited and it is very difficult to develop new technologies based on what people really need and what is best for the environment. Capitalist science is inhibited by its need to keep profit (exchange value rather than use value) as the bottom line.

We have shown examples of how science has been inhibited in one farming family: through economic and social/political constraints, which are reinforced by an educational system that, despite its stated aims, actually teaches people to stop asking questions, to stop doing science, and to become deskilled. Pre-capitalist science was always limited due to lack of resources, lack of communication, and the riskiness of cultivation. But even with the introduction of some capitalist science, technology, and economy, these constraints are being maintained (if not magnified) in the family.

9.1.3 Tradition vs Science

We have mentioned how philosophers of science and researchers in science education have debated whether there is some kind of 'science' or 'knowledge' which is different from 'modern western knowledge'. Multiculturalists have claimed that there is 'indigenous knowledge' or 'traditional knowledge'. Others have insisted that 'science' is universal.

Often what is being compared are 'bodies of knowledge'. But when we look carefully at these so-called 'bodies of knowledge', both in mainstream science and on the farm in Rudravali, what do we see? We do not see anything that is sufficiently static or well-defined to be called a 'body of knowledge'. Neither do we see such a thing in capitalist science. What we actually see are not things, but very dynamic processes that are not separate from a process of 'constructing knowledge' or, as we prefer to say, doing science.

The dynamism and the process of questioning and investigating do not receive adequate emphasis when we use the word 'knowledge'. By focussing on the 'knowledge', science education tends to be aimed at trying to transfer knowledge rather than trying to figure out how to understand the world, how others have tried to understand it, how and why we choose particular questions for investigation, and for whom we do science. These are important aspects of science that need more explicit attention in education.

Therefore, in this study, we have not been focussing on different technologies, concepts, or beliefs as things in themselves. Instead, we are interested in the question of how the family develops and learns to produce crops, technologies and methods of cultivation.

We claim the family has been engaging in a practice of doing science which is fairly universal. But the technologies, concepts, and beliefs that have resulted from this process depend on their environment and historical circumstances, and may be distinctive to their culture.

We are not arguing that either the way they do science or their results are producing a culture that is worth preserving just for the sake of preserving tradition or maintaining diverse cultures. Cultures keep changing, and people of each culture need to try to figure out how and why we should change, and how we can work through education and science to advocate, encourage, and demand necessary changes. People will make more progress in developing their own cultures of cultivation if they are not prevented from doing more science.

Perhaps the problem of defining science is itself dialectical: there is a universal way of doing science everywhere, but there is not a universal way of doing science anywhere. What we mean is that the very definition of science anywhere is that it is a network of variable aspects, with variable numbers and orders of aspects. Actually, the interconnections between the aspects make it impossible to even distinguish the aspects from each other or see any discernible order. And this variable process may be occurring everywhere. That said, there may be some generalisable differences between the sets of aspects that are included in the way science is done in different places. We have already mentioned several aspects that were less prevalent in Rudravali and more prevalent in our experience of capitalist science.

We do not see any reason why we should try to preserve these differences between how science is done in different places. Actually, we should try to encourage a more HDM form of science everywhere, and we should try to do away with doing capitalist science for profit.

9.2   Implications and Suggestions for Education

We will now return to the questions that motivated our study of the cultivation of cultivation, as mentioned in the Introduction (Chap. 1). We cannot reach general answers to these broad questions, but we can summarise what our small case study may imply.

We conclude that hunger and poverty are only partially due to the centuries of stagnation, backwardness, and lack of scientific advancement that have been thought to characterise what we called 'traditional' agriculture. Within the practice of agriculture, we suspect that there have always been (to varying extents) dialectical conflicts between the doing of HDM science for people and various forces that prevent science from advancing in ways that will help people to live better lives. We saw some evidence for this in our brief review of the historical development of paddy cultivation. And we saw evidence for it in our study of the family in Rudravali. We saw that there are some very negative effects of the attempts to 'modernise' and industrialise agriculture, as was begun by the 'Green Revolution' in the 1960's, and continued through the imposition of 'neoliberal' policies.

However, we also conclude that the present agricultural and environmental problems are not simply due to modern science and technology. We lay the blame on the particular types of science and technology that are characteristic of capitalist development.

In some ways, paddy cultivation, and the present way of life on the farm are improvements over the kinds of cultivation and ways of life that existed one or two hundred years ago. Even capitalist cultivation and agribusiness, to the extent that it exists, is an improvement in many ways. It is neither possible nor desirable to go back to the old way of life of a few hundred years ago. It was neither in balance with nature, nor humane.

However, in many ways, the old village way of life was less alienating, and on the farm in Rudravali we did find less alienation of people from nature, from each other, and from their own work. The increase in alienation that comes with capitalist development is probably the main reason why past village life seems so appealing.

We do not see any reason why capitalist development and capitalist science are inevitable. In Rudravali we have seen evidence that a more HDM kind of science is possible. This is a form of science that would stimulate, and would be stimulated by, an anti-capitalist form of development. It would provide ways for people to figure out how to solve the present agricultural crisis.

9.2.1 Do we Need Indigenous Science and Technology?

Some people have opposed the promotion of scientific temper in India because they claim that it imposes a 'western' scientific method (as reviewed by Meera Nanda 2003). They advocate indigenous science and technology instead. But rather than seeing science as a subculture of 'western' culture, we say that it needs to be reclaimed and recast by its indigenous practitioners throughout the world.

The role played by agricultural workers in learning and doing science should be acknowledged and supported as a form of authentic science education that takes place on farms. We support an education in which, rather than focussing on teaching the 'The Body of Knowledge' (which is defined by hegemonic powers as particular lists of commodified concepts), we need to encourage people to become more engaged in the process of science in school and throughout their everyday lives. This is what scientific temper is all about.

We agree with Arjun Joshi in his attack on pseudoscience, black magic, and the dispension and glorification of easy answers which dodge sceptical scrutiny:

> The consequences of scientific illiteracy are far more dangerous in our times than in any time that has come before. ...It is enormously easier to present, in an appealing way, the wisdom distilled from centuries of patient and collective interrogation of Nature than to detail the messy distillation apparatus. The method of science, as stodgy and grumpy as it may seem, is far more important than the findings of science. (Joshi 2016)

One of the main problems with formal science education at present – and also with the way professional science is sometimes done – is the de-emphasis on the process of questioning, and the disinclination to question authority as well as the disinclination to question one's own answers.

One of the groups that has advocated indigenous science and technology and attacked efforts to spread scientific temper in India was a group called 'Patriotic and People Oriented Science and Technology' (PPST). Their approach has been extended to efforts to use 'science' (actually pseudoscience) to validate religious beliefs and myths and prove the antiquity of various technologies in India. This even includes claims by prominent politicians and 'scientists' that more than a thousand years ago Indians travelled into outer-space, transplanted an elephant's head onto a human, and made atom bombs. (One of the questions that is too infrequently asked by either these cultural traditionalists or professional scientists is why an atom bomb is a scientific advance to be proud of.) There has been a proliferation of various sorts of 'scientists' who, rather than doing evidence-based research, rely on an appeal to authority, an appeal to a mystical 'inner knowing', or an appeal for nationalism (or an appeal against 'anti-nationalism').

The need to support the struggle for scientific temper has intensified as attacks on it increase. Narendra Dabholkar, Govind Pansare, and MM Kalburgi were murdered because they were rationalists and activists for scientific temper, and stood up against black magic and superstition. So far, no one has been convicted for these murders.

Ordinary people face many constraints that make it difficult for them to engage in the process of questioning, which is an essential part of doing science. Certain politicians have threatened people and tried to lead people to think that it is anti-national and undesirable for people to question their rulers, authorities, and official government policies. It appears to be easier than ever for politicians to proclaim the truth and validity of various statements they make without any mention of evidence, and even in the face of obvious counter-evidence. If people are afraid to ask questions, politicians can say anything without fear. They can create their own unquestioned 'body of knowledge' that is far removed from the doing of science. These politicians have found that it is expedient to solidify their power using this anti-science approach. They use it to appeal to Hindu fundamentalist sentiments, and also to suppress progressive and left leaning activism. Rather than using religion to question or oppose science, they try to use religion to validate science 'facts', and also try to use the 'body of knowledge' of science to validate religion. They do not even question or speak out against 'western modern science'.

Considering science as a process rather than a 'body of knowledge', we question whether it is actually western. We agree with researchers who say that other cultures are threatened by 'western' culture – but the threat we are here concerned with is the hegemonic refusal of one culture to recognise, support, or in some cases even allow other cultures to engage in what we see as a universal process of doing science throughout our lives. People who support multiculturalism may say that they do not object to 'modern western science', they just think other ways of knowing also need to be supported since they are equally valid. However, they may actually end up denying certain marginalised groups of people access to science. The reasons for this are embedded in the structure of capitalism and the role of science within capitalism. Science has been misappropriated and needs to be rescued. Perhaps educationists can play some role in the rescue.

However, we do not advocate trying to return to some kind of pre-capitalist science. We do not romanticise past village life, or deny the need for the development of new technologies and improved agriculture. But we do need to ask: better for whom? We advocate creating a new way of doing science that will be more widely available to all people, in which people will ask questions and do investigations in order to work for a more just and egalitarian world, rather than doing science for the profit of capital. What we need is to develop a new, non-capitalist way of living, working, and doing HDM science. In the present circumstances, capitalist science is hegemonic and it will not be easy to replace it with HDM science.

But we are convinced that HDM science is necessary in order to produce necessary technological development. As the farmers told us, they need technological change in order to make their work easier. We have to support them in this need. But in order to do so, we have to realise that technological change should not be driven primarily by the need to produce commodities or increase profit.

In order to solve the agricultural crisis, we need to ask the most basic questions, and investigate and understand the reasons for the crisis. The real question is: What kind of development is needed – and development for whom? Is modern capitalist agriculture necessary or desirable for the farming family? We do not think so. There are better alternatives.

The reason for the crisis is that farmers are, increasingly, not getting the true value of their labour – because this is the nature of any form of capitalism as it develops. The only way out of the agricultural crisis is to remove the profit motive as the driving force: remove capitalism. Further corporatisation of agriculture will only make things worse.

Rather than doing capitalist science for its exchange value, we need to do HDM science for its use value. Agriculture must be driven by the rational (and emotional) need to supply food, clothing, and shelter for all. Science must be done for this purpose, and the science of ordinary farmers/peasants must no longer be suppressed and prevented from developing. Rather, capitalist science, which is done for profit, needs to be suppressed, uncommodified, and redesigned. People must stop becoming alienated from science in order to further cultivate cultivation.

One of the main reasons that the family in Rudravali does not do science to a greater extent is that they do not have the resources and money, and their risks are too great. The entire capitalist structure needs to be changed in order to allow the farmers to get the true value for their crops. This can only happen through collective action. Forming unions and local cooperatives could be a beginning. At present these do not exist in Rudravali. Learning about such efforts made by other communities, and communication with them, would be useful. Perhaps through community action and cooperation farmers could also get some relief from backbreaking labour, get more free time, and more time to engage in the process of demanding and creating systemic change.

9.2.2 Suggestions for Future School Education

Education is intimately connected to the social structure in which it exists. In many ways it serves the interests of those in power, but often does so in ways that are not very obvious. In order to consider how education needs to be improved, we need to look at the objectives of education. We need to ask how science education supports the status quo and how it encourages the meaningful questioning and investigating that is needed to encourage creative involvement in making the world a better place.

At present, Indian school education is hardly concerned with trying to help students understand the political economic system in which the world is immersed.

Textbook writers usually shy away from using the word, 'capitalism' or discussing how capitalism develops (Haydock 2015). However, a draft of an NCERT textbook on economic development for Class X, contained the following explanation of "Why are some countries developed and some others underdeveloped or less developed?":

> In general, economic development is a process that begins when at least four interrelated things happen. First, people start saving more out of their income. This means that more money is available which can be used for productive investment. What is saved is used to start a factory, open a shop, build a road, run a business, build an irrigation scheme etc. After some time, because of these investments, your income increases even further. A developed country like USA or Japan saves a substantial portion of its income.
>
> Second, markets for goods produced expand either within the country or outside....
>
> Third, technological progress leads to creation of manufacturing industries and mass production of industrial commodities. ...
>
> Fourth, countries also start using more energy and natural resources to run the machines and factories. Coal contributed significantly to Britain's industrial revolution in the 18th century. It could also exploit natural resources of countries of Asia and Africa through colonialism. Utilisation of huge coal deposits in the Lorraine valley and electricity powered Germany's move towards becoming an industrial country.
>
> An underdeveloped country today understandably desires to follow a similar path. ...
> (Anonymous, 2005)

What is most disturbing about this text is the utter lack of any mention of the class structure of society. It says that 'people' start saving more – the language is ambiguous, but it seems to imply that people are all people or people in general. This interpretation is reinforced by the passive voice used in the following two sentences. But then, there is a jump to 'you': "your income increases even further". Could this 'you' be the reader – the individual student, or some general set of people? (Could my income increase?) No, apparently not, because the last sentence of this paragraph is about countries, implying that the talk all along has been about a country, not an individual or some subset of people in a country. This fits very well into the kind of discourse that discusses 'our' GDP or what 'we' (the nation) does or thinks, as wishful efforts to define or invent the existence of a monolithic nationalism. As we continue reading, it becomes more clear that the text is about countries. Countries which 'start using more energy and natural resources...' Britain is used as an example of a country – a country as a monolithic, singular identity. This leads to the conclusion that an undeveloped country has a monolithic, unanimous desire. But what is this desire? To become a colonial power? To become developed? To become developed into what must be a capitalist country, even though it is not called capitalist, and apparently does not have even a hint of class division (or any division)?

In the unlikely chance that you are a literate peasant reading this textbook passage, how do you make sense of it relative to your own life and view of society? Do you feel personally defective and guilty for not having saved more? Or do you think the majority of farmers are naturally defective in this regard? Or in case you have invested something back into agriculture, do you wonder why your karma deviated from the textbook case? And do you agree that colonialism is wonderful because it allows the country with colonial power to develop?

This entire passage was deleted from the textbook before it was published. But it may not have been a bad idea to include at least the latter part of it, because it actually is inadvertently spelling out one of the main reasons why the peasant is still a peasant: despite it's purported GDP, India is not a sufficiently imperialist power and there are insufficient places for peasants to be displaced to. As Utsa Patnaik points out,

> It is usually taken for granted that today's developing nations will follow the same or similar trajectories of development as today's advanced nations did, with a reduction in the share of the primary sector both in the nation's output and in its employment. Labour and population would face displacement from agriculture, but ... displaced populations would be re-absorbed into [secondary and tertiary] sectors along with a fast rate of urbanisation of the population. (Patnaik 2016)

As she says, according to this point of view, the displacement of small peasant producers from the land is considered to be an unavoidable part of capitalist accumulation, and "The population subsisting on agriculture should largely shift away to other more-paying activities ..., and the corporate sector should enter agriculture directly to raise the technological level."

This reminds Karen of what an elite 15-year-old student of hers once told her after visiting a village school (she had complained that he did not need to eat so much chips and chocolate):

> "I NEED chips and chocolates – the food is so bad here! ... It's not MY problem if they [the children in the village school] don't have chocolates – the parents should take care of their children they don't work hard enough there's too many of them that's why they're poor! If they had more brains they would work in a bank – they don't use their brains – if they were educated they wouldn't BE farmers – the farmers in US make a lot of money – the rich don't depend on the poor – what kind of economics is that?"

When Karen tried to tell him something about where wealth comes from, he told her, "I was just doing economics up until 5 minutes ago – that's not what WE learn!" In hindsight, his blatant honesty is somewhat refreshing – he obviously was not indoctrinated enough to know that he should not say (or know that he thinks!) such things.

Actually, what he was saying was almost in-line with something the family in Rudravali told us: they do not want their children to continue in agriculture. They would be quite happy if their children would study hard in school and get jobs in banks – as Kalpana actually did!

But, on the other hand, the family also says that their work is vital. They are cultivators, and without cultivation, people would have no food. Their demand is for a kind of cultivation that is more humane, that is not so tiring, and that does not rob them of the true value of the paddy they produce. What kind of a world do we live in if this is not easily achievable? How could it be considered to be inevitable that the family leaves cultivation, loses their land, and becomes unemployed in order that we can have mechanised corporate cultivation (which is apparently what our leaders mean when they say 'sustainable development')?

How many human beings actually desire to follow a path similar to the capitalist path that Britain followed (even if it was possible)? That path does not lead to prosperity for all – it leads to an ever-increasing gap between the rich and the poor.

At present, children are not learning how to do farming through school education. And the family in Rudravali said that they do not need or want schools to teach farming. They have convinced us that they are correct. We now agree, and we argue that in the present context, schools should not try to teach children how to do agriculture. People such as the farming family in Rudravali are already becoming very skilful farmers through the practice of doing farming in the fields. They are learning and teaching using pedagogies that are much better than those available to overworked school teachers confined to overcrowded classrooms.

One thing that is missing from formal school education is the study of agricultural development and its historical, political and economic connection to capitalist development. What is needed is a historical dialectical materialist approach which can surpass the facile assumptions that people shift (or do not shift) out of agriculture due to their individual subjective wishes or intelligence. Common misconceptions, such as those mentioned above, need to be critiqued.

Rather than teaching how to cultivate, school education should rather focus on teaching something about cultivation in a manner that is integrated with natural and social sciences, literature, mathematics, arts, etc. But we are against making students memorise 'facts', lists of crops and the names of practices or exhortations to use (or not to use) new seeds, 'organic' methods, machinery, artificial fertilisers, weedicides, or pesticides. The main objective of agricultural education in schools should be to encourage students and teachers to do science: to question, examine and understand the present agricultural crisis, its interdependencies, its history, and its systemic causes, so that people continue to search for systemic solutions. This study of the crisis is not a body of knowledge or a list of facts. It is an exploration – a continual process of doing HDM science: questioning and observing, investigating, and analysing material evidence.

Education will be successful if it encourages students to ask the types of questions and look for the sorts of answers that upset the status quo. This may even be called "anti-national". But actually, it is anti-capitalist – because it involves questioning capitalism rather than taking it as an inevitable 'natural' way of living. It is actually the present mainstream type of education that subordinates the national interest – and human/nature interest – to global capital.

It is useful for people to ask basic questions and investigate how the underlying social – political – economic structure affects the way people cultivate. By so doing, we come to realise that the problems of poor people are not just due to the greed and cruelty of a few oppressors. Therefore, the problems will not be overcome by teaching rich people to be more caring to poor people. As we discussed, the rich and powerful are compelled to behave in certain ways due to the coercive laws of competition and the need to increase profits. If they do otherwise, they will lose the competition.

Although we (the researchers) were not very successful in learning to cultivate paddy, we did learn a lot during this study. We learned how difficult cultivation is – back-breaking for the body, and complex for the mind. We were surprised that the cultivation of cultivation was not occurring as we had expected. When we began this research project, we thought our ideology was pro-farmer. However, the analysis of our actions and interactions reveals how in some ways we were actually just supporting capitalism without understanding the problems farmers face. We initially advocated what we thought was a pro-environment ideology for organic farming. We thought our research would lead us to develop materials and methods to teach farmers to do experiments and develop their own seeds. We thought the family needed a better education in order to do better cultivation. We now see that to some extent the family was already doing 'organic' farming, experimenting, and selecting seeds. The reason they were not doing these things to a greater extent was not that they did not know how to do so, or did not appreciate the advantages, but that they faced systemic constraints. We initially advocated an innovative, learn-by-doing kind of formal science education that would teach children how to farm. Now we see that the farmers were already doing science and using teaching/learning methods in their fields that were better than anything we might do in a school. We were further surprised to find evidence that in many ways, their science was HDM science. We had expected that the farmers may be more apt to ask authorities for solutions to their problems – but instead we observed them figuring out solutions for themselves. The farm at first appeared to us to be quite traditional, and we expected that unchanging methods of cultivation were being handed down from authoritative elders. We did not expect to find that Smita was playing such an important role in doing cultivation, and in doing science. We were surprised that the family did not rely more on rituals and that religion was not more closely intertwined with their work in the field.

Moreover, the farming family did not see a need for their children to go to school to learn skills on how to farm. They were demanding education to get out of agriculture. We had thought that the parents would be hoping their children would take up cultivation. We were surprised to learn that that was actually the last thing they wanted. They did not want their children to study agriculture in school. The form of formal education we initially advocated would probably just end up deskilling and sorting children out of job opportunities. What we learned is that instead of indoctrinating farmers to conform to a somewhat modified form of commodified capitalist farming, we instead need to allow people to investigate and understand their basic conflicts and work collectively to invent their own systemic solutions.

Agricultural development is desired and needed, but how can it be brought about? What kind of development will it be? And for whom will it be? Will development occur by teaching farmers skills, modern methods, precision, and abstract principals? Will it occur by teaching them to be concerned with global issues, and to stop respecting elders, to read and write in non-local languages, and to be more individualistic (counteracting the traditional ways of Table 2.1)? We doubt that agricultural development can or should be brought about by intentionally teaching these things.

The reason the farming family does not 'develop' or modernise is not that they have an old-fashioned, traditional, conservative mind-set, do not want modern methods, or do not know how to do science. The family never told us that they wanted to preserve tradition for the sake of tradition or for maintaining cultural diversity. It is because they cannot afford to purchase modern technology and cannot afford to take more risks. Also, rather than having any essential desire to stick to their ancestors' methods, they may not believe a new method will be better without seeing the evidence and testing it out. It is not because they are uneducated or unskilled. As we observed, much more skill is required to use a pair of oxen for ploughing than to drive a tractor. Transplanting and harvesting by hand require much more skill than pushing buttons on an automatic transplanting machine or a combine harvester. Making a reed basket by hand requires much more skill than dumping some plastic pellets into a machine and pushing a few buttons to make a plastic basket. Teaching farmers modern skills is not the way to modernise agriculture, bring about development, and 'raise the GDP'. Modern capitalist agriculture does not require more workers with more skills, but fewer workers who have fewer skills and who are less questioning and more ready to follow authority.

Therefore, we do not advocate skill-development programmes. We have discussed how the main function of these programmes is to create a certification system that serves to sift people out of opportunities for better jobs. They serve to increase inequality. We think that skills should rather be learned while they are being practiced. However, manual skills should not be separated from work with the mind, either in education or in jobs. Rather than envisioning a world of increasing division of people doing manual labour from people doing the work with the mind, we envision a world in which technological advancement through HDM science will allow all people to engage in creative labour that integrates work with the hands and mind – and also provides people with more free time.

An integration between work with the hands and work with the mind is needed in order to address both the agricultural crisis and the environmental crisis. In relation to these problems there has been a long-standing exhortation to educate students to 'think globally and act locally'. Noel Gough has critiqued this exhortation, questioning what it means to think globally (Gough 2003). He argues that 'thinking globally', under present circumstances, is subject to differential international power relations. He discusses whether it may also be necessary to 'think locally and act globally'. We agree with him that 'western' science concepts, understandings, and technology should not be privileged, although at present they are. In other words, that (or any other) 'body of knowledge' should not be privileged. They should all be questioned: the questioning is what should at times be 'privileged', depending on **what** questioning and **for whom** it is being done. We assert that thinking (with the mind) and acting (with the hands) need to be integrated at both local and un-local levels. When we say 'thinking', we do not refer to an abstract 'body of knowledge'. We refer to interdependently (and dialectically) thinking, questioning, investigating, and acting based on local (as well as un-local) material and historical contexts. We do not think it is necessarily problematic to 'privilege' this way of doing HDM science, as we call it, by allowing or encouraging students to practice it in all schools. Maybe it is already fairly universally practiced to some extent in various forms.

It is imperative to support school education for all. We need to oppose the privatisation and commodification of education. We saw the problems that the high cost of education was causing for the farming family. It is only through government schools that education is more widely available and responsive to the needs of people rather than being concerned with profit. We need to demand that education is made free for all, from anganwadi (pre-school care) and kindergarten through to the highest post graduate levels. Private schools, colleges, and universities should be banned. Higher education should be available for all, including non-english speaking people living in villages.

However, education should be valued for its use to all people. We heard how Bhimraj and Smita were telling us that at present education is only valued for its exchange value in securing better-paying jobs to a few. If the government provided massive, well-paying employment programmes in rural areas, this would at least help people survive the immediate crisis so that they could improve their education in meaningful ways, rather than just trying to get certification for a job that is practically non-existent. At present there are government efforts for employment coming under the Mahatma Gandhi National Rural Employment Guarantee Act (MGNREGA), but this is woefully inadequate.

There is much that we can and should do to improve school education and make it more meaningful for the needs of people who are engaged in agricultural work. Formal education is needed for general literacy for all. We saw that deficiencies in communication are one of the reasons that the family on the farm was not engaging in more science. Learning and getting more experience in reading and writing, integrated with oral communication, would help. Access to libraries and internet are also needed, and should be provided at school as well as in community centres and at home. Problems regarding choices of the medium of instruction and the languages to be taught also have to be addressed. Schools should be integrated into the community, with opportunities for all sorts of people to get full-time or part-time employment as teachers, and work with students to address local/global problems using local/global means of communication. We have seen the pedagogy used by the adults in the farming family, and we think this is the kind of teaching that could be done in schools.

Schools should also play important roles in allowing and encouraging students to develop social relationships in which they work cooperatively and creatively to do meaningful HDM science. This does not mean that students should be required to engage in doing HDM science in relation to every topic in a prescribed syllabus. It does not mean that teachers should assign questions for the students to investigate. We believe that questioning, doing science, and practicing scientific temper is transferable across domains. It is done by necessity, as and when the authentic conflicts arise in real life. But real-life conflicts need not be segregated out of schools.

At present, students hardly get chances to engage in conversations with each other, ask questions, or do investigations as part of the main classroom discourse. Schools are mainly concerned with trying to teach students to be quiet and obedient. This has been a longstanding criticism that has been made by educationists and policy makers. We realise that there are many systemic reasons why these problems nevertheless persist. We must demand improvements in the educational system, as well as an end to capitalism itself. Education, as it is now, is functioning to decrease actual learning. Increasing capitalism means decreasing the doing of actual science. Capitalism has proved itself to be unsustainable. Improved and meaningful education will facilitate the necessary systemic change.

Meaningful education is very difficult under the present inequitable economic system. It can only be attained through considerable struggle by students and teachers – or as subversive activities carried on at the peripheries, in the cracks, and by whispering at the back of the class. We are optimistic that it is possible and will help people to ask the sorts of questions they need to ask in order to recognise and understand the problems they face and cooperate to work collectively towards systemic solutions.

References

Anonymous, 2015. Unpublished draft of a social science textbook in India.

Gough, Noel. 2003. Thinking globally in environmental education: Some implications for internationalizing curriculum inquiry. In *International Handbook of Curriculum Research*, ed. William F. Pinar, 53–72. New York: Lawrence Erlbaum Associates.

Haydock, Karen. 2015. Stated and unstated aims of NCERT social science textbooks. *Economic and Political Weekly* L (17): 109–119.

Joshi, Arjun. 2016. Science on television. *Economic & Political Weekly* 51 (22): 5.

Kosambi, D.D. 1952. Imperialism and peace, science and freedom. *Monthly Review* 4: 200–205.

Nanda, Meera. 2003. *Prophets Facing Backward: Postmodern Critiques of Science and Hindu Nationalism in India*. New Brunswick: Rutgers University Press.

Patnaik, Utsa. 2016. Capitalist trajectories of global interdependence and welfare outcomes: The lessons of history for the present, Ch.4. In *Critical Perspectives on Agrarian Transition: India in the Global Debate*, ed. B.B. Mohanty. Basingstoke: Taylor & Francis Ltd.

Afterword

What happened after this book was written?[1]

At the beginning of 2020, the very definition of what it means to be Indian was being questioned. The government had been making new citizenship laws and adding requirements for identity certification. The rights of various marginalised people were becoming increasingly restricted. In addition to certification for education, skills, and jobs, people were being required to have certification to continue residing in India.

At the beginning of 2020, economic problems were increasing for most people in India. As we discussed (in Sect. 7.1.2), due to the escalating agricultural crisis, people were increasingly being forced to look for non-agricultural employment wherever they could get it. Unemployment was reaching new heights. In many farming families, at least one member of the family was forced to go and find work in construction, as a driver or security guard, or do some other form of manual, daily wage labour. Many men (and some women) had to go and live in distant cities to find work. But because the wages were too low, they usually had to remain separated from their families, sleeping under the stairs, in a loft shelf, or floor of a shop or factory, or packed into a rented room with barely enough space for each person to stretch out (or sometimes sleeping in shifts).[2] There was usually no running water or toilet. Workers had to pay to wait in queues to use overcrowded public toilets, and collect water at certain times from public taps – or sometimes they had to pay for water. They cooked their own food, or ate at roadside stalls. Whatever money they managed to save, they sent back to their villages so that their families could survive.

---

[1] The writing was basically completed, except for editing, by the end of 2019. The last time the researchers visited Rudravali was on 6 November 2019.

[2] The people who fed us at the Homi Bhabha Centre for Science Education were poorly paid contract workers who were sleeping on the tables in our canteen, until they were forced to rent a room outside (one room without plumbing for half a dozen people).

This way of life was occurring because, as inhumane as it was, it was better than remaining in the village without jobs, and without even enough food.

Then the corona came. The central government imposed an enforced 'lockdown' in order to prevent the virus from spreading. But, incredibly, they prohibited people from working or going out in the streets without making any provision for people to have food, water, shelter, jobs, or wages. Even after several months of lockdown, they had hardly provided any 'fiscal stimulus' (as governments in some other countries had done), neither by providing direct income transfers and free rations nor by creating jobs for infrastructure development. As of September 2020, there have been no meaningful income transfers, and even food subsidies were late and were not available to many hungry people who did not possess the required documents (people need 'certification' not only for skill but also for hunger). It was an impossible nightmare that was so unbelievable that it appeared to be a joke: a crime being carried out by some power with an inhumane sense of humour. (Capitalism gone amok? Or capitalism status quo?) The Prime Minister actually told people to go out and bang their empty plates, rather than making sure they had food!

The central government should have immediately organised massive testing, contact tracing, and isolation of people who were suspected of having the virus. They should have acted quickly to nationalise all healthcare, hire and train healthcare workers at all levels, and build additional hospitals and facilities for isolation. In a city like Mumbai, about one third of all flats are lying vacant because they are being bought and sold as investments, rather than as places for people to live. People who were homeless or who were crowded into inhumane spaces should have immediately been allowed to live in these empty flats – places that some of them had constructed with their own hands. Massive employment programmes should have been set up throughout the country so that people could work safely, doing essential work that would relieve poor people from poverty. Food that the government was storing in godowns (warehouses) should have been immediately given back to all people who needed it. Emergency funds should have been given to all. None of these sorts of solutions to the pandemic are impossible. What is impossible is that, so far, this has not happened. Instead, corporations are trying to make profits off of the pandemic. Throughout the world, governments are paying the heads of corporations to buy stocks in their own companies, to keep financial capital spinning upwards, without producing anything. The Government of India is doing away with labour laws that came into being after years of struggle by workers. Bypassing the Parliament, the central government passed ordinances in order to further privatise the buying and selling of agricultural produce and to legalise contract farming, which facilitates corporatisation of agriculture. Farmers across the country are protesting these moves, which are bound to escalate the agricultural crisis. The government is also barrelling ahead on the road to privatising education and making it increasingly inequitable (e.g. through online education for the privileged).

Who can believe the claims that physical distancing is not possible in India? Not possible because the present living conditions are the naturalised, unavoidable, unquestionable living conditions? Who can believe the claims that things have to be the way they are because 'India is a poor country'? It is actually the reverse – the people of India are poor because things are the way they are: a few billionaires are making grotesque profits.[3]

We mentioned (in Sect. 6.4) how peasants give up their land only as a very last resort, and even when people are forced to go to the city to look for jobs outside of agriculture, they need to preserve the option of going back to their villages in case things do not work out. This is exactly what happened during the coronavirus pandemic: hundreds of thousands of daily wage labourers in cities were forced to go back to their villages because they had no food, shelter, or money in the cities. Since all trains, buses, and other means of transport were shut down, people had to (illegally) walk hundreds of kilometres. But the situation in the villages has been worse than ever. As of this writing (in early September 2020), lockdowns are being eased, even without evidence that there is a decrease in the number of corona cases and deaths. Therefore, people are returning to cities to look for jobs. Prospects for farmers are bleaker than ever. For example, although the price of milk in cities has remained constant, many people can no longer afford it, and the farmers producing the milk are selling it for less than half the price they got previously.

So where does all this leave us and the family on the farm in Rudravali?

When the national pandemic lockdown began in the last week of March 2020, we were physically isolated from each other. We were also socially isolated from each other if we did not have adequate access to communication technologies. Therefore, it was difficult for the researchers to find out what was going on in Rudravali, and it was difficult for the family to find out what was going on outside the village. Throughout mainstream and social media, the spread of misinformation, the general lack of scientific temper (Sect. 7.2.8), and the disinclination to ask or investigate the most important questions made it difficult for all of us to understand what was happening and why it was happening. For example, the mainstream media kept stating, with 'scientific' authority, that so far, xxx number of people in India, or in some part of India, have had Covid-19. But they almost always failed to even hint that we did not really know how many people had the virus, because very little testing was being done. They reported that there are no cases in villages, but they failed to mention that no one was testing people in villages to see whether they had the virus. They reported that yyy number of people have so far died from coronavirus, but they failed to mention that many other people who died were not tested for coronavirus and were not counted. They failed to report that in many places, especially in villages, deaths (from any cause) may not be accurately reported. 'Educated' people are not asking the relevant questions, or doing investigations. Rather than doing science, too many people are easily trusting and believing selected 'authorities', or making claims without evidence.

People live their lives according to their class, caste, and other kyriarchal constraints. These constraints have become more obvious during the corona pandemic, and it has become more obvious that marginalised people suffer the most, and for those with more privilege, suffering is less.

---

[3] This enormous wealth actually belongs to the people, including the family in Rudravali.

In February 2020, Abhijit, who was working on his PhD thesis, had gone on a student exchange program to the Czech Republic. He managed to complete his project, despite restrictions due to the pandemic, since much of his work could be done online. Although he was supposed to return in May, due to the travel restrictions he could not return until September. While he was away, his uncle in Mumbai passed away due to Covid-19.

On 10 May, Gurinder came down with a fever, which continued for a few days. In his building, a doctor and some members of the doctor's family had tested positive for Covid-19, and were in hospital. The entire building was sealed off, without even allowing anyone to go out to buy food and medicine, as before. Although it seemed likely that Gurinder had contracted the virus, he was not tested, and recovered after self-isolation at home. However, this happened in the same week that he was to present his online PhD thesis defence, which had to be postponed by a month.

Kalpana has a government job in a bank, which is classified as an essential service. Therefore, she has had to continue working at the bank throughout the lockdown.

Karen has not stepped out of the door of her building in Mumbai for the last 6 months, and orders food deliveries online. She has managed to continue working almost as before, since she has been working from home for the last year or so.

On 17 May, with the stringent lockdown still in effect in Mumbai and Raigad District, Kalpana got to know that Namdev had passed away on 12 May. The stated cause was heart attack. He did not have a Covid-19 test, but it is possible that his heart attack was related to the virus. He was in his mid-60s, was physically very fit, and had been doing much of the work on the farm. The farm was right on the main highway, and exposure to the virus was not unlikely.

Those of us who were in Mumbai did not succeed in contacting the family by phone. The family had very transitory phone service, even in the best of times. We were all worried about how the family would survive without Namdev. They were already complaining that the work was too much. They wanted the boys, Pratik and Pranay, to get a good education and get good jobs out of agriculture. But now, Pratik was the same age that Bhimraj was when he left school and began doing cultivation full time. What would Pratik do? Was there any choice, or any hope? Just 6 months ago it had seemed like a chance for a whole new world may be opening up for him, through education.

Then, on 3 June, a major cyclone, Nisarga, struck Maharashtra, and Raigad District was the worst hit. The areas around Mumbai and Raigad District generally do not experience dangerous cyclones. However, based on media reports and warnings by the Indian Meteorological Department, we were expecting the worst. Such unusual weather was being blamed on climate change. But in Mumbai, all we experienced, peering out of our (hopefully) safe locked-down locations onto the almost empty streets, was howling winds, waves of blowing rain, and some tin roofs being blown off nearby terraces. The mainstream media focussed on Mumbai and reported that there was not much damage, and they quickly stopped reporting much about the cyclone.

However, the cyclone had gone directly over Raigad District, and directly over Rudravali. Most of the houses in Rudravali were damaged, although there were no deaths in the village. Many of the large trees were uprooted: mangos, palms, guavas, lemons, …. The cowshed and the tile roof of the family's house were destroyed. Electricity in the region was down for at least a month. Therefore, phones were also not operating. After some delay, the government has given some compensation for the damage. Fortunately, the cyclone hit just before the beginning of the paddy cultivation season.

One of the things we have been discussing in this book is the gap between work with the hands and the mind. We have seen evidence of how this widening gap affects the family in Rudravali: in education, in science, in the spread of capitalist development, and in the development of paddy cultivation. The gap between the mind and the hands is even extending to the constitution and law enforcement. Capital is taking on a life of its own. But democracy, equality, and freedom are becoming theoretical. They may be preached, but in practice, they are diminishing. We have seen how the members of the family are compelled to spend their waking hours engaging in the struggle to satisfy their basic daily needs. Where is their freedom? Where is their equality? Where are their democratic rights? Is education helping them to attain their needs and desires, or even to voice them? We have glimpsed a little of their struggle to learn and to do historical dialectical materialist science. When will they win this struggle and be free to cultivate a sort of cultivation that is egalitarian, useful, and truly valuable?

Index

Printed in the United States
by Baker & Taylor Publisher Services